Reinforced Concrete Design

Jonsson Engineering Center. (Courtesy Rennselear Polytechnic Institute)

Reinforced

Concrete

Design

LEONARD SPIEGEL, P.E.

Hudson Valley Community College

GEORGE F. LIMBRUNNER, P.E.

Hudson Valley Community College

Prentice-Hall, Inc., Englewood Cliffs, N.J. 07632

Library of Congress Cataloging in Publication Data

Spiegel, Leonard.
   Reinforced concrete design.

   Includes bibliographical references and index.
   1.–Reinforced concrete construction.
I.–Limbrunner, George F., joint author. II.–Title.
TA683.2.S58   624′.1834   79-24807
ISBN 0-13-771659-1

Cover design by A Good Thing, Inc.
Manufacturing buyer: Gordon Osborne

Printed in the United States of America

10 9 8 7 6 5 4 3 2 1

PRENTICE-HALL INTERNATIONAL, INC., *London*
PRENTICE-HALL OF AUSTRALIA PTY. LIMITED, *Sydney*
PRENTICE-HALL OF CANADA, LTD., *Toronto*
PRENTICE-HALL OF INDIA PRIVATE LIMITED, *New Delhi*
PRENTICE-HALL OF JAPAN, INC., *Tokyo*
PRENTICE-HALL OF SOUTHEAST ASIA PTE. LTD., *Singapore*
WHITEHALL BOOKS LIMITED, *Wellington, New Zealand*

Contents

v

3 Reinforced Concrete Beams: T-Beams and Doubly Reinforced Beams   63

4 Shear in Beams   110

5 Development, Splices, and Simple-Span Bar Cutoffs

6 Continuous Construction Design Considerations

7 Serviceability

8 Walls

Preface

The primary objective of this book is to furnish the reader with a basic understanding of the strength and behavior of reinforced concrete members and simple concrete structural systems.

With relevant reinforced concrete research and literature increasing at an exponential rate, it is the intent of this book to translate this vast amount of information and data into an integrated source that reflects the latest information available. It is not intended to be a comprehensive theoretical treatise of the subject, because it is our contention that such a document could easily obscure the fundamentals that we strive to emphasize in the engineering technology programs. In addition, we are of the opinion that adequate comprehensive books on reinforced concrete design do exist for those who seek the theoretical background, the research studies, and more rigorous applications.

Throughout the book the specifications of the *Building Code Requirements for Reinforced Concrete* (ACI 318-77) have been used as a basis of design utilizing the ultimate strength method, which since 1971 has been referred to as the *strength method*. Because the ACI Code serves as a design standard in the United States, the authors strongly recommend that the Code be utilized as a companion publication to this book.

The text content is primarily an elementary, noncalculus, practical approach to the design and analysis of reinforced concrete structural members utilizing numerous example problems and a step-by-step solution format. In addition, chapters are furnished, utilizing a conceptual approach, on such topics as prestressed concrete, composite construction, and detailing of reinforced concrete members. The metric system (SI), which is still alien to the reinforced concrete design and construction field in the United States, is introduced in a comprehensive manner utilizing several example problems.

The book has been thoroughly classroom-tested in our engineering technology programs and should serve as a valuable design guide and source for technologists, technicians, and engineering and architectural students. Additionally, it will aid architects and engineers preparing for state licensing examinations for professional registration.

The authors wish to express their appreciation to the many people who assisted in the preparation of this book. We are indebted particularly to Virginia van Dyck, who typed the major portion of the manuscript. We are especially indebted to our wives, Beverly Spiegel and Jane Limbrunner, for their patience, encouragement, and support, and it is to them that we affectionately dedicate this book.

LEONARD SPIEGEL, P.E.
GEORGE F. LIMBRUNNER, P.E.

Reinforced Concrete Design

1

Materials and Mechanics of Bending

1–1 CONCRETE

Concrete consists principally of a mixture of cement and fine and coarse aggregates (sand, gravel, crushed rock, and/or other materials) to which water has been added as a necessary ingredient for the chemical reaction of curing. The bulk of the mixture consists of the fine and coarse aggregates. The resulting concrete strength and durability are a function of the proportions of the mix as well as other factors, such as the concrete placing, finishing, and curing history.

The compressive strength of concrete is relatively high. However, it is a relatively brittle material, the tensile strength of which is small compared with its compressive strength. Hence, steel reinforcing rods (which have high tensile and compressive strength) are used in combination with the concrete; the steel will resist the tension and the concrete the compression. *Reinforced concrete* is the result of this combination of steel and concrete. In many instances, steel and concrete are positioned in members so that they both resist compression.

1–2 THE ACI BUILDING CODE

The design and construction of reinforced concrete buildings is controlled by the "Building Code Requirements for Reinforced Concrete" (ACI 318-77) of the American Concrete Institute.[1] The use of the term "Code" in this text refers to the ACI Code unless otherwise stipulated. Previous editions of the code, to which reference is occasionally made, appeared in 1963 (ACI 318-63) and 1971 (ACI 318-71). The code itself has no legal status. However, it has been incorporated into the building codes of almost all states and municipalities throughout the United States. When so incorporated, it has official sanction, becomes a legal document, and is part of the law controlling reinforced concrete design and construction in a particular area.

1–3 CEMENT AND WATER

Structural concrete utilizes, almost exclusively, hydraulic cement. With this cement water is necessary for the chemical reaction of *hydration*. In the process of hydration, the cement sets and bonds the fresh concrete into one mass. *Portland cement*, which originated in England, is undoubtedly the most common form. Portland cement consists chiefly of calcium and aluminum silicates. The raw materials are limestones, which provide calcium oxide (CaO), and clays or shales, which furnish silicon dioxide (SiO_2) and aluminum oxide (Al_2O_3). Following processing, cement is marketed in bulk or in 94-lb (1-ft^3) bags.

In fresh concrete, the ratio of the amount of water to the amount of cement, by weight, is termed the *water/cement ratio*. This ratio can also be expressed in terms of gallons of water per bag of cement. For complete hydration of the cement in a mix, a water/cement ratio of 0.35–0.40 (4–4½ gal/bag) is required. To increase the *workability* of the concrete (the ease with which it can be mixed, handled, and placed), higher water/cement ratios are normally used.

1–4 AGGREGATES

In ordinary structural concretes the aggregates occupy approximately 70–75% of the volume of the hardened mass. Gradation of aggregate size to produce close packing is desirable because, in general, the more densely the aggregate can be packed, the better are the strength and durability.

Aggregates are classified as fine or coarse. *Fine aggregate* is generally sand and may be categorized as consisting of particles that will pass a No. 4 sieve (four openings per linear inch). A *coarse aggregate* consists of particles that would be retained on a No. 4 sieve. The maximum size of coarse aggregate in reinforced concrete is governed by various ACI Code requirements, but obviously the concrete must be capable of being placed into forms as well as between adjacent reinforcing bars.

1-5 CONCRETE IN COMPRESSION

The theory and techniques relative to the design and proportioning of concrete mixes, as well as the placing, finishing, and curing of concrete, are outside the scope of this book and are adequately discussed in many other publications.[2-5] Field testing, quality control, and inspection are also adequately covered elsewhere. This is not to imply that these are of less importance in overall concrete construction technology but only to reiterate that the objective of this publication is to deal with the design and analysis of reinforced concrete members.

We are primarily concerned with how a reinforced concrete member behaves when subjected to load. It is generally accepted that the behavior of a reinforced concrete member under load depends on the stress–strain relationship of the materials, as well as the type of stress to which it is subjected. With concrete used principally in compression, the compressive stress–strain curve is of primary interest.

The compressive strength of concrete is denoted f'_c and is assigned the units *pounds per square inch* (psi). Frequently, for calculations, f'_c is used with the units *kips per square inch* (ksi).

A test that has been standardized by the American Society for Testing and Materials (ASTM)[6] is used to determine the compressive strength (f'_c) of concrete. The test involves compression loading to failure of a specimen cylinder of concrete. The compressive strength so determined is the highest compressive stress to which the specimen is subjected. Note in Fig. 1–1 that f'_c is not the stress that exists in the specimen at failure but that which occurs at a strain of about 0.002.

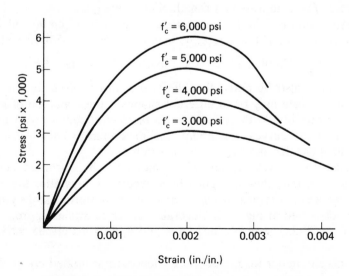

FIGURE 1-1   Typical stress–strain curves for concrete

Currently, 28-day concrete strengths (f'_c) range from 2,500 to 9,000 psi, with 3,000–4,000 psi being common for reinforced concrete structures and 5,000–6,000 psi being common for prestressed concrete members. Higher-strength concretes have been achieved under laboratory conditions. The curves shown in Fig. 1–1 represent the result of compression tests on 28-day standard cylinders for varying design mixes.

A review of the stress–strain curves for different-strength concretes reveals that generally the maximum compressive strength is achieved at a unit strain of approximately 0.002 in./in. Stress then decreases, accompanied by additional strain. Higher-strength concretes are more brittle and will fracture at a lower maximum strain than will the lower-strength concretes. The initial slope of the curve varies unlike that of steel and only approximates a straight line. Since the slope of the curve is the modulus of elasticity, we observe that there exists no true modulus of elasticity for any given-strength concrete, since stress is only approximately proportional to strain. It may also be observed that the slope of the initial portion of the curve (if it is approximately a straight line) varies with different-strength concretes. Even if we assume a straight-line portion, the modulus of elasticity is different for different-strength concretes.

The ACI Code, Section 8.5.1, provides the accepted empirical expression for *modulus of elasticity*:

$$E_c = w_c^{1.5} 33 \sqrt{f'_c}$$

where E_c = modulus of elasticity of concrete in compression (psi)
   w_c = unit weight of concrete (lb/ft^3)
   f'_c = compressive strength of concrete (psi)
This expression is valid for concretes having w_c between 90 and 155 lb/ft^3. For normal-weight concrete, w_c may be taken as 145 lb/ft^3 and

$$E_c = 57,000 \sqrt{f'_c} \qquad \text{(see Table A–5 for values of } E_c)$$

It should also be noted that the stress–strain curve for the same-strength concrete may be of different shapes if the condition of loading varies appreciably. With different *rates of strain* (loading) we will have different shape curves. Generally, the maximum strength of a given concrete is smaller at slower rates of strain.

Concrete strength varies with time, and the specified concrete strength is usually that strength which occurs 28 days after the placing of the concrete. A typical strength–time curve for normal stone concrete may be observed in Fig. 1–2. Generally, concrete attains approximately 70% of its 28-day strength in 7 days and approximately 85–90% in 14 days.

Concrete, under load, exhibits a phenomenon termed *creep*. This is

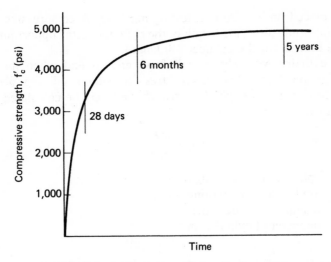

FIGURE 1-2 Strength–time relationship for concrete

the property by which concrete continues to deform (or strain) over long periods of time while under constant load. Creep occurs at a decreasing rate over a period of time and may cease after several years. Generally, high-strength concretes exhibit less creep than do lower-strength concretes. The magnitude of the creep deformations is proportional to the magnitude of the applied load as well as to the length of time of the load application.

1-6 CONCRETE IN TENSION

Tensile and compressive strengths of concrete are not proportional, and an increase in compressive strength is accompanied by an appreciably smaller percentage increase in tensile strength. For ordinary stone concrete, a rough estimate of tensile strength is that it is approximately 9% of the compressive strength.

The true tensile strength of concrete is difficult to determine. One common approach has been to use the *modulus of rupture* (which is the bending tensile stress of a plain concrete test beam at failure) as a measure of tensile strength. The *split-cylinder test* has also been used to determine the concrete tensile strength and is generally accepted as a good measure of the true tensile strength. A reasonable estimate of the split-cylinder strength is 6–7 times $\sqrt{f'_c}$ for ordinary stone concrete, and an average value often used is $6.7\sqrt{f'_c}$.

The split-cylinder test utilizes a standard 6-in.-diameter 12-in.-long cylinder placed on its side in a testing machine. A compressive load is applied uniformly along the length of the cylinder, with support furnished along the bottom for the cylinder full length.

The cylinder will split in half from end to end when its tensile strength is reached. This tensile stress at which splitting occurs is referred to as the *split-cylinder strength* and may be calculated by the following expression:

$$f_t = \frac{2P}{\pi LD}$$

where f_t = splitting tensile strength (psi)
   P = applied load at splitting (lb)
   L = length of cylinder (in.)
   D = diameter of cylinder (in.)

1-7 REINFORCING STEEL

Concrete cannot withstand very much tensile stress without cracking; therefore, tensile reinforcement must be embedded in the concrete to overcome this deficiency. In the United States this reinforcement is in the form of steel bars or welded wire fabric composed of steel wires. In other countries, however, many approaches have been tried in the search for an economical reinforcement for concrete. Principal among these are the fiber-reinforced concretes, where the reinforcement is obtained through the use of short fibers of steel or other materials, such as fiberglass. The specifications for steel reinforcement published by the American Society for Testing and Materials (ASTM) are generally accepted for the steel used in reinforced concrete construction in the United States.

All steel bars used for reinforcing are round, deformed bars with some form of patterned ribbed projections rolled onto their surfaces. The patterns vary depending on the producer, but all patterns conform to ASTM specifications. The bars vary in designation from No. 3 through No. 11, with two additional bars, No. 14 and No. 18.

For bars No. 3 through No. 8, the designation represents the bar diameter in eighths of an inch. The No. 9, No. 10, and No. 11 bars have diameters which provide areas equal to 1-in. square bars, 1⅛-in. square bars, and 1¼-in. square bars, respectively. The No. 14 and No. 18 bars correspond to 1½-in. square bars and 2-in. square bars, respectively, and are generally only available by special order.

ASTM specifications require that producers roll onto the bar the following information: a symbol indicating the producer, a number indicating the size of the bar, a symbol indicating the type of steel from which the bar was rolled, and a number to indicate grade 60 steel. The grade indicates the minimum specified yield stress in ksi. For instance, a grade 60 steel bar has a minimum specified yield stress of 60 ksi. No symbol indicating grade is rolled onto grade 40 or 50 steel bars.

Reinforcing bars are usually made from newly manufactured steel (billet steel). Bars rerolled from railroad rails (rail steel) and used axles (axle steel) are relatively uncommon. Steel types and ASTM specification numbers for bars are tabulated in Table A–1 in Appendix A. As may be observed in the tabulation, some ASTM designations include more than one grade of steel. For example, ASTM A615, which is new billet steel, is available in grades 40 and 60. ASTM A706, low alloy steel, which is intended primarily for welding, is available in only one grade. Tables A–2 and A–3 contain useful information on cross-sectional areas of bars.

The most useful physical properties of reinforcing steel for reinforced concrete design calculations are yield stress (f_y) and modulus of elasticity. A typical stress–strain diagram for reinforcing steel is shown in Fig. 1–3.

The yield stress (or yield point) of steel is determined through procedures governed by ASTM standards. For practical purposes, the yield stress may be thought of as that stress at which the steel exhibits increasing strain with no increase in stress. The yield stress of the steel will usually be one of the known (or given) quantities in a reinforced concrete design or analysis problem. See Table A–1 for the range of f_y.

The modulus of elasticity of carbon reinforcing steel (the slope of the stress–strain curve in the elastic region) varies over a very small range and has been adopted as 29,000,000 psi (ACI Code, Section 8.5.2).

Welded wire fabric (commonly called *mesh*) is another type of

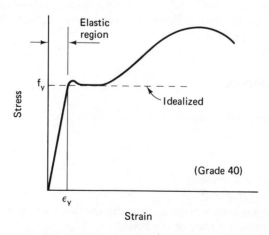

FIGURE 1-3   Typical stress–strain diagram for reinforcing steel

reinforcement. It consists of cold-drawn wire in orthogonal patterns, square or rectangular, resistance-welded at all intersections. It may be supplied in either rolls or sheets, depending on wire size. Normally, fabric with wire diameters larger than about ¼ in. is available only in sheets.

Two types of welded wire fabric are available: smooth (ASTM A185) with a specified minimum yield strength of 65,000 psi, and deformed (ASTM A497) with a specified minimum yield strength of 70,000 psi. The deformed wire is normally more expensive but its use may be justified because of its higher strength and increased bond with concrete.

A rational method of designating wire sizes has been adopted by the wire industry to replace the formerly used gage system. Smooth wires are described by the letter W followed by a number equal to 100 times the cross-sectional area of the wire in square inches. Deformed wire sizes are similarly described, but the letter D is used. Thus a W9 wire has an area of 0.090 in.^{2} and a D8 wire has an area of 0.080 in.^{2}. A W8 wire has the same cross-sectional area as the D8 but is smooth rather than deformed. Intervals between full numbers are given by decimals, for example W9.5 and D8.3.

Generally, the fabric is indicated by the symbol WWF followed by spacings first of longitudinal wires and then of transverse wires and last by the sizes of longitudinal and transverse wires. Thus WWF6×12-W8.2×W4.4 indicates a smooth welded wire fabric with 6 in. longitudinal spacing, 12 in. transverse spacing, and a cross-sectional area equal to 0.082 in.^{2} for the longitudinal wires and 0.044 in.^{2} for the transverse wires.

Additional information about welded wire fabric, as well as tables relating size number with wire diameter, area, and weight, may be obtained through the Wire Reinforcement Institute[7] or the Concrete Reinforcing Steel Institute.[8]

1–8 BEAMS—GENERAL

In a homogenous elastic beam, the flexure formula

$$f = \frac{Mc}{I}$$

where f = calculated bending stress
   M = applied moment
   c = distance from neutral axis to the outside tension or compression fiber of the beam
   I = moment of inertia of the cross section about the neutral axis

may be used to calculate bending stress. By rearranging the flexure formula, the maximum moment that may be applied to the beam cross section, called the *resisting moment*, may be found:

$$M_R = \frac{F_b I}{c}$$

where M_R = resisting moment

F_b = allowable bending stress

This procedure is straightforward for a beam of known cross section for which the moment of inertia can easily be found. However, for a reinforced concrete beam, the use of the flexure formula presents many complications, since the beam is not homogeneous and concrete does not behave elastically over its full range of strength.

The *internal couple* concept of beam strength is more general and may be applied to homogeneous or nonhomogeneous beams having linear (straight-line) or nonlinear stress distributions. For reinforced concrete beams, it has the advantage of using the basic resistance pattern that is found in the beam.

As an introduction to the internal couple method, the following three different analysis problems of homogeneous timber beams are offered. The problems utilize both approaches as a means of comparison: the internal couple approach and the flexure formula approach.

EXAMPLE 1–1

A homogeneous timber beam of cross section shown is used on a simple span of 20 ft. It carries a total load (which includes the weight of the beam) of 1,200 lb/ft. Using the nominal dimensions shown:

a. Find the maximum bending stress using the internal couple method.

b. Check part (a) using the flexure formula.

Solution. From the load diagram [Fig. 1–4(a)], the maximum moment may be found:

$$M_{max} = \frac{w\ell^2}{8} = \frac{1,200(20)^2}{8}$$

$$= 60,000 \text{ ft-lb}$$

a. Figure 1-4(c) depicts a free body diagram of the left half of the beam as it is acted upon by the left reaction, the load, and the internal couple. The maximum bending stress may be found as follows:

   1. By inspection, the neutral axis (N.A.) is at midheight. The stress and strain vary as a straight line from zero at the neutral axis (which is also the centroidal axis) to a maximum at the outer fiber. The area above the neutral axis is in compression and the area below is in tension. These are

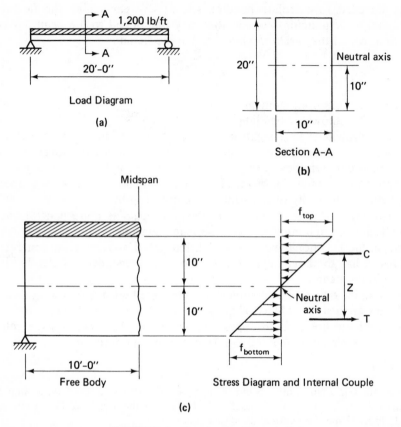

FIGURE 1-4 Sketches for Example 1-1

generally called *bending stresses*, since they result from the bending behavior of the member.

2. *C* represents the resultant compressive force or summation of all the unit compressive stresses above the neutral axis.

3. *T* represents the resultant tensile force or summation of all the unit tensile stresses below the neutral axis.

4. *C* and *T* each act at the centroid of their respective triangles of stress distribution. Therefore, $Z = 13.33$ in.

5. *C* must equal *T* since $\Sigma H = 0$. The two forces *C* and *T* act together to form the internal couple (or internal resisting moment) of magnitude *CZ* or *TZ*.

6. The internal resisting moment must equal the bending moment due to the external loads at any section. Therefore,

$$M = CZ \quad \text{or} \quad TZ$$
$$60,000(12) = C(13.33)$$

from which

$$C = 54,014 \text{ lb} = T$$

7. $C =$ (area of triangle of stress) (width of beam), or

$$C = 54,014 = \frac{1}{2}(10)(f_{top})(10)$$

Solving for f_{top} yields

$$f_{top} = 1,080.0 \text{ psi} = f_{bottom}$$

The beam is satisfactory assuming that the timber allowable bending stress is greater than 1,080 psi.

b. Check part (a) using the flexure formula

$$f = \frac{Mc}{I}$$

where

$$I = \frac{bh^3}{12} = \frac{10(20)^3}{12} = 6,667 \text{ in.}^4$$

$$f = \frac{60,000(12)(10)}{6,667} = 1,080 \text{ psi} \qquad \text{(OK)}$$

The preceding example is based on elastic theory and assumes the following: (1) a plane section before bending remains a plane section after bending (the variation in strain throughout the depth of the member is linear from zero at the neutral axis); and (2) the modulus of elasticity is constant; therefore, stress is proportional to strain (within the proportional limit) and the stress distribution throughout the depth of the beam is also linear from zero at the neutral axis to a maximum at the outer fibers.

The internal couple approach may also be used to find the moment strength (resisting moment) of a beam.

EXAMPLE 1–2

Find the resisting moment (M_R) of a timber beam having a rectangular cross section as shown in Fig. 1-5(a). The allowable bending stress is 1.2 ksi (1,200 psi). Use the nominal dimensions shown. Assume the linear stress distribution shown in Fig. 1–5(b).

The resultant tensile and compressive forces are

$$C = T = \frac{1}{2}(1.2)(3)(2) = 3.6 \text{ kips}$$

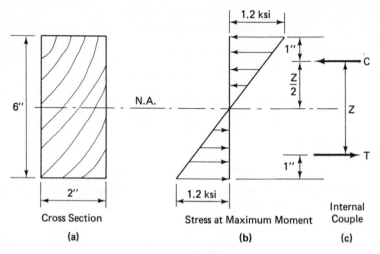

6″

N.A.

2″

Cross Section

(a)

1.2 ksi

1″

$\frac{Z}{2}$

1″

1.2 ksi

Stress at Maximum Moment

(b)

C

Z

T

Internal
Couple

(c)

FIGURE 1-5 Sketch for Example 1–2

$$\frac{Z}{2} = \frac{2(3)}{3} = 2 \text{ in.}$$

$$M_R = CZ = TZ = 3.6(4) = 14.4 \text{ in.-kips}$$

Check the foregoing using the flexure formula:

$$f = \frac{Mc}{I} \quad \text{and} \quad M_R = \frac{F_b I}{c}$$

$$I = \frac{bh^3}{12} = \frac{2(6)^3}{12} = 36 \text{ in.}^4$$

$$M_R = \frac{1.2(36)}{3} = 14.4 \text{ in.-kips} \tag{OK}$$

The internal couple may also be used to analyze irregularly shaped cross sections, although for homogeneous beams it is more cumbersome than is the use of the flexure formula.

EXAMPLE 1–3

Find the resisting moment (M_R) for the T-shaped timber member shown in Fig. 1–6(a). The allowable bending stress is 1.0 ksi. Assume a linear stress distribution. Use the nominal dimensions shown.

The neutral axis must be located so that the strain and stress diagrams may be defined.

$$\bar{y} = \frac{\Sigma(Ay)}{\Sigma A} = \frac{5.0(6.5) + 12(3)}{17} = 4.03 \text{ in.}$$

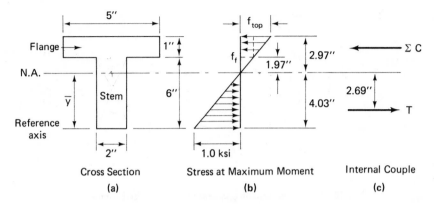

FIGURE 1-6 Sketch for Example 1-3

Since the bottom of the cross section is farther from the neutral axis, it will reach allowable stress first. The stresses may then be represented as in Fig. 1–6(b). By similar triangles,

$$f_{top} = \frac{2.97}{4.03}(1.0) = 0.737 \text{ ksi}$$

The stress at bottom of flange (f_f) is

$$f_f = \frac{1.97}{4.03}(1.0) = 0.489 \text{ ksi}$$

The tensile force can be evaluated and located in Fig. 1–6(c) as follows:

$$T = \text{stress} \times \text{area} = \frac{1.0}{2}(4.03)(2.0) = 4.03 \text{ kips}$$

T is located $\frac{2}{3}(4.03) = 2.69$ in. below the neutral axis. The compressive force will be broken up into components, because of the irregular area, as shown in Fig. 1–7.

Referring to both Figs. 1–6 and 1–7, the component internal compressive forces, component internal couples, and M_R may now be evaluated.

Component forces:

$$C_1 = 0.489(5)(1) = 2.445 \text{ kips}$$

$$C_2 = \frac{0.248}{2}(5)(1) = 0.620 \text{ kip}$$

$$C_3 = \frac{0.489}{2}(1.97)(2) = 0.963 \text{ kip}$$

Moment arm distance from component compressive force to T:

$$Z_1 = 2.69 + 1.97 + 0.5 = 5.16 \text{ in.}$$

$$Z_2 = 2.69 + 1.97 + \frac{2}{3}(1) = 5.33 \text{ in.}$$

13

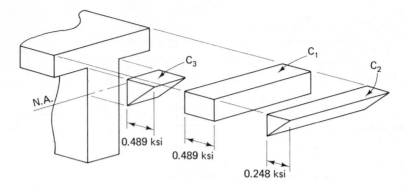

FIGURE 1-7 Component compressive forces for Example 1–3

$$Z_3 = 2.69 + \tfrac{2}{3}(1.97) = 4.00 \text{ in.}$$

Component internal couple = (force) × (moment arm):

$$M_{R1} = 2.445(5.16) = 12.62 \text{ in.-kips}$$

$$M_{R2} = 0.620(5.33) = 3.30 \text{ in.-kips}$$

$$M_{R3} = 0.963(4.00) = 3.85 \text{ in.-kips}$$

Summation of compressive force components:

$$C = C_1 + C_2 + C_3 = 4.028 \text{ kips} \qquad (\text{checks with } T = 4.03 \text{ kips})$$

Summation of internal couple components:

$$M_R = M_{R1} + M_{R2} + M_{R3} = 19.77 \text{ ft-kips}$$

Check by the flexure formula.

$$I = I_{cg} + Ad^2 = \tfrac{1}{12}(2)(6)^3 + \tfrac{1}{12}(5)(1)^3 + 12(1.03)^2 + 5(2.47)^2 = 79.65 \text{ in.}^4$$

$$M_R = \frac{F_b I}{c} = \frac{1.0(79.65)}{4.03} = 19.76 \text{ in.-kips} \approx 19.77 \text{ in.-kips} \qquad (\text{OK})$$

As previously mentioned, reinforced concrete beams are not homogeneous, since they consist of two materials, steel and concrete. The flexure formula is not directly applicable. Therefore, the basic tool used is the internal couple method.

PROBLEMS

1–1. The unit weight of normal reinforced concrete is commonly assumed to be 150 lb/ft³. Find the weight per lineal foot (lb/ft) for a reinforced concrete beam which:

14

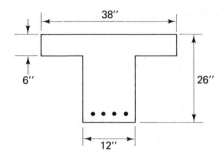

FIGURE P1-1

    a. has a cross section 16 in. wide and 28 in. deep
    b. has a cross section as shown in the accompanying diagram.

1-2. a. Find the resisting moment (M_R) of the rectangular timber beam using the internal couple method.

$$F_b = 1,000 \text{ psi}$$

    b. Check part (a.) by using the flexure formula.

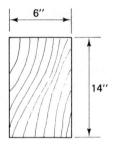

FIGURE P1-2

1-3. A timber beam of cross section shown is on a simple span of 12 ft. It carries a dead load (which includes the weight of the beam) of 0.7 kip/ft. There is a concentrated load of 6 kips located at midspan.

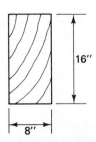

FIGURE P1-3
15

   a. Using the internal couple method, find the maximum bending stress.
   b. Check part (a.) by using the flexure formula.

1–4. A timber beam of cross section shown is on a simple span of 12 ft. It carries a total load (which includes the weight of the beam) of 1,200 lb/ft. (Use nominal dimensions.)

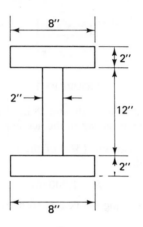

FIGURE P1-4

   a. Find the maximum bending stress using the internal couple method.
   b. Check part (a.) using the flexure formula.
   c. Find the magnitude of the total compressive force.
   d. Using the internal couple method, find the resisting moment if the allowable bending stress is 1,600 psi.

REFERENCES

1. "Building Code Requirements for Reinforced Concrete" (ACI 318–77), The American Concrete Institute, P. O. Box 19150, Redford Station, Detroit, Michigan 48219.

2. ACI Committee 211, "Recommended Practice for Selecting Proportions for Normal and Heavyweight Concrete" (ACI 211.1–74), The American Concrete Institute, P.O. Box 19150, Redford Station, Detroit, Michigan 48219. 1975.

3. George Earl Troxell, Harner E. Davis, and Joe W. Kelly, *Composition and Properties of Concrete*, 2nd ed. New York: McGraw-Hill Book Company, 1968.

4. *Design and Control of Concrete Mixtures*, 11th ed., Engineering Bulletin of the Portland Cement Association, Chicago, 1968.

5. Joseph J. Waddell, ed., *Concrete Construction Handbook*, 2nd ed. New York: McGraw-Hill Book Company, 1974.

6. "ASTM Standards," American Society for Testing and Materials, 1916 Race Street, Philadelphia, Pennsylvania 19103.

7. Wire Reinforcement Institute, 7900 West Park Drive, McLean, Virginia 22101.

8. Concrete Reinforcing Steel Institute, 180 North LaSalle Street, Chicago, Illinois 60601.

2

Rectangular Reinforced Concrete Beams and Slabs: Tension Steel Only

2–1 INTRODUCTION

When a beam is subjected to bending moments (also termed *flexure*), bending strains are produced. Under positive moment (as normally defined), compressive strains are produced in the top of the beam and tensile strains are produced in the bottom. These *strains* produce *stresses* in the beam—compression in the top and tension in the bottom. Bending members must, therefore, be able to resist both tensile and compressive stresses.

For a concrete flexural member (beam, wall, slab, etc.) to have any significant load-carrying capacity, its basic inability to resist tensile stresses must be overcome. By embedding reinforcement (usually steel bars) in the tension zones, we create a *reinforced concrete* member. Members composed of these materials, when properly designed and constructed, perform very adequately when subjected to flexure.

Initially, we will consider simply supported single-span beams, which, since they carry only positive moment (tension in the bottom), will be reinforced with steel placed near the bottom of the beam.

2–2 METHOD OF ANALYSIS AND DESIGN

In the timber beam problems in Chapter 1, we assumed both a straight-line strain distribution and straight-line stress distribution from the neutral axis to the outer fibers. This, in effect, stated that stress is proportional to strain. This is generally valid only up to a point called the *proportional limit*. With structural steel normally used in construction, the proportional limit and the yield stress have nearly the same value, and an allowable bending stress is determined by applying a factor of safety to the yield stress. With timber the determination of an allowable bending stress is less straightforward, but it may be thought of as some fraction of the breaking bending stress. Utilizing the allowable bending stresses and the assumed linear stress–strain relationship, both analysis and design of timber and structural steel members are performed in a manner similar to that used in the previous problems.

This approach, despite the fact that a reinforced concrete beam is a nonhomogeneous member, was also considered to be valid for concrete design for many years and was known as the working stress design (W.S.D.) method.

The basic assumptions for the *working stress design method* may be summarized as follows: (1) a plane section before bending remains a plane section after bending; (2) Hooke's law (i.e., stress is proportional to strain) applies to both the steel and concrete; (3) the tensile strength of concrete is zero and the reinforcing steel carries all the tension; and (4) the bond between the concrete and the steel is perfect, so that no slip occurs.

Based on these assumptions the flexure formula was still used even though the beam was nonhomogeneous. This was accomplished by theoretically transforming one material into another based on the ratio of the concrete and steel moduli of elasticity.

In the last 25 years, a more modern and realistic approach has been developed based on the fact, among others, that at some point in the loading, the proportional stress–strain relationship for the compressive concrete ceased to exist. This approach was termed the *ultimate strength design* (U.S.D.) *method*. Both methods were fully recognized and acceptable to use in the 1963 ACI Building Code. In the 1971 ACI Building Code, for all practical purposes, the ultimate strength method became the sole design and analysis technique, although W.S.D. was still allowed as an alternative method. In addition, the ultimate strength method designation was changed to *strength method* by the 1971 ACI Building Code.

The assumptions for the strength design method are similar to those itemized for the W.S.D. method, with one notable exception. Research has indicated that the compressive concrete stress is approximately proportional to strain only up to moderate loads. With an increase in load, the approximate proportionality ceases to exist and the compressive stress diagram takes a shape similar to the concrete compressive stress–strain curve of Fig. 1-1. Additional assumptions for strength design are discussed in Section 2-4.

The difference in approach of the two methods may be briefly expressed at this point. In the working stress method, actual working loads, now called *service loads*, were utilized and a member was designed or analyzed based on an allowable compressive bending stress (generally $0.45f'_c$) and a compressive stress pattern that was assumed to vary linearly from zero at the neutral axis. In the modern strength method, service loads are amplified using load factors. Members are then designed so that their practical strength at failure, which is somewhat less than the true strength at failure, is sufficient to resist the amplified loads. The strength at failure is still commonly called the *ultimate strength*, and the load at or near failure is commonly called the *ultimate load*. The stress pattern assumed for strength design is such that predicted strengths are in substantial agreement with test results.

2–3 BEHAVIOR UNDER LOAD

Prior to the discussion of the strength method, let us review the behavior of a long-span rectangular reinforced concrete beam as the load on the beam increases from zero to that magnitude which would cause failure.

At very small loads, assuming that the concrete has not cracked, both concrete and steel will resist the tension and concrete alone will resist the compression. The stress distribution will be as shown in Fig. 2-1. Note that stresses vary linearly with distance from zero at the neutral axis and are, for all practical purposes, proportional to strains.

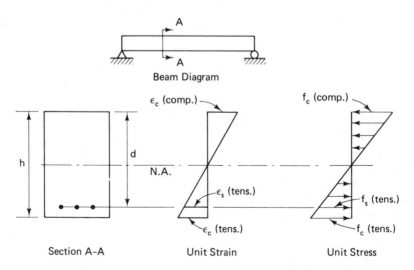

FIGURE 2-1   Flexural behavior at very small loads

This will be the case when stresses are low (below the modulus of rupture).

At moderate loads, the tensile strength of the concrete will be exceeded and the concrete will crack (hairline cracks) in the manner shown in Fig. 2-2. Since the concrete cannot transmit any tension across a crack, the steel bars will then resist the entire tension. The stress distribution at or near a cracked section then becomes as shown in Fig. 2-2. This is approximately valid up to a concrete stress f_c of about $f_c'/2$.

The concrete compressive stress is still assumed to be proportional to the concrete strain.

With further load increase, the compressive strains and stresses will increase; however, they will cease to be proportional, and some nonlinear stress curve will result on the compression side of the beam. This stress curve above the neutral axis will be essentially the same shape as the concrete stress–strain curve (see Fig. 1-1). The stress and strain distribution which exists at or near the ultimate load may be observed in Fig. 2-3. Eventually, the ultimate capacity of the beam will be reached and the beam will fail. The actual mechanism of the failure will be discussed later in this chapter.

At this point, the reader may well recognize that the process of attaining the ultimate capacity of a member is quite irreversible. The member has cracked and deflected significantly; the steel has yielded and will not return to its original length. If other members in the structure have similarly reached their ultimate capacitites, the structure itself is probably crumbling and in a state of ruin or semiruin, even though it may not have completely collapsed. Naturally, although we cannot

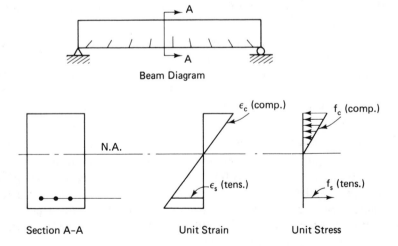

FIGURE 2-2 Flexural behavior at moderate loads

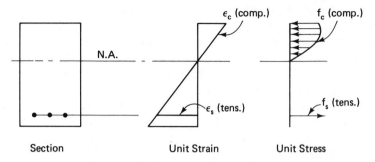

Section   Unit Strain   Unit Stress

FIGURE 2-3   Flexural behavior near ultimate load

ensure that this state will never be reached, factors are introduced to create the commonly accepted margins of safety. Nevertheless, the ultimate capacities of members are, at present, the basis for reinforced concrete analysis and design.

In this text, it is in such a context that we will speak of failures of members.

2–4 STRENGTH DESIGN METHOD ASSUMPTIONS

The development of the strength design approach depends on the following basic assumptions:

1. A plane section before bending remains a plane section after bending. That is, the strain throughout the depth of the member varies linearly from zero at the neutral axis. Tests have shown this assumption to be essentially correct.

2. Stresses and strains are approximately proportional only up to moderate loads (assuming that the concrete stress does not exceed approximately $f'_c/2$). When the load is increased and approaches an ultimate load, stresses and strains are no longer proportional. Hence the variation in concrete stress is no longer linear.

3. In calculating the ultimate moment capacity of a beam, the tensile strength of the concrete is neglected.

4. The maximum usable concrete compressive strain at the extreme fiber will be assumed equal to 0.003. This value is based on extensive testing which indicated that the flexural concrete strain at failure for rectangular beams generally ranges from 0.003 to 0.004 in./in. Hence the assumption that the concrete is about to crush when the maximum strain reaches 0.003 is slightly conservative.

5. The steel is assumed to be uniformly strained to that strain which exists at the level of the centroid of the steel. Further, if the strain in the steel (ϵ_s) is less than the yield strain of the steel (ϵ_y), the stress in the steel will be $E_s\epsilon_s$. This assumes that for stresses less than f_y, the steel stress is proportional to strain. For strains equal to or greater than ϵ_y, the stress in the reinforcement will be considered independent of strain and equal to f_y.

6. The bond between the steel and concrete is perfect and no slip occurs.

Assumptions 4 and 5 constitute what may be termed "code criteria" with respect to failure. The true ultimate strength of a member will be somewhat greater than that computed using these assumptions. However, the strength method of design and analysis of the ACI Code is based on these criteria and consequently is our basis for bending member design and analysis.

2-5 FLEXURAL STRENGTH OF RECTANGULAR BEAMS

Based on the assumptions previously stated, we can now examine the strains, stresses, and forces that exist in a reinforced concrete beam which is subjected to its *ultimate moment*, that is, the moment that exists just prior to the failure of the beam. (This moment may also be called the beam's *ultimate flexural strength*.) In Fig. 2-4 the assumed beam has a width b, an effective depth d, and is reinforced with a steel area of A_s. (A_s is the total *cross-sectional area* of tension steel present.)

Based on the preceding assumptions, it is possible that the beam may be loaded to the point where the maximum concrete compressive bending strain equals 0.003 in./in. and the tensile steel unit stress equals its yield stress (as a limit). It is also possible that the steel strain that occurs when the concrete strain reaches 0.003 in./in. may be in excess of the steel yield strain ϵ_y. When this occurs, it implies a specific mode of failure which will be discussed further in a later section.

As previously stated, the compressive stress distribution above the neutral axis for a flexural member is similar to the concrete compressive stress–strain curve as depicted in Fig. 1-1. As may be observed in Fig. 2-4, the ultimate compressive stress f'_c does not occur at the outer fiber. Neither is the shape of the curve the same for different-strength concretes. Actually, the magnitudes of the compressive concrete stresses are defined by some irregular curve, which could vary not only from concrete to concrete but also from beam to beam. Present theories

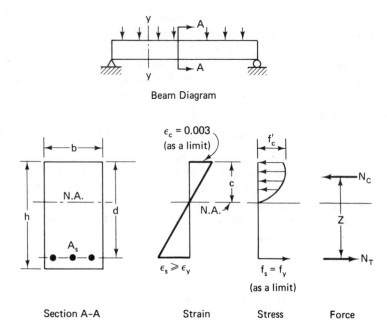

FIGURE 2-4 Beam subjected to ultimate moment

accept the fact that at ultimate moment, concrete stresses and strains are no longer proportional above the neutral axis but vary from zero at the neutral axis to a maximum at or near the extreme outside fiber.

The flexural strength or resisting moment of a rectangular beam is created by the development of these internal stresses, which in turn may be represented as internal forces. As may be observed in Fig. 2-4, N_C represents a theoretical internal resultant compressive force, which in effect constitutes the total internal compression above the neutral axis. N_T represents a theoretical internal resultant tensile force, which in effect constitutes the total internal tension below the neutral axis.

These two forces, which are parallel, equal and opposite, and separated by a distance Z, constitute an internal resisting couple whose maximum value may be termed the flexural strength or *resisting moment* of a bending member.

As a limit, this internal resisting moment must be capable of resisting the actual design bending moment induced by the applied loads. Consequently, if we wish to design a beam for a prescribed loading condition, we must arrange its concrete dimensions and the steel reinforcements so that it is capable of developing a resisting moment at least equal to the maximum bending moment induced by the loads.

The determination of this internal resisting moment is quite complex because of the shape of the compressive stress diagram above the neutral axis. Not only is N_C difficult to evaluate, but its location relative to the tensile steel is difficult to establish. However, since the internal resisting moment is actually a function of the magnitude of N_C and Z, it is not really necessary to know the exact shape of the compressive stress distribution above the neutral axis. To determine the internal resisting moment, it is only necessary to know (1) the total resultant compressive force N_C in the concrete, and (2) its location from the outer compressive fiber (from which the distance Z may be established).

These two values may easily be established by replacing the unknown complex compressive stress distribution by a fictitious one of simple geometrical shape provided that the fictitious distribution results in the same total compressive force N_C applied at the same location as in the actual distribution when it is at the point of failure.

2–6 EQUIVALENT STRESS DISTRIBUTION

For purposes of simplification and practical application, a fictitious but equivalent rectangular concrete stress distribution was proposed by Whitney[1] and subsequently adopted by the ACI Code (Sections 10.2.6 and 10.2.7). The ACI Code also stipulates that other compressive stress distribution shapes may be used provided that the results are in substantial agreement with comprehensive test results. However, because of the simplicity of the rectangular shape, it has become the more widely used fictitious stress distribution for design purposes.

With respect to this equivalent stress distribution as shown in Fig. 2-5, the average stress intensity is taken as $0.85f'_c$ and is assumed to act over the upper area of the beam cross section defined by the width b and a depth of a. The magnitude of a may be determined by

$$a = \beta_1 c$$

where c = distance from the outer compressive fiber to the neutral axis

β_1 = a constant which is a function of the strength of the concrete. The value of β_1 is taken as 0.85 for $f'_c \leq 4,000$ psi and is reduced continuously at a rate of 0.05 for each 1,000 psi of strength in excess of 4,000 psi; β_1 must not be taken less than 0.65

It is in no way maintained that the compressive stresses are actually distributed in this most unlikely manner. It is maintained, however, that

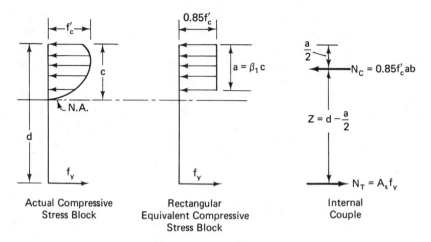

FIGURE 2-5 Equivalent stress block for strength design/analysis

this equivalent rectangular distribution gives results close to those of the complex actual stress distribution. An isometric view of the accepted internal relationships may be observed in Fig. 2-6.

Using the equivalent stress distribution in combination with the strength design assumptions, we may now determine the ideal flexural strength M_n of rectangular reinforced concrete beams which are reinforced for tension only.

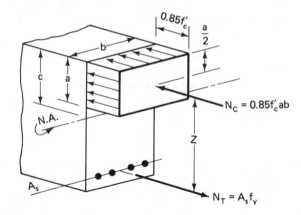

FIGURE 2-6 Equivalent stress block for strength design/analysis

EXAMPLE 2–1

Determine M_n for a beam of cross section shown in Fig. 2-7. $f'_c = 4,000$ psi. Assume A615 grade 60 steel.

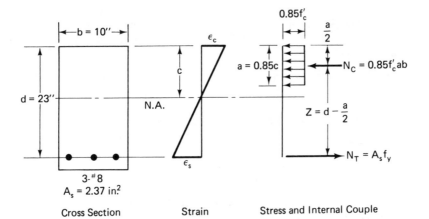

3-#8
A_s = 2.37 in.²

Cross Section   Strain   Stress and Internal Couple

FIGURE 2-7 Sketch for Example 2-1

1. We will assume that f_y exists in the steel, subject to later check. By $\Sigma H = 0$,

$$N_C = N_T$$

$$(0.85f'_c)ab = A_s f_y$$

$$a = \frac{A_s f_y}{0.85f'_c b} = \frac{2.37(60)}{0.85(4)(10)} = 4.18 \text{ in.}$$

This is the depth of the stress block that must exist if there is to be horizontal equilibrium.

2. Calculate M_n

$$M_n = N_C(d - \frac{a}{2}) \quad \text{or} \quad N_T(d - \frac{a}{2})$$

Based on the concrete,

$$M_n = (0.85f'_c)ab(d - \frac{a}{2})$$

$$= 0.85(4.0)(4.18)(10)\left(23 - \frac{4.18}{2}\right)$$

$$= 2,970 \text{ in.-kips}$$

$$\frac{2,970}{12} = 247 \text{ ft-kips}$$

or, based on the steel,

$$M_n = A_s f_y(d - \frac{a}{2})$$

$$= 2.37(60)(20.9)$$

$$= 2,970 \text{ in.-kips}$$

27

3. In the foregoing computations, the assumption was made that the steel reached its yield strain (and therefore its yield stress) before the concrete reached its "ultimate" (by definition) strain of 0.003. This assumption will now be checked.

Referring to Fig. 2-8, we may locate the neutral axis as follows:

$$a = \beta_1 c \quad \text{(ACI Code, Section 10.2.7)}$$

$$\beta_1 = 0.85 \text{ since } f_c' = 4,000 \text{ psi}$$

Therefore,

$$c = \frac{a}{0.85} = \frac{4.18}{0.85} = 4.92 \text{ in.}$$

By similar triangles in the strain diagram, we may find the strain in the steel when the concrete strain is 0.003:

$$\frac{0.003}{c} = \frac{\epsilon_s}{d - c}$$

Then

$$\epsilon_s = \frac{d - c}{c}(0.003) = \frac{23 - 4.92}{4.92}(0.003)$$

$$= 0.011 \text{ in./in.}$$

The strain at which the steel yields (ϵ_y) may be determined from the basic definition of modulus of elasticity, $E = \text{stress/strain}$.

$$\epsilon_y = \frac{f_y}{E_s} = \frac{60,000}{29,000,000} = 0.00207 \text{ in./in.} \quad \text{(see Table A-1)}$$

This represents the strain in the steel when the stress first reaches 60,000 psi.

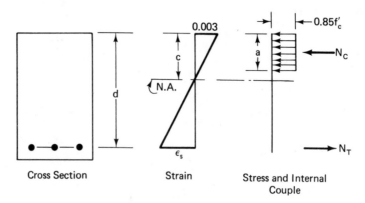

Cross Section   Strain   Stress and Internal Couple

FIGURE 2-8   Steel strain check

Since the actual strain in the steel (0.011) is greater than the yield strain (0.00207), the steel reaches its yield stress before the concrete reaches its strain of 0.003 and the assumption that the stress in the steel is equal to the yield stress was correct. (See Assumption No. 5 in Section 2–4.)

2–7 BALANCED, OVERREINFORCED, AND UNDERREINFORCED BEAMS

In Fig. 2-8, the *ratio* between the steel strain and the maximum concrete strain is fixed once the neutral axis is established. The location of the neutral axis will vary based on the amount of tension steel in the cross section, since the stress block is just deep enough to ensure that the resultant compressive force is equal to the resultant tensile force ($\Sigma H = 0$). If more tension bars are added to the bottom of a reinforced concrete cross section, the depth of the compressive stress block will be greater, and therefore the neutral axis will be lower. Referring again to Fig. 2-8, if there were just enough steel to put the neutral axis at a location where the yield strain in the steel and the maximum concrete strain of 0.003 existed at the same time, the cross section would be said to be *balanced* (see the ACI Code, Section 10.3.2). The amount of steel required to create this condition is large, and balanced beams are not generally economical. However, the balanced condition is the dividing line between two distinct types of reinforced concrete beams which are characterized by their failure modes. If a beam has more steel than is required to create the balanced condition, the beam is said to be *overreinforced*. The additional steel will cause the neutral axis to be low (see Fig. 2-9). This will, in turn, cause the concrete to reach a strain of 0.003 *before* the steel yields. Should more moment (and therefore strain) be applied to the beam cross section, failure will be initiated by a sudden crushing of the concrete.

If a beam has less steel than is required to create the balanced condition, it is said to be *under reinforced*. The neutral axis will be higher than the balanced neutral axis (refer to Fig. 2-9), and the steel will reach its yield strain (and therefore its yield stress) before the concrete reaches a strain of 0.003. A slight additional load will cause the steel to stretch a considerable amount. The strains in the concrete and the steel continue to increase. However, the tensile force *is not* increasing. Since the compressive force cannot increase ($\Sigma H = 0$) and since the concrete *stress is* increasing, the area under compression must be decreasing and the neutral axis must rise. This process continues until the reduced area fails in compression as a secondary effect. This failure due to yielding is a gradual one, with the beam showing greatly increased deflection after

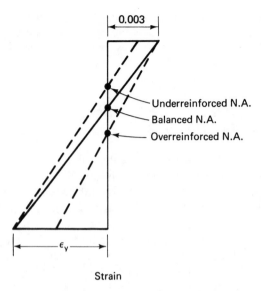

0.003

Underreinforced N.A.
Balanced N.A.
Overreinforced N.A.

ϵ_y

Strain

FIGURE 2-9 Variation in neutral-axis locations

"running time"

the steel reaches the yield point; hence there is adequate warning of impending failure. The steel, being ductile, will not actually pull apart even at failure of the beam.

2–8 REINFORCEMENT RATIO LIMITATIONS AND GUIDELINES

While failure due to yielding of the steel is a gradual one, with adequate warning of collapse, failure due to crushing of the concrete is sudden and without warning. The first type of failure is preferred, and it is ensured by restrictions of the ACI Code. The Code stipulates, in Section 10.3.3, that the amount of tensile steel must not exceed 0.75 times the amount of steel that would produce balanced conditions. This is an attempt to assure a ductile failure which is produced by yielding of the steel as compared to the brittle type of failure which occurs when the failure is in compression. This may be expressed as $A_s \leq 0.75 A_{sb}$.

The code puts this stipulation in terms of a *reinforcement ratio*— sometimes called the *steel ratio*. This is a ratio of A_s to the product of beam width b and effective depth d and is denoted by the Greek lowercase rho (ρ):

$$\rho = \frac{A_s}{bd}$$

30

The maximum permissible reinforcement ratio, according to the Code, is 0.75 times that steel ratio which would produce the balanced condition (ρ_b). This may be expressed as:

$$\rho_{max} = 0.75\,\rho_b$$

The use of the reinforcement ratio concept will be convenient in later calculations.

EXAMPLE 2–2

In Example 2-1, determine the amount of steel required to create the balanced condition. $d = 23$ in., $b = 10$ in. From Table A-1; $\epsilon_y = 0.00207$ in./in. By definition of the balanced condition, the strain diagram must be as shown in Fig. 2-10.

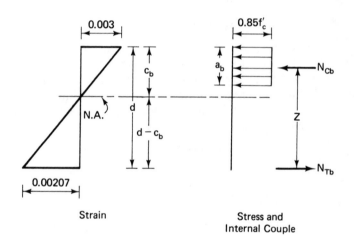

FIGURE 2-10   Sketch for Example 2–2

1.
$$\frac{c_b}{0.003} = \frac{d - c_b}{0.00207}$$

$$0.00207\,c_b = 0.003\,(23 - c_b)$$

$$0.00207\,c_b + 0.003c_b = 0.069$$

$$c_b = \frac{0.069}{0.00507} = 13.6 \text{ in.}$$

2.
$$a_b = \beta_1 c_b = 0.85\,(13.6) = 11.6 \text{ in.}$$

3.
$$N_{Cb} = (0.85f'_c)\,a_b b = 0.85\,(4.0)\,(11.6)\,(10)$$

$$N_{Cb} = 393 \text{kips}$$

$$N_{Tb} = A_{sb} f_y$$

$$N_{Tb} = N_{Cb}$$

4. Therefore,

$$\text{required } A_{sb} = \frac{N_{Tb}}{f_y} = \frac{N_{Cb}}{f_y}$$

$$= \frac{393}{60} = 6.56 \text{ in.}^2$$

Comparing the balanced area of steel required with the actual amount of steel present (3 No. 8 bars = 2.37 in.²), we see that this is an *underreinforced* beam which is controlled by the steel yielding.

Check the code requirement for a ductile-type beam (ACI Code, Section 10.3.3):

$$0.75 A_{sb} = 0.75 (6.56 \text{ in.}^2) = 4.92 \text{ in.}^2 > 2.37 \text{ in.}^2 \qquad \text{(OK)}$$

It may be noted again, at this point, that heavily reinforced beams are less efficient than their more lightly reinforced counterparts. One structural reason for this is that, for a given beam size, an increase in A_s is accompanied by a decrease in the lever arm of the internal couple ($Z = d - a/2$). This may be illustrated by doubling the tension steel for the beam of Example 2-1 and recalculating M_n:

$$A_s = 2(2.37) = 4.74 \text{ in.}^2 \qquad (100\% \text{ increase})$$

$$a = \frac{4.74 (60)}{0.85 (4)(10)} = 8.36 \text{ in.}$$

$$M_n = 0.85 (4)(8.36)(10)\left(23 - \frac{8.36}{2}\right) = 446 \text{ ft-kips}$$

For the beam of Example 2-1, M_n was 247 ft.-kips; therefore

$$\frac{446 - 247}{247} \times 100 = 80\% \text{ increase}$$

Instead of utilizing the preceding laborious technique for determining the balanced steel, a general expression for that reinforcement ratio which would produce the balanced condition (ρ_b) may be derived as shown in Fig. 2–11.

The neutral axis for the *balanced condition* is located utilizing similar triangles.

$$\frac{c_b}{0.003} = \frac{d}{0.003 + \dfrac{f_y}{E_s}}$$

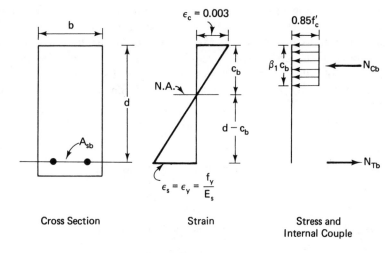

| Cross Section | Strain | Stress and Internal Couple |

FIGURE 2-11 The balanced condition

Reorganizing and substituting, with $E_s = 29{,}000{,}000$ psi, yields

$$c_b = \frac{0.003}{0.003 + f_y/29{,}000{,}000}(d)$$

$$= \frac{87{,}000}{87{,}000 + f_y}(d) \qquad (2\text{-}1)$$

And, since $N_{Cb} = N_{Tb}$ ($\Sigma H = 0$),

$$(0.85f_c')\beta_1 ab = A_{sb}f_y$$

$$c_b = \frac{A_{sb}f_y}{(0.85f_c')\,\beta_1 b}$$

Since $A_{sb} = \rho_b bd$,

$$c_b = \frac{\rho_b bd f_y}{(0.85f_c')\,\beta_1 b} = \frac{\rho_b d f_y}{(0.85f_c')\beta_1} \qquad (2\text{-}2)$$

Equations (2-1) and (2-2) may now be equated and solved for ρ_b.

$$\rho_b = \frac{0.85f_c'\,\beta_1}{f_y}\left(\frac{87{,}000}{f_y + 87{,}000}\right) \qquad (2\text{-}3)$$

Note that in Eq. (2-3), units are pounds and inches.

ρ_b can easily be calculated for different combinations of f_c' and f_y and subsequently tabulated. However, the upper limit for the reinforcement ratio set by the ACI Code is $0.75\rho_b$. Table A-4 gives values of $0.75\rho_b$ for various combinations of steel yield stress and concrete strength. This table may be used for quick reference to determine if a beam is acceptable under the present ACI Code with respect to ductility.

Hence the concept of a balanced section serves as a datum or reference as to whether a design or analysis conforms to the specific mode of failure as prescribed by the ACI Code. Where the area of steel is in excess of the area of steel required for the balanced condition, a brittle form of failure will occur. Where the area of steel is less than the area of steel required for the balanced condition, a ductile mode of failure will occur.

EXAMPLE 2–3

Re-check Example 2-1 for ductility using $0.75\rho_b$ as a limit.

$$\rho = \frac{A_s}{bd} = \frac{2.37}{10\,(23)} = 0.0103$$

From Table A-4, for $f_y = 60,000$ psi and $f'_c = 4,000$ psi,

$$\rho_{max} = 0.75\rho_b = 0.0214$$

$$0.0103 < 0.0214 \qquad\qquad\qquad \text{(OK)}$$

The Code requirement may also be expressed as $A_{s(max)} = 0.75A_{sb}$. A_{sb} was calculated in Example 2-2.

$$A_{s\,(max)} = 0.75\,(6.56) = 4.92 \text{ in.}^2 > 2.37 \text{ in.}^2 \qquad\qquad \text{(OK)}$$

The ACI Code, Section 10.5.1, also places a *lower limit* on the reinforcement ratio:

$$\rho_{min} = \frac{200}{f_y}$$

This lower limit guards against sudden failure by essentially ensuring that a beam with a very small amount of tensile reinforcement has a greater moment strength as a reinforced concrete section than that of the corresponding plain concrete section computed from its modulus of rupture. This limitation does not apply to slabs of uniform thickness. Minimum reinforcement in slabs is governed by the required shrinkage and temperature steel as outlined in the ACI Code, Section 7.12.

2–9 STRENGTH REQUIREMENTS

The basic criterion for strength design may be expressed as follows:

Strength furnished ⩾ strength required

All members and all sections of members must be proportioned to meet this criterion.

The required strength may be expressed in terms of design loads or their related moments, shears, and forces. Design loads may be defined as service loads multiplied by the appropriate load factors. (When the word "design" is used as an adjective throughout the ACI Code, it indicates that load factors are included.) The subscript u will be used to indicate design loads, moments, shears and forces.

For example, the *design load* $w_u = 1.4 w_{DL} + 1.7 w_{LL}$ and the required or *design moment* strength for dead and live load is $M_u = 1.4 M_{DL} + 1.7 M_{LL}$. This is, in effect, a safety provision. The load factors attempt to assess the possibility that prescribed service loads may be exceeded. Obviously, a specified live load is more apt to be exceeded than a dead load, which is largely fixed by the weight of the structure. With regard to the weight of the structure, recall that the unit weight of reinforced concrete is taken as 150 lb/ft³. The ultimate strength of the members must take into account the total of all service loads, each multiplied by its respective load factor. Load factors are different in magnitude for dead load, live load, wind, or earthquake loading.

For dead and live loads, the ACI Code, Section 9.2, specifies that design loads, design shears, and design moments be obtained from service loads by using the relation

$$U = 1.4D + 1.7L \qquad \text{ACI} \quad \text{Eq. (9-1)}$$

where U is any of the design strengths, D the service dead load, and L the service live load. Load factors for other types of loading, as well as combinations of loads, may be obtained from Section 9.2 of the ACI Code.

A second safety provision provided in the ACI Code, Section 9.3, is the reduction of the theoretical capacity of a structural element by a *strength reduction factor* ϕ.

The ϕ *factor* provides for the possibility that small adverse variations in material strengths, workmanship, and dimensions, although within acceptable tolerance and limits of good practice, may combine to result in undercapacity. In effect, the ϕ factor, when multiplied by the theoretical ideal strength of a member, will furnish us with a practical and dependable strength which is obviously less than the ideal strength.

In assigning strength reduction factors, the degree of ductility and the importance of the member are considered as well as the degree of accuracy with which the strength of the member can be established. The ACI Code, Section 9.3, provides for these variables by using the following ϕ factors:

Bending	= 0.90
Shear and torsion	= 0.85
Compression members (spirally reinforced)	= 0.75
Compression members (tied)	= 0.70
Bearing on concrete	= 0.70
Bending in plain concrete	= 0.65

Hence we can state that the *usable* moment strength (or capacity) M_R will equal the *ideal* moment strength M_n multiplied by the ϕ factor:

$$M_R = \phi M_n$$

2-10 RECTANGULAR BEAM ANALYSIS FOR MOMENT (TENSION REINFORCEMENT ONLY)

The *flexural analysis problem* is characterized by knowing precisely what comprises the cross section of a beam. That is, the following things are *known*: tension bar size and number (or A_s), beam width (b), effective depth (d) or total depth (h), f'_c, and f_y. To be found, basically, is the beam strength, although this may be manifested in various ways: find M_R, or check adequacy of the given beam, or find an allowable load that the beam can carry. The *flexural design problem*, on the other hand, requires the determination of one or more of the dimensions of the cross section and/or the determination of the main tension steel to use. It will be important to recognize the differences between these two types of problems since the methods of solution are different.

EXAMPLE 2–4

Determine if the beam shown in Fig. 2–12 is adequate as governed by the ACI Code (318–77).

Service loads:

Uniform dead load = 0.8 kip/ft   (excludes beam weight)

Uniform live load = 0.8 kip/ft

Concentrated live load = 12 kips   (located at midspan)

Assume $f'_c = 4,000$ psi and A615 grade 60 billet steel ($f_y = 60,000$ psi).

A logical approach to this type of problem is to compare the practical moment strength (M_R) with the applied design moment resulting from the *factored* loads. This latter moment will be noted M_u. If the beam is adequate for the moment, $M_R \geqslant M_u$. The procedure outlined here for M_R is summarized in Section 2–11.

a. Determination of M_R

1. Given:

$$f_y = 60,000 \text{ psi}$$
$$b = 12 \text{ in.}$$
$$d = 17.5 \text{ in.}$$
$$A_s = 4.00 \text{ in.}^2 \text{ (Table A–2)}$$
$$f'_c = 4,000 \text{ psi}$$

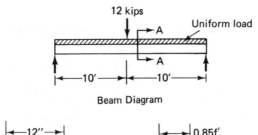

Beam Diagram

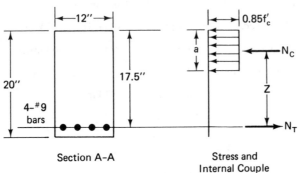

Section A-A

Stress and
Internal Couple

FIGURE 2-12   Sketch for Example 2-4

2. To be found: M_R.

3.
$$\rho = \frac{A_s}{bd} = \frac{4.00}{12(17.5)} = 0.0190$$

4. From Table A–4, $0.75\rho_b = 0.0214$. $0.0190 < 0.0214$. Therefore, failure by yielding of steel is assured.

$$\rho_{min} = \frac{200}{f_y} = \frac{200}{60,000} = 0.0033$$

$$0.0191 > 0.0033 \qquad\qquad \text{(OK)}$$

5.
$$a = \frac{A_s f_y}{(0.85 f_c')b}$$

$$a = \frac{4.00(60)}{0.85(4)(12)} = 5.88 \text{ in.}$$

6.
$$Z = d - \frac{a}{2} = 17.5 - \frac{5.88}{2} = 14.6 \text{ in.}$$

7. Based on steel: $M_n = A_s f_y Z$

$$= 4.00(60)(14.6) = 3,490 \text{ in.-kips}$$

$$\frac{3,490}{12} = 291 \text{ ft-kips}$$

8. $M_R = \phi M_n = 0.9(291) = 262$ ft-kips

b. Find M_u

Since the given service dead load excluded the beam weight, it will now be calculated:

$$\text{Beam weight} = \frac{20(12)}{144}(0.150) = 0.25 \text{ kip/ft}$$

$$\text{Given applied dead load} = 0.80 \text{ kip/ft}$$

$$\text{Total service uniform dead load} = 1.05 \text{ kips/ft}$$

$$\text{Factored uniform dead load} = 1.05(1.4) = 1.47 \text{ kips/ft}$$

$$\text{Factored uniform live load} = 0.8(1.7) = 1.36 \text{ kips/ft}$$

$$\text{Factored concentrated live load} = 12(1.7) = 20.4 \text{ kips}$$

$$w_u = 1.47 + 1.36 = 2.83 \text{ kips/ft}$$

$$P_u = 20.4 \text{ kips}$$

$$M_u = \frac{w_u \ell^2}{8} + \frac{P_u \ell}{4}$$

$$= \frac{2.83(20)^2}{8} + \frac{20.4(20)}{4}$$

$$= 243.5 \text{ ft-kips} < 262 \text{ ft-kips}$$

Therefore, the beam is adequate.

The present ACI Code (1977) has placed the upper limit of $0.75\rho_b$ on the reinforcement ratio. The older working stress design codes did not restrict ρ in this manner, and therefore overreinforced beams do exist in structures. Although these are not acceptable within the philosophy of the 1977 Code, they have been usable and their moment capacity can logically be found by using no more steel than that represented by $0.75\rho_b$. Essentially, this approach ignores the strength of any steel that is in excess of 75% of the amount of steel which would produce the balanced condition.

EXAMPLE 2–5

Find M_R for the beam of cross section shown in Fig. 2–13. Steel is billet steel, A615 grade 40. Concrete strength is 3,000 psi.

1. Given:

$$f_y = 40,000 \text{ psi}$$

$$f'_c = 3,000 \text{ psi} \qquad d = 20 \text{ in.}$$

$$b = 11 \text{ in.} \qquad A_s = 7.62 \text{ in.}^2$$

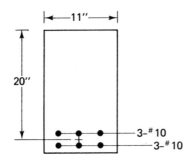

FIGURE 2-13 Sketch for Example 2–5

2. To be found: M_R.

3.
$$\rho = \frac{A_s}{bd} = \frac{7.62}{11(20)} = 0.0346$$

4. $0.75\rho_b = 0.0278 = \rho_{max}$

  Since $\rho > 0.75\rho_b$, we will use $\rho = 0.0278$ as a maximum. Therefore,

$$\text{Effective } A_s = \rho bd = 0.0278(11)(20)$$
$$= 6.12 \text{ in.}^2$$

5.
$$a = \frac{A_s f_y}{0.85 f'_c\ b} = \frac{6.12(40)}{0.85(3)(11)} = 8.73 \text{ in.}$$

6.
$$Z = d - \frac{a}{2} = 20 - \frac{8.73}{2} = 15.6 \text{ in.}$$

7. $M_n = A_s f_y Z = 6.12(40)(15.6) = 3,820 \text{ in.-kips}$

8.
$$M_R = \phi M_n = (0.9)\frac{3,820}{12} = 286 \text{ ft-kips}$$

2–11 SUMMARY OF PROCEDURE FOR RECTANGULAR BEAM ANALYSIS FOR M_R (TENSION REINFORCEMENT ONLY)

1. List the known quantities. Use a sketch.

2. Determine what is to be found. (An "analysis" may require any of the following to be found: M_R, M_n, allowable service live load or dead load.)

3. Calculate the reinforcement ratio.

$$\rho = \frac{A_s}{bd}$$

39

4. Compare actual ρ with $0.75\rho_b$ and ρ_{min} to determine if the beam cross section is acceptable.

5. Calculate the depth of the (rectangular) compressive stress block.

$$a = \frac{A_s f_y}{(0.85 f_c')b}$$

6. Calculate the internal couple lever arm length.

$$Z = d - \frac{a}{2}$$

7. Calculate the ideal resisting moment M_n.

$$M_n = N_t Z = A_s f_y Z$$

or

$$M_n = N_c Z = 0.85 f_c' \, abZ$$

8. $M_R = \phi M_n$

A flow diagram of this procedure is presented in Appendix B.

2-12 SLABS—INTRODUCTION

Slabs constitute a specialized category of bending members and are utilized in both structural steel and reinforced concrete structures. Probably the most basic and common type of slab is termed a *one-way slab*. A one-way slab may be described as a structural reinforced concrete slab supported on two opposite sides so that the bending occurs in one direction only, that is, perpendicular to the supported edges.

If a slab is supported along all four edges, it may be designated as a *two-way slab*, with bending occurring in two directions perpendicular to each other. However, if the ratio of the lengths of the two perpendicular sides is in excess of 2, the slab may be assumed to act as a one-way slab with bending primarily occurring in the short direction.

A specific type of two-way slab is categorized as a *flat slab*. A flat slab was defined by the 1963 ACI Building Code (ACI 318–63) as *a concrete slab reinforced in two or more directions, generally without beams or girders to transfer the loads to supporting members*. The slab, then, could be considered to be supported on a grid of shallow beams which are themselves integral with, and have the same depth as, the slab. The columns tend to punch upward through the slab, resulting in a high shearing stress along with inclined slab cracking. (This "punching shear" is considered later in our discussion on column footings.) Thus it is common to both thicken the slab in the vicinity of the column, utilizing

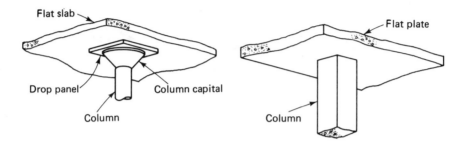

FIGURE 2-14   Reinforced concrete slabs

a *drop panel*, and at the same time enlarge the top of the column in the shape of an inverted frustrum called a *column capital* (see Fig. 2–14). Another type of two-way slab is a *flat plate*. This is similar to the flat slab, without the drop panels and column capitals; hence it is a slab of constant thickness supported directly on columns. Generally, the flat plate is utilized where spans are smaller and loads lighter than those requiring a flat slab design (see Fig. 2–14).

2–13 ONE-WAY SLABS—ANALYSIS FOR MOMENT

In our discussion of slabs, we will be primarily concerned with one-way slabs. Such a slab is assumed to be a rectangular beam with a width $b = 12$ in. When loaded with a uniformly distributed load, the slab deflects so that it has curvature, and therefore bending moment, in only one direction. Hence the slab is analyzed and designed as though it were composed of 12-in.-wide segments placed side by side with a total beam depth equal to the slab thickness. With a width of 12 in. the uniformly distributed load, generally specified in pounds per square foot (psf) for buildings, automatically becomes the load per linear foot (lb/ft) for the design of the slab (see Fig. 2–15).

In the one-way slab the main reinforcement for bending is placed perpendicular to the supports. Since analysis and design will be done for a typical 12-in.-wide segment, it will be necessary to specify the amount of steel in that segment. Reinforcing steel in slabs is normally specified by bar size and center-to-center spacing, and the amount of steel considered to exist in the 12-in.-wide typical segment is an *average amount*. Table A-3 is provided to facilitate this determination. For example, if a slab is reinforced with No. 7 bars spaced 15 in. apart (center to center), a typical 12-in.-wide segment of the slab would contain an average steel area of 0.48 in.^{2}. This is sometimes denoted 0.48 in.^{2}/ft.

In addition, the ACI Code stipulates that reinforcement for shrinkage

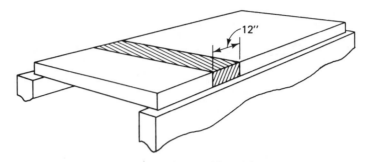

FIGURE 2-15 One-way slab

and temperature stresses normal to the principal reinforcement must be provided in structural floor and roof slabs where the principal reinforcement extends in one direction only. The ACI Code further states that for grade 40 or 50 deformed bars, the minimum area of such steel must be $A_s = 0.0020bh$, and for grade 60 deformed bars the minimum area must be $A_s = 0.0018bh$, where b = width of member (12 in. for slabs) and h = total slab thickness. The ACI Code also stipulates that in structural slabs of uniform thickness, the minimum amount of reinforcement in the direction of the span (principal reinforcement) must not be less than that required for shrinkage and temperature reinforcement (ACI Code, Section 10.5.3).

The principal reinforcement may not be spaced farther apart than three times the slab thickness nor more than 18 inches. Shrinkage and temperature reinforcement may not be spaced farther apart than five times the slab thickness nor more than 18 inches.

The required thickness of a one-way slab may depend on the bending, deflection, and/or shear strength requirements. The ACI Code imposes span/depth criteria in an effort to prevent excessive deflection, which might adversely affect the strength or performance of the structure at service loads. Table 9.5(a) of the ACI Code establishes minimum thicknesses for beams and one-way slabs in terms of fractions of the span length. These may be used for members not supporting or attached to construction likely to be damaged by large deflections. If the member supports such construction, deflections must be calculated. For members of lesser thickness than that indicated in the table, deflection should be computed and if satisfactory the member may be used. Further, the tabular values are for use with nonprestressed reinforced concrete members made with normal-weight concrete and grade 60 reinforcement. If a different grade of reinforcement is used, the tabular values must be multiplied by a factor equal to

$$0.4 + \frac{f_y}{100,000}$$

As an example, for simply supported solid one-way slabs of normal-weight concrete and grade 60 steel, the minimum thickness required when deflections are not computed equals $\ell/20$, where ℓ is the span length of the slab. Deflections receive further treatment in Chapter 7.

According to the ACI Code, Section 7.7.1, the concrete protection for reinforcement in slabs must be not less than ¾ in. for surfaces not exposed directly to the weather or in contact with the ground. This is applicable for No. 11 and smaller bars. For surfaces exposed to the weather or in contact with the ground, the minimum concrete protection for reinforcement is 2 in. for No. 6 through No. 18 bars and 1½ in. for No. 5 and smaller bars. If a slab is cast against and permanently exposed to the ground, the minimum concrete cover for all reinforcement is 3 in.

EXAMPLE 2–6

A one-way structural interior slab having a cross section shown is to span 12 ft. The steel is A615 grade 40. The concrete strength is 3,000 psi and the cover is ¾ in. Determine the service live load (psf) which the slab can support.

From Fig. 2-16, the bars in this slab are No. 5 bars spaced 7 in. apart. This is sometimes denoted "7 in. o.c." meaning 7 in. center-to-center distance. They are perpendicular to the supports.

1. Given: $A_s = 0.53$ in.^{2}/ft (Table A–3)

$f'_c = 3,000$ psi, $f_y = 40,000$ psi

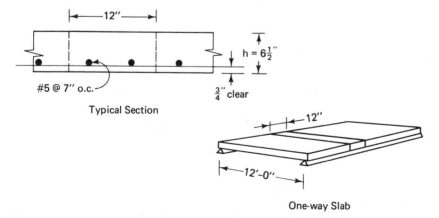

FIGURE 2-16 Sketch for Example 2–6

$$b = 12 \text{ in.}$$

$$d = 6.5 - 0.75 - \frac{0.625}{2} = 5.44 \text{ in.}$$

2. Find: M_R and permissible service live load.

3.
$$\rho = \frac{A_s}{bd} = \frac{0.53}{12(5.44)} = 0.0081$$

4. From Table A-4, $\rho_{max} = 0.75\rho_b = 0.0278$.

$$0.0081 < 0.0278 \tag{OK}$$

The minimum A_s is that required for shrinkage and temperature steel. Therefore, the minimum A_s will be checked in place of a ρ_{min} check.

Minimim $A_s = 0.0020bh = 0.0020(12)(6.5) = 0.16$ in.²/ft

$$0.53 > 0.16 \tag{OK}$$

5.
$$a = \frac{A_s f_y}{(0.85f_c')b} = \frac{0.53(40)}{0.85(3)(12)} = 0.69 \text{ in.}$$

6
$$Z = d - \frac{a}{2} = 5.44 - \frac{0.69}{2} = 5.10 \text{ in.}$$

7. $M_n = A_s f_y Z = 0.53(40)(5.10) = 108$ in.-kips   (per foot of slab width)

8.
$$M_R = 0.9M_n = 0.9(108) = \frac{97.2}{12} = 8.1 \text{ ft-kips}$$

The service live load that the slab can support will be found next. The slab supports dead load (its own weight) and live load (to be found). Find the dead load moment. (The notation M_u is used to denote moment resulting from *factored* applied loads.)

$$M_{u\,(DL)} = \frac{1.4w\ell^2}{8}$$

$$w_{DL} = \text{slab weight} = \frac{6.5(12)}{144}(0.150) = 0.081 \text{ kip/ft}$$

$$M_{u\,(DL)} = \frac{1.4(0.081)(12)^2}{8} = 2.04 \text{ ft-kips}$$

Therefore, the M_R available to resist the live load is

$$8.10 - 2.04 = 6.06 \text{ ft-kips}$$

$$M_{u\,(LL)} = 1.7\frac{w\ell^2}{8} = 6.06 \text{ ft-kips}$$

$$w_{LL} = \frac{8(6.06)}{1.7(12)^2} = 0.198 \text{ kip/ft}$$

Since the segment is 12 in. wide, this is equivalent to 198 psf. It will be noted that the procedure for finding M_R for a one-way slab is identical to that for a beam. See Section 2–11 for a summary of this procedure.

2–14 RECTANGULAR BEAM DESIGN FOR MOMENT (TENSION REINFORCEMENT ONLY)

In the design of rectangular sections for moment, with f'_c and f_y usually prescribed, *three basic quantities* are to be determined: beam width, beam depth, and steel area. It should be recognized that there is a large multitude of combinations of these three quantities that will satisfy the moment strength required in a particular application. Theoretically, a wide, shallow beam may have the same M_R as a narrow, deep beam. It must also be recognized that practical considerations and code restraints will affect the final choices of these quantities. There is no easy way to determine the *best* cross section, since economy depends on much more than simply the volume of concrete and amount of steel in a beam.

Utilizing the internal relationships previously developed for beam analysis, modifications may be made whereby the design process may be simplified. We have previously developed the analysis expression for ultimate strength of a rectangular beam with tension reinforcement only:

$$M_R = \phi N_C Z = \phi N_T Z$$

and

$$M_R = \phi(0.85 f'_c) ba\left(d - \frac{a}{2} \right)$$

where

$$a = \frac{A_s f_y}{(0.85 f'_c) b}$$

The use of these formulas will now be simplified through the development of certain constants which will subsequently be tabulated.

$$\rho = \frac{A_s}{bd} \qquad \text{therefore, } A_s = \rho b d$$

$$a = \frac{A_s f_y}{(0.85 f'_c) b} = \frac{\rho b d f_y}{(0.85 f'_c) b} = \frac{\rho d f_y}{0.85 f'_c}$$

Arbitrarily define ω:

$$\omega = \rho \frac{f_y}{f'_c}$$

Then

$$a = \frac{\omega d}{0.85}$$

Substitute into the M_R expression:

$$M_R = \phi(0.85 f'_c)(b)\frac{\omega d}{0.85}\left[d - \frac{\omega d}{2(0.85)} \right]$$

Simplify and rearrange:

$$M_R = \phi b d^2 f'_c \omega(1 - 0.59\omega)$$

Arbitrarily define $\bar{k}$:

$$\bar{k} = f'_c \omega(1 - 0.59\omega)$$

The term $\bar{k}$ is sometimes called the *coefficient of resistance*. It varies with ρ, f'_c, and f_y. Tables A–6 through A–10 give the value of $\bar{k}$ in ksi for values of ρ and various combinations of f'_c and f_y. The maximum ρ tabulated in each table is 0.75 ρ_b. The general analysis expression for M_R may now be written as

$$M_R = \phi b d^2 \bar{k}$$

In this form, it may readily be used for design purposes. Since $\bar{k}$ is in units of ksi and b and d are in inches, M_R will come out in in.-kips. If the desired M_R is to be in ft-kips, the expression may be written as

$$M_R = \frac{\phi b d^2}{12} \bar{k}$$

These expressions may also be used to simplify rectangular beam *analysis*. As an example, let us recompute the resisting moment M_R for Example 2–4.

1.
$$\rho = \frac{A_s}{bd} = \frac{4.0}{12(17.5)} = 0.0190$$

2. From Table A–9, $\bar{k} = 0.9489$ (ksi).

$$M_R = \frac{\phi b d^2 \bar{k}}{12} = 0.9(12)\frac{(17.5)^2}{12}(0.9489)$$

$$= 262 \text{ ft-kips}$$

As may be observed, this analysis approach is shorter and simpler than that of Example 2–4 and is equally acceptable. The choice of which approach to use is entirely up to the reader.

Utilizing the newly developed expressions and relationships for design problems, the first example will be one where the cross section (b and h) is known by either practical and/or architectural considerations, leaving the selection of the reinforcing bars as the only unknown.

EXAMPLE 2–7

Design a rectangular beam to carry a service dead load moment of 50 ft-kips (which includes the moment due to the weight of the beam) and a service live load moment of 100 ft-kips. Architectural considerations require the beam width to be 10 in. and the total depth (h) to be 25 in. Use f'_c = 3,000 psi and f_y = 60,000 psi.

Of the three basic quantities to be found, two are specified in this problem and the solution for the required steel area is direct. The procedure outlined here is summarized in Section 2–15.

1. Total design moment:

$$M_u = 1.4 M_{DL} + 1.7 M_{LL}$$

$$= 1.4(50) + 1.7(100)$$

$$= 240 \text{ ft-kips}$$

2. Estimate d to be equal to $h - 3$ in.

$$d = 25 - 3 = 22 \text{ in.}$$

Since $M_R = \phi bd^2 \overline{k}$ and since M_R must $= M_u$ as a lower limit, the expression may be written as $M_u = \phi bd^2 \overline{k}$ and

$$\text{Required } \overline{k} = \frac{M_u}{\phi bd^2} = \frac{240(12)}{0.9(10)(22)^2} = 0.66 \text{ ksi}$$

3. From Table A-7, $\overline{k}$ of 0.6607 ksi will be provided if the steel ratio ρ = 0.0130. Therefore, required ρ = 0.013. From Table A–4 (or A–7),

$$0.75\rho_b = 0.0161 = \rho_{\max}$$

$$\rho_{\min} = \frac{200}{f_y} = 0.0033$$

4.
$$\text{Required } A_s = \rho bd$$

$$= 0.013(10)(22) = 2.86 \text{ in.}^2$$

5. Theoretically, any bar or combination of bars that provides at least 2.86 in.² of steel area will satisfy the design requirements. Preferably, not less than two bars should be used. The bars should be of the same diameter and placed in one layer whenever possible.

The ACI Code specifies minimum clearance and cover requirements for steel. The clear space between bars in a single layer must not be less than:

a. The bar diameter or 1 in. (ACI Code, Section 7.6.1).

b. 1⅓ × maximum aggregate size (ACI Code, Section 3.3.3).

Further, should multiple layers of bars be necessary, a 1-in. minimum clear distance is required between layers (ACI Code, Section 7.6.2).

Cover requirements are stated in the ACI Code, Section 7.7.1, for cast-

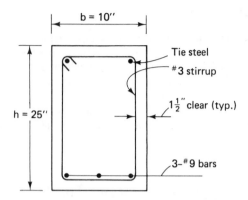

FIGURE 2-17 Design sketch for Example 2–7

in-place concrete. This listing is extensive. However, for beams, girders, and columns not exposed to the weather or in contact with the ground, the minimum concrete cover on any steel is 1½ in.

Table A-11 combines spacing and cover requirements into a tabulation of minimum beam widths for multiples of various bars. The assumptions are stated. It should be noted that No. 3 stirrups are assumed. Stirrups are a special form of reinforcement primarily for shear stress purposes and will be discussed in a later chapter. One type of stirrup may be observed in Fig. 2–17.

If we now consider the bar selection based on the foregoing required A_s (2.86 in.²), the following combinations may be considered:

$$3 \text{ No. 9 bars:} \quad A_s = 3.00 \text{ in.}^2$$
$$4 \text{ No. 8 bars:} \quad A_s = 3.16 \text{ in.}^2$$
$$5 \text{ No. 7 bars:} \quad A_s = 3.00 \text{ in.}^2$$

A review of Table A-11 indicates that the only acceptable combination is 3 No. 9 bars. The other two bar groups will not fit into a 10-in.-wide beam in one layer. The width of beam required for 3 No. 9 bars is 9½ in., which obviously is satisfactory.

At this point one should check the actual effective depth d and compare it with the estimated d.

$$\text{Actual } d = 25 - 1.5 - 0.38 - \frac{1.128}{2} = 22.6 \text{ in.}$$

This is slightly in excess of the estimated d and is therefore conservative (on the safe side). Because of the small difference, no revision is either suggested or required.

6. Generally, concrete dimensions should be to the ½ in. The final design sketch should show the following (see Fig. 2-17):

a. Beam width.

b. Total beam depth.

c. Main reinforcement size and number of bars.

d. Cover on reinforcement.

e. Stirrup size.

A second type of design problem is presented in Example 2–8. This may be categorized as a *free design* because of the *three* unknown variables: beam width, beam depth, and area of reinforcing steel. There are, therefore, a large number of combinations of these variables which will theoretically solve the problem. However, we do have some idea as to the required and/or desired relationships between these unknowns.

We know that the steel ratio ρ must be greater than ρ_{min} and less than ρ_{max}, which, in effect, establishes the range of the acceptable amount of steel. Table A-4 contains recommended values of ρ to use for design purposes. These are recommended maximum values. Try not to use larger ρ values. If larger values are used, a smaller concrete section will result with potential deflection problems.

These ρ values are based on the 1963 ACI Code (Part II), which stipulated that

$$\text{where } \rho > \frac{0.18f'_c}{f_y}, \text{ deflection must be checked}$$

$$\text{where } \rho < \frac{0.18f'_c}{f_y}, \text{ deflections need not be checked}$$

Experience and judgment developed over the years have also established a range of acceptable and economical depth/width ratios for rectangular beams. The extremes of the d/b ratio should be

$$1.0 \leqslant \frac{d}{b} \leqslant 3.0$$

However, the desirable d/b ratio preferably should be between 1.5 and 2.2. These relationships furnish us with an approach to the problem.

EXAMPLE 2–8

Design a simply supported rectangular reinforced concrete beam with tension steel only, to carry a service dead load of 0.9 kip/ft and a service live load of 2.0 kips/ft. (The dead load does not include the weight of the beam.) The span is 18 ft. Assume No. 3 stirrups. Use $f'_c = 3,000$ psi and $f_y = 40,000$ psi. See Fig. 2-18.

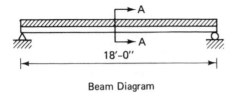

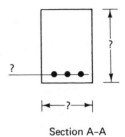

Section A-A

FIGURE 2-18 Sketch for Example 2–8

1. Find the applied design moment M_u, temporarily neglecting the beam weight, which will be included at a later step.

$$w_u = 1.4w_{DL} + 1.7w_{LL}$$

$$= 1.4(0.9) + 1.7(2.0) = 4.66 \text{ kips/ft}$$

Note that this is the factored design load,

$$M_u = \frac{w_u\ell^2}{8} = \frac{4.66(18)^2}{8} = 189 \text{ ft-kips}$$

2. Assume a value for ρ. Use $\rho = 0.0135$ (see Table A-4).

3. From Table A-6, $\bar{k} = 0.4828$ ksi.

4. At this point we have two unknowns, b and d, which can be established using two different approaches. One approach is to assume b and then solve for d. This approach is practical and simple, since many practical and architectural considerations usually establish b within relatively narrow limits.

   Assuming that $b = 11$ in. and utilizing the relationship $M_u = \phi bd^2\bar{k}$,

$$\text{Required } d = \sqrt{\frac{M_u}{\phi b\bar{k}}} = \sqrt{\frac{189(12)}{0.9(11)(0.4828)}} = 21.8 \text{ in.}$$

   A check of the d/b ratio gives $21.8/11 = 1.98$, which is a reasonable value.

5. At this point, the beam weight may be estimated. Realizing that the total design moment will increase, the final beam size may be estimated to be about 11 in. × 26 in. for purposes of calculating its weight.

$$\text{Beam dead load} = \frac{11(26)}{144}(0.150) = 0.298 \text{ kip/ft}$$

6. Additional M_u due to beam weight:

$$M_u = 1.4\frac{0.298(18)^2}{8} = 17 \text{ ft-kips}$$

$$\text{Total } M_u = 189 + 17 = 206 \text{ ft-kips}$$

7. Using the same ρ, $\bar{k}$, and b as previously, compute the new required d.

$$\text{Required } d = \sqrt{\frac{M_u}{\phi b k}} = \sqrt{\frac{206(12)}{0.9(11)(0.4828)}} \quad 22.7 \text{ in.}$$

A check of the d/b ratio gives $22.7/11 = 2.06$, which is reasonable.

8. Required $A_s = \rho b d$

$$= 0.0135(11)(22.7) = 3.37 \text{ in.}^2$$

9. From Table A-2, select 3 No. 10 bars. $A_s = 3.81$ in.^{2}. From Table A-11,

$$\text{Minimum required } b = 10.5 \text{ in.} \qquad \text{(OK)}$$

10. Determine the total beam depth h.

$$\text{Required } h = 22.7 + \frac{1.27}{2} + 0.38 + 1.5 = 25.2 \text{ in.}$$

Use $h = 25.5$ in. The actual effective depth d may now be checked.

$$d = 25.5 - 1.5 - 0.38 - \frac{1.27}{2} = 23 \text{ in.} > 22.7 \text{ in.} \qquad \text{(OK)}$$

11. A design sketch is shown in Fig. 2-19. This design is not necessarily the best. The final ρ is

$$\frac{3.81}{11(23)} = 0.015$$

and the final d/b ratio is

$$\frac{23}{11} = 2.09$$

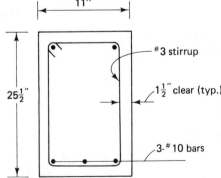

FIGURE 2-19 Design sketch for Example 2–8

The design moment was based on a section 11 in. × 26 in., where the designed beam turned out to be 11 in. × 25½in. Small modifications in proportions could very well make more efficient use of the steel and concrete provided.

The percentage of overdesign could be determined (not necessary, however) by analyzing the designed cross section and comparing strength with applied moment.

A second approach to this problem would be to establish a desired d/b ratio and then mathematically solve for b and d. For example, if we establish a desirable d/b ratio of 2.0, then $d = 2b$. Again utilizing the relationship $M_u = \phi bd^2 \bar{k}$,

$$\text{Required } bd^2 = \frac{M_u}{\phi \bar{k}} = \frac{189(12)}{0.9(0.4828)} = 5220 \text{ in.}^3$$

We can substitute for d as follows:

$$\text{Required } bd^2 = 5{,}220 \text{ in.}^3$$

$$b(2b)^2 = 5{,}220 \text{ in.}^3$$

$$b^3 = \frac{5{,}220}{4} = 1305 \text{ in.}^3$$

$$\text{Required } b = \sqrt[3]{1{,}305} = 10.93 \text{ in.}$$

Hence, assuming that $b = 11$ in., the required effective depth d can then be calculated in a manner similar to that previously used.

2–15   SUMMARY OF PROCEDURE FOR RECTANGULAR REINFORCED CONCRETE BEAM DESIGN FOR MOMENT (TENSION REINFORCEMENT ONLY)

a. Cross section (b and h) known; find the required A$_s$

1. Convert the service loads or moments to design M_u (include the beam weight).

2. Based on knowing h, estimate d by using the relationship $d = h - 3$ in. and calculate the required $\bar{k}$.

$$\text{Required } \bar{k} = \frac{M_u}{\phi bd^2}$$

3. From Tables A6–A10, find the required steel ratio ρ.

4. Compute the required A_s.

$$\text{Required } A_s = \rho bd$$

5. Select the bars. Check to see if the bars can fit into the beam in one layer (preferable). Check the actual effective depth and compare with the assumed

effective depth. If the actual effective depth is slightly in excess of the assumed effective depth, the design will be slightly conservative (on the safe side). If the actual effective depth is less than the assumed effective depth, the design is on the unconservative side and should be revised.

6. Sketch the design. (See Example 2–7, step 6, for a discussion.)

b. Design for cross section and required A_s

1. Convert the service loads or moments to design moment M_u. An estimated beam weight may be included in the dead load if desired. Be sure to apply the load factor to this additional dead load.

2. Select a desired steel ratio ρ. See Table A-4 for recommended values. (Use the ρ values from Table A-4 unless a small cross section or decreased steel is desired.)

$$\rho_{min} \leqslant \rho \leqslant \rho_{max}$$

3. From Tables A6–A10, find $\bar{k}$.

4. Assume b and compute the d required.

$$\text{Required } d = \sqrt{\frac{M_u}{\phi b \bar{k}}}$$

If the d/b ratio is reasonable (1.5–2.2), use these values for the beam.

5. Estimate h and compute the beam weight. Compare with the estimated beam weight if an estimated beam weight was included.

6. Revise design M_u to include the moment due to the beam's own weight using the latest weight determined.

7. Using b and $\bar{k}$ previously determined along with the new total design M_u, find the new required d.

$$\text{Required } d = \sqrt{\frac{M_u}{\phi b \bar{k}}}$$

Check to see if the d/b ratio is reasonable.

8. Find the required A_s.

$$\text{Required } A_s = \rho b d$$

9. Select the bars and check to see if the bars can fit into a beam of width b in one layer (preferable).

10. Establish the final h, rounding this upward to the next ½ in. This will make the actual effective depth greater than the design effective depth and the design will be slightly conservative (on the safe side).

11. Sketch the design. (See Example 2–7, step 6, for a discussion.)

A flow diagram of this procedure is presented in Appendix B.

2–16 DESIGN OF ONE-WAY SLABS FOR MOMENT (TENSION REINFORCEMENT ONLY)

As higher-strength steel and concrete have become available for use in reinforced concrete members, the sizes of the members have decreased. *Deflections* of members are affected very little by material strength but are affected greatly by the size of a cross section and its related moment of inertia. Therefore, deflections will be larger for a member of high-strength materials than would be the deflections for the same member fabricated from lower-strength materials, since the latter member will be essentially larger in cross-sectional area. Deflections will be discussed in detail in Chapter 7. As discussed previously, one method that the ACI Code allows in order to limit adverse deflections is the use of a minimum thickness (see Section 2-13). A slab that meets the minimum thickness requirement must still be designed for flexure. However, deflections need not be calculated or checked unless the slab supports or is attached to construction likely to be damaged by large deflections.

The following example will illustrate the use of the ACI minimum thickness for one-way slabs. With respect to the span length to be used for design, the ACI Code, Section 8.7.1, recommends for beams and slabs not integral with supports:

$$\text{Span length} = \text{clear span} + \text{depth of member}$$

but not to exceed the distance between centers of supports. For design purposes, we will use the distance between centers of supports, since the slab thickness is not yet determined.

EXAMPLE 2–9

Design a simple span one-way slab to carry a uniformly distributed live load of 325 psf. The span is 10 ft (center to center of supports). Use $f'_c = 3{,}000$ psi and $f_y = 40{,}000$ psi. Conform to the ACI minimum thickness requirement and make the slab thickness to a ¼-in. increment.

Determine the required minimum h and use this to estimate the slab dead weight.

1. From ACI Table 9.5(a),

$$\text{Minimum } h = \frac{\ell}{20}\left(0.4 + \frac{40{,}000}{100{,}000}\right) = \frac{10(12)}{20}(0.8) = 4.8 \text{ in.}$$

Round to the next higher ¼ in. Try $h = 5$ in. and design a 12-in.-wide segment.

2. Determine the slab weight dead load.

$$\frac{5(12)}{144}(0.150) = 0.0625 \text{ ksf}$$

The total design load is
$$w_u = 1.4 w_{DL} + 1.7 w_{LL}$$
$$= 1.4(0.0625) + 1.7(0.325)$$
$$= 0.64 \text{ kip/ft}$$

3. Determine the design moment.

$$M_u = \frac{w_u \ell^2}{8} = \frac{0.64(10)^2}{8} = 8.0 \text{ ft-kips}$$

4. Establish the approximate d. Assuming No. 6 bars and minimum concrete cover on the bars of ¾ in.,

$$\text{Assumed } d = 5.0 - 0.75 - 0.375 = 3.88 \text{ in.}$$

5. Determine the required $\bar{k}$.

$$\bar{k} = \frac{M_u}{\phi b d^2}$$
$$= \frac{8.0(12)}{0.9(12)(3.88)^2} = 0.590 \text{ ksi}$$

6. From Table A-6, for a required $\bar{k} = 0.590$, the required $\rho = 0.0171$. Check ρ_{max}.

$$\rho_{max} = 0.0278 > 0.0171 \qquad \text{(OK)}$$

Use $\rho = 0.0171$.

7. Required $A_s = \rho b d = 0.0171(12)(3.88) = 0.80 \text{ in.}^2/\text{ft}$

8. Select the main steel (from Table A-3). Select No. 6 bars at 6½ in. o.c. ($A_s = 0.81 \text{ in.}^2$). The assumption on bar size was satisfactory. The code requirements for maximum spacing have been discussed in Section 2–13. Minimum spacing of bars in slabs, practically, should not be less than 4 inches, although the ACI Code allows bars to be placed closer together as discussed in Example 2–7. Check the maximum spacing (ACI Code, Section 7.6.5).

$$\text{Maximum spacing} = 3h \quad \text{or} \quad 18 \text{ in.}$$
$$3h = 3(5) = 15 \text{ in.}$$
$$6½ \text{ in.} < 15 \text{ in.} \qquad \text{(OK)}$$

Therefore, use No. 6 bars at 6½ in. o.c.

9. Select shrinkage and temperature reinforcement (ACI Code, Section 7.12).

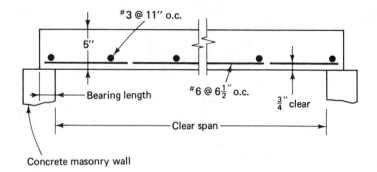

FIGURE 2-20   Design sketch for Example 2-9

Required $A_s = 0.0020bh$

$$= 0.0020(12)(5) = 0.12 \text{ in.}^2/\text{ft.}$$

Select No. 3 bars at 11 in. o.c. ($A_s = 0.12$ in.²) or No. 4 bars at 20 in. o.c. ($A_s = 0.12$ in.²).

$$\text{Maximum spacing} = 5h \quad \text{or} \quad 18 \text{ in.}$$

Therefore, use No. 3 bars at 11 in. o.c.

10. The main steel area must exceed the area required for shrinkage and temperature steel (ACI Code, Section 10.5.3).

$$0.81 \text{ in.}^2 > 0.12 \text{ in.}^2 \qquad \text{(OK)}$$

11. A design sketch is shown in Fig. 2-20.

2-17 SUMMARY OF PROCEDURE FOR DESIGN OF ONE-WAY SLABS FOR MOMENT (TO SATISFY ACI MINIMUM h)

1. Compute the minimum h based on the ACI Code, Table 9.5(a). Round off h to the next higher ¼ in. and use.

2. Compute the slab weight and compute w_u (total design load).

3. Compute the design moment M_u.

4. Calculate an assumed effective depth d (assuming No. 6 bars and ¾ in. cover) by using the relationship

$$d = h - 1.12 \text{ in.}$$

5. Calculate the required $\bar{k}$.

$$\text{Required } \bar{k} = \frac{M_u}{\phi bd^2}$$

6. From Tables A6–A10, find the required steel ratio ρ. (This must be less than ρ_{max}. If the required $\rho > \rho_{max}$, the slab must be made thicker.)

7. Compute the required A_s.

$$\text{Required } A_s = \rho bd$$

8. Select the main steel (Table A-3). Check with maximum spacing $3h$ or 18 in. Check the assumption of step 4.

9. Select shrinkage and temperature steel as per the ACI Code.

$$\text{Required } A_s = 0.0020bh \quad \text{(grade 40 and 50 steel)}$$

$$\text{Required } A_s = 0.0018bh \quad \text{(grade 60 steel)}$$

Check with maximum spacing $5h$ or 18 in.

10. The main steel area *cannot* be less than the area of steel required for shrinkage and temperature.

11. Sketch the design.

2–18 SLABS ON GROUND

The previous discussion primarily concerns itself with a structural slab. Another category of slabs, generally used as a floor, may be termed a *slab on grade* or *slab on ground*. As the name implies, it is a slab that is supported throughout its entire area by some form of subgrade. The design of such a slab is significantly different from the design of a structural slab.

The theoretical design of a slab on grade would involve not only the loading and slab itself but also the elastic behavior of the subgrade. Such a design or analysis is highly laborious and involved and is based on assumed conditions whose existence may be questionable. Such a design is generally not warranted.

The usual design procedure is an arbitrary selection based on an engineer's past experience or the use of recommended values obtained from tables such as those presented in the *CRSI Design Handbook*, Vol. II[2] (1965 edition), which is based on the 1963 ACI Code, or other design aids.[3]

A slab for the very lightest function should be not less than 4 in.

thick. This and other slabs may be empirically selected from Table 2–1, which is reprinted from the *CRSI Design Handbook*. These slab thicknesses and associated steel reinforcement suggestions are recommended *minimums*.

In addition to the CRSI recommendations, the Wire Reinforcement Institute (WRI) recommends that where one layer of welded wire fabric is used, the placement should be as follows:

4-in. slab: middle of slab

5-in. slab: 2 in. below top surface

6-in. slab: 2 in. below top surface

TABLE 2–1
**Slab Thicknesses and Steel Recommended for Slabs on Ground for
Various Uniformly Distributed Loads**

Uniform Loads (PSF)*	Minimum Slab Thickness (in.)	Reinforcement†
Less than 100	4	One layer 6 × 6-W1.4 × W1.4 welded wire fabric minimum for ideal conditions, 6 × 6-W2.0 × W2.0 for average conditions
100–200	5	One layer 6 × 6-W2.0 × W2.0 welded wire fabric or one layer 6 × 6-W2.9 × W2.9
Not over 400–500	6	One layer 6 × 6-W2.9 × W2.9 welded wire fabric or one layer 6 × 6-W4.0 × W4.0
600–800	6	Two layers 6 × 6-W2.9 × W2.9 welded wire fabric or two layers 6 × 6-W4.0 × W4.0
1,500‡	7	Two mats of bars (one top, one bottom) each of No. 4 bars at 12 in. o.c. each way
2,500‡	8	Two mats of bars (one top, one bottom) each of No. 5 bars at 12 in. o.c. each way
3,000–3,500‡	9	Two mats of bars (one top, one bottom) each of No. 5 bars at 8–12 in. o.c. each way

* Fill material and compaction should be equivalent to that of ordinary highway practice. If laboratory control of compaction is available, the load capacities can be increased in the ratio of the actual compaction coefficient k to 100.

† Place the first layer of reinforcement 2 in. below the top of the slab and the second layer 2 in. up from the bottom of the slab.

‡ For loads in excess of 1,500 psf, the subsoil condition should be investigated with extra care.

PROBLEMS

The following problems concern tension reinforcing only. For beams, assume 1½ in. cover and No. 3 stirrups. For slabs, assume ¾ in. cover.

2–1. The beam of cross section shown has $f'_c = 4,000$ psi. The bars are A615 grade 40. Find the ideal moment strength (M_n) if:

    a. $d = 14$ in.
    b. $d = 21$ in.
    c. $d = 28$ in.

Compare the results in a tabulation.

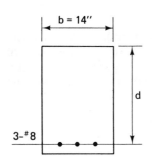

FIGURE P2-1

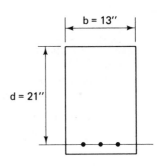

FIGURE P2-2

2–2. Calculate the ideal moment capacity (M_n) for the beam of cross section shown. $f'_c = 4,000$ psi and $f_y = 60,000$ psi. Reinforcement is:

    a. 3 No. 8 bars.
    b. 3 No. 9 bars.
    c. 3 No. 10 bars.

Compare the results in a tabulation.

2–3. Calculate the ideal moment capacity (M_n) for the beam of cross section shown. The steel is A615 grade 40. $b = 12$ in.; 4 No. 9 bars.

    a. $f'_c = 3,000$ psi
    b. $f'_c = 4,000$ psi
    c. $f'_c = 5,000$ psi

Compare the results in a tabulation.

2–4. Calculate the resisting moment (M_R) for the beam of cross section shown. $b = 13$ in.; 4 No. 8 bars.

    a. $f'_c = 4,000$ psi, $f_y = 40,000$ psi
    b. $f'_c = 4,000$ psi, $f_y = 60,000$ psi

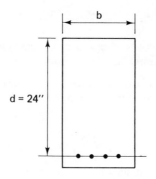

FIGURE P2-3 and P2-4

2-5. A reinforced concrete beam having the cross section shown is on a simple span of 28 ft. It carries uniform service loads of 3.60 kips/ft live load and 2.20 kips/ft dead load in addition to its own weight. Check the adequacy of the beam with respect to moment. Check the steel ratio to ensure that it is within the limitations of the ACI Code. $f'_c = 3,000$ psi and $f_y = 40,000$ psi.

  a. Reinforcing is 6 No 10 bars. NG

  b. Reinforcing is 6 No. 11 bars. ok

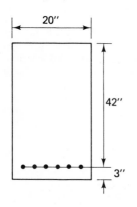

FIGURE P2-5

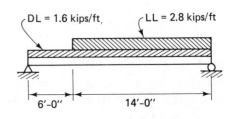

FIGURE P2-6

2-6. A rectangular reinforced concrete beam carries *service loads* on a span of 20 ft as shown. The dead load of 1.6 kips/ft does not include the beam weight. $f'_c = 3,000$ psi and $f_y = 60,000$ psi. $b = 14.5$ in., $h = 24$ in., and reinforcing is 5 No. 9 bars. Determine whether the beam is adequate with respect to moment.

2-7. The one-way slab shown spans 12 ft from center of support to center of support. Calculate M_R and determine the *service live load* (psf) that the slab may carry. (Assume that the only dead load is the weight of the slab.) $f'_c = 3,000$ psi and $f_y = 40,000$ psi.

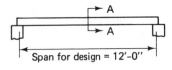

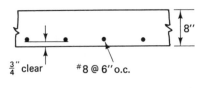

Span for design = 12'-0"

$\frac{3}{4}''$ clear   #8 @ 6"o.c.

Section A-A

FIGURE P2-7

2–8. A simply supported rectangular reinforced concrete beam is to span 22 ft and is to carry uniform service loads of 1.6 kips/ft dead load and 1.4 kips/ft live load. The assumed dead load includes an estimated beam weight. Use A615 grade 40 steel and $f'_c = 3,000$ psi. Use the recommended ρ from Table A-4. Make the beam width 15 in. and keep the overall depth (h) to full inches. Assume No. 3 stirrups. Check the adequacy of the beam you design by comparing M_u with M_R. Sketch your design.

2–9. Design a rectangular reinforced concrete beam to resist a total design moment M_u of 133 ft-kips. (This includes the moment due to beam weight.) Architectural considerations require that the width (b) be 11½ in. and the overall depth (h) to be 23 in. $f'_c = 3,000$ psi and $f_y = 40,000$ psi. Sketch your design.

2–10. Repeat Problem 2–9 with $M_u = 400$ ft-kips, $b = 16$ in., $h = 28$ in., $f'_c = 4,000$ psi, and $f_y = 40,000$ psi.

2–11. For the beam designed in Problem 2–10, if the main reinforcement were incorrectly placed so that the actual effective depth were 24 in., would the beam be adequate? Check by comparing M_u with the M_R resulting from the beam using actual steel, actual b, and $d = 24$ in.

2–12. Design a rectangular reinforced concrete beam for a simple span of 30 ft. The beam is to carry uniform *service* loads of 1.0 kip/ft dead load and 2.0 kips/ft live load. The assumed dead load does not include the weight of the beam. Because of column sizes, the beam width should not exceed 16 in. Use $f'_c = 3,000$ psi and $f_y = 60,000$ psi. Sketch your design.

2–13. Design a rectangular reinforced concrete beam for a simple span of 32 ft. Uniform service loads are 1.5 kips/ft dead load and 2.0 kips/ft live load. The assumed dead load does not include the weight of the beam. The width of the beam is limited to 18 in. Use $f'_c = 3,000$ psi and $f_y = 40,000$ psi. Sketch your design.

2–14. Design a rectangular reinforced concrete beam for a simple span of 40 ft. Uniform service loads are 0.8 kip/ft dead load (does not include weight of beam) and 1.4 kips/ft live load. Use $f'_c = 4,000$ psi and $f_y = 60,000$ psi. Sketch your design.

2–15. Design a simply supported rectangular reinforced concrete beam for the span and *service loads* shown. The uniform service dead load shown does not include the beam weight. $f'_c = 3,000$ psi and $f_y = 60,000$ psi.

61

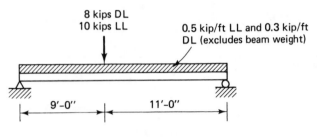

8 kips DL
10 kips LL

0.5 kip/ft LL and 0.3 kip/ft
DL (excludes beam weight)

9'-0''   11'-0''

FIGURE P2-15

2–16. Design a simply supported one-way reinforced concrete floor slab to span 8 ft and carry a service live load of 300 psf in addition to its own weight. $f'_c =$ 3,000 psi and $f_y = 40,000$ psi. Sketch your design.

#2.17.

2–17. Design a simply supported one-way reinforced concrete floor slab to span 10 ft, carry a service live load of 175 psf, and a service dead load of 25 psf in addition to its own weight. $f'_c = 3,000$ psi and $f_y = 40,000$ psi. Make the slab thickness to ½ in.

    a. Design the slab for the ACI Code minimum thickness.
    b. Design the thinnest possible slab allowed by the ACI Code.

2–18. Design the simply supported one-way reinforced concrete slab as shown. The service live load is 200 psf. $f'_c = 3,000$ psi and $f_y = 60,000$ psi. Sketch your design.

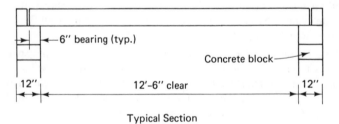

6'' bearing (typ.)

Concrete block

12''   12'-6'' clear   12''

Typical Section

FIGURE P2-18

REFERENCES

1. Charles S. Whitney, "Plastic Theory of Reinforced Concrete Design," *Trans. ASCE*, Vol. 68, 1942.

2. *CRSI Design Handbook*, Vol. II, 1965, Concrete Reinforcing Steel Institute, 180 North LaSalle Street, Chicago, Illinois 60601.

3. Ralph E. Spears, *Concrete Floors on Ground*, Portland Cement Association, 1978

3

Reinforced Concrete Beams: T-Beams and Doubly Reinforced Beams

3-1 T-BEAMS—INTRODUCTION

Floors and roofs in reinforced concrete buildings may be composed of slabs which are supported so that loads are carried to columns and then to the building foundation. These, as previously discussed, are termed flat slabs or flat plates. The span of such a slab cannot become very large before its own dead weight causes it to become uneconomical. Many types of systems have been devised to allow greater spans without the problem of excessive weight.

One such system, which may be termed a *beam and girder system*, is composed of a slab on supporting reinforced concrete beams, which in turn frame into reinforced concrete girders and/or columns. The girders, in turn, frame into the supporting columns. In such a system, the beams and girders are commonly placed monolithically with the slab. Systems other than the monolithic system do exist, and these may make use of some precast and some cast-in-place concrete. These are generally of a proprietary nature. The typical monolithic system is shown in Fig. 3-1. The beams are commonly spaced so that they intersect the girders at the midpoint, third points, or quarter points, as shown in Fig. 3-2.

In the analysis and design of such floor and roof systems, it is common practice to assume that the monolithically poured slab and supporting beam interact as a unit in resisting *positive bending moment*. As shown in Fig. 3-3, the slab becomes the compression flange, and the supporting beam becomes the web or stem. The interacting flange and stem produce the cross section having the typical T-shape from which

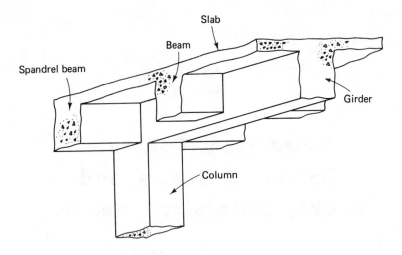

FIGURE 3-1   Beam and girder floor system

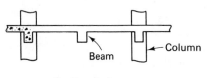

Section A-A

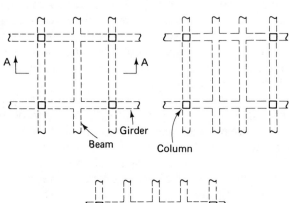

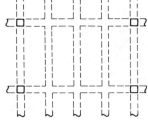

FIGURE 3-2   Common beam and girder layouts

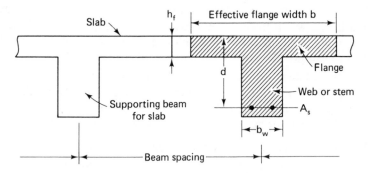

FIGURE 3-3   T-beam as part of a floor system

the T-beam gets its name. It should be noted that the slab, which comprises the T-beam flange, must itself be designed to span across the supporting beams. Therefore, the slab behaves as a bending member acting in two directions. It should also be noted that should the T-beam cross section be subjected to *negative bending moment*, the slab at the top of the stem will be in tension while the bottom of the stem is in compression. It will be seen that this situation will occur at interior supports of continuous beams, which will be discussed later.

To simplify the complex two-way behavior of the flange, the ACI Code, for design and analysis purposes, has established criteria whereby the flange, when acting together with the web, will have a limited width that may be considered effective in resisting applied moment. This effective flange width for symmetrical shapes will always be equal to or less than the beam spacing (see Fig. 3-3).

3-2 T-BEAM ANALYSIS

For purposes of analysis and design, the ACI Code, Section 8.10, has established limits on the effective flange width as follows:

1. The effective flange width must not exceed one-fourth of the span length of the beam, and the effective overhanging slab width on each side of the web must not exceed eight times the thickness of the slab nor one-half of the clear distance to the next beam.

In other words, the lesser of the three values will control and the effective flange width must not exceed:

    a. One-fourth of the span length.

    b. $b_w + 16h_f$.

    c. Center-to-center spacing of beams.

2. For beams having a flange on one side only, the effective overhanging flange width must not exceed one-twelfth of the span length of the beam nor six times the slab thickness, nor one-half of the clear distance to the next beam.

3. For isolated beams in which the T-shape is used only for the purpose of providing additional compressive area, the flange thickness must not be less than one-half of the width of the web, and the total flange width must not be more than four times the web width.

The ductility requirements for T-beams are similar to rectangular beams. The maximum steel ratio ρ shall not exceed $0.75\rho_b$. However, this $0.75\rho_b$ is not the same value as that tabulated for rectangular beams because of the T-shaped compressive area. A variation of the rectangular beam approach for $0.75\rho_b$ may be observed in Section 3-3. The minimum steel ratio is determined as previously:

$$\rho_{\min} = \frac{200}{f_y}$$

However, the *actual* steel ratio which is used to compare against the minimum steel ratio is determined by using the *web width* (b_w), not the effective flange width (b) (see the ACI Code, Section 10.5.1). The foregoing applies where the stem is in tension.

Because of the relatively large compression area available in the flange of the T-beam, the resisting moment capacity is usually limited by the yielding of the tensile steel. It is, therefore, both customary and safe to assume that the tensile steel will yield before the concrete reaches its ultimate strain and crushes. The total tensile force N_T, at the ultimate condition, may then be found by

$$N_T = A_s f_y$$

To proceed with the analysis, the shape of the compressive stress block must be defined. As in our previous analyses, the total compressive force N_C must be equal to the total tensile force N_T. The shape of the stress block must be compatible with the area in compression. Two conditions may exist: the stress block may be completely within the flange or it may cover the flange and extend into the web. These two conditions will result in what we will term, respectively, a *rectangular T-beam* and a *true T-beam*. The basic difference between the two, in addition to the shape of the stress block, is that the rectangular T-beam with effective flange width b is analyzed in the same way as is a rectangular beam of width b, while the analysis of the true T-beam must consider the T-shaped stress block.

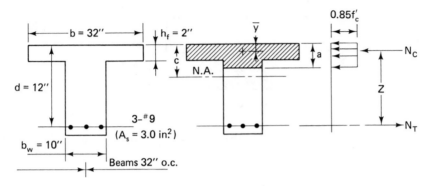

FIGURE 3-4 Sketch for Example 3-1

EXAMPLE 3-1

The T-beam shown in Fig. 3-4 is part of a floor system. Determine the dependable resisting moment strength M_R if f_y = 60,000 psi (A615 grade 60) and f'_c = 3,000 psi.

1. Since the span length is not given, determine the effective flange width in terms of the flange thickness and beam spacing.

$$b_w + 16h_f = 10 + 16(2) = 42 \text{ in.}$$

Beam spacing = 32 in. o.c.   Use b = 32 in.

2. Assume that the steel yields and find N_T.

$$N_T = A_s f_y = 3.00(60,000) = 180,000 \text{ lb}$$

3. The flange, if fully stessed to $0.85f'_c$, would produce a total compressive force of

$$N_C = (0.85f'_c) h_f b$$

$$= 0.85(3,000)(2)(32) = 163,000 \text{ lb}$$

4. Since 180,000 > 163,000, the stress block must extend below the flange far enough to provide the remaining compression.

$$180,000 - 163,000 = 17,000 \text{ lb}$$

Hence the stress block extends below the flange and the analysis is one for a true T-beam.

5. The remaining compression ($N_T - N_C$) may be obtained by the additional web area.

$$N_T - N_C = (0.85f'_c) b_w (a - h_f)$$

67

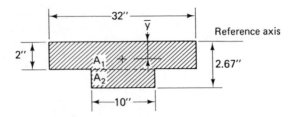

FIGURE 3-5 T-beam compressive area

Solving for a, we obtain

$$a = \frac{N_T - N_C}{(0.85f'_c)b_w} + h_f = \frac{17,000}{0.85(3,000)(10)} + 2$$

$$= 2.67 \text{ in.}$$

6. Check ρ_{min}.

$$\rho_{min} = \frac{200}{f_y} = \frac{200}{60,000} = 0.0033$$

$$\text{Actual } \rho = \frac{A_s}{b_w d} = \frac{3.00}{10(12)} = 0.025 > 0.0033 \qquad \text{(OK)}$$

7. To calculate the magnitude of the internal couple, it is necessary to know the lever-arm distance between N_C and N_T. The location of N_T is assumed to be at the centroid of the steel area and we will locate N_C at the centroid of the compression area (see Fig. 3-5). Using a reference axis at the top of the section, the centroid may be located a distance $\overline{y}$ below the reference axis, as follows:

$$\overline{y} = \frac{\Sigma(Ay)}{\Sigma A}$$

$$A_1 = 32(2) = 64 \text{ in.}^2 \qquad A_2 = 10(0.67) = 6.7 \text{ in.}^2$$

$$\overline{y} = \frac{64(1) + 6.7(2 + 0.34)}{64 + 6.7} = 1.13 \text{ in.}$$

This locates N_C. Therefore,

$$Z = d - \overline{y}$$

$$= 12 - 1.13 = 10.87 \text{ in.}$$

8. The nominal (or ideal) internal resisting moment may be found:

$$M_n = N_T Z = \frac{180,000(10.87)}{12,000} = 163 \text{ ft-kips}$$

from which the dependable resisting moment is

$$M_R = \phi M_n = 0.9(163) = 147 \text{ ft-kips}$$

9. We will now check our assumption of a ductile failure (yielding of steel). In T-beams this is most conveniently accomplished by comparing the actual tension steel area with 75% of the amount of steel required to produce the balanced condition $(0.75 A_{sb})$. The location of the balanced neutral axis is found as follows:

$$c_b = \frac{87,000}{f_y + 87,000}(d) \quad \text{(see Section 2–8)}$$

$$= \frac{87,000}{60,000 + 87,000}(12) = 7.10 \text{ in.}$$

Using the rectangular beam relationship $a = 0.85c$, which is approximate for T-beams,

$$a_b = 0.85(7.10) = 6.035 \text{ in.}$$

Then the total *balanced* compression force N_{Cb} must equal

$$N_{Cb} = 0.85 f'_c \left[b(h_f) + b_w(a_b - h_f) \right]$$

$$= 0.85(3,000) \left[32(2) + 10(6.035 - 2) \right]$$

$$= 266,093 \text{ lb} = N_{Tb}$$

Also, since $N_{Tb} = A_{sb} f_y$,

$$A_{sb} = \frac{266,093}{60,000} = 4.43 \text{ in.}^2$$

which is the area of steel which makes this beam a balanced beam.

$$A_{s(\text{max})} = 0.75 A_{sb}$$

$$= 0.75(4.43) = 3.32 \text{ in.}^2$$

$$3.32 \text{ in.}^2 > 3.0 \text{ in.}^2 \qquad \text{(OK)}$$

3–3 DEVELOPMENT OF T-BEAM PERMISSIBLE A_s

As may be observed, the preceding check in step 9 is based on the following relationships:

1. $c_b = \dfrac{87,000}{f_y + 87,000}(d)$

2. $a_b = 0.85 c_b$ (where $\beta_1 = 0.85$)

3. $N_{Cb} = 0.85 f'_c \left[b(h_f) + b_w(a_b - h_f) \right]$

4. $N_{Cb} = N_{Tb} = A_{sb} f_y$

5. $A_{s\,(max)} = 0.75 A_{sb}$

Combining these expressions and solving for $A_{s\,(max)}$, we arrive at the following expression:

$$A_{s\,(max)} = \frac{0.75}{f_y} N_{Cb}$$

$$= \frac{0.75}{f_y}(0.85f'_c)\left\{ bh_f + \left[\beta_1 \frac{87,000(d)}{87,000 + f_y} - h_f \right] b_w \right\}$$

$$= \frac{0.638}{f_y} f'_c h_f \left\{ b + b_w \left[\frac{\beta_1}{h_f}\left(\frac{87,000(d)}{87,000 + f_y} \right) - 1 \right] \right\}$$

Substituting for various combinations of f'_c and f_y, $A_{s\,(max)}$ expressions result which are listed in Table 3-1. Using these expressions, let us now calculate the $A_{s\,(max)}$ allowed by the Code for Example 3-1:

$$A_{s\,(max)} = 0.0319 h_f \left[b + b_w \frac{0.503(d)}{h_f} - 1 \right]$$

$$= 0.0319(2)\left[32 + 10 \frac{0.503(12)}{2} - 1 \right]$$

$$= 0.0638(32 + 20.18)$$

$$= 3.33 \text{ in.}^2 \approx 3.32 \text{ in.}^2$$

This represents the maximum permissible area of steel allowed for ductility considerations based on the ACI Code. Since this exceeds the actual furnished A_s (3.33 > 3.00), the ductile failure is assured and the code criteria are satisfied.

TABLE 3–1
Expressions for $A_{s\,(max)}$

f'_c(psi)	f_y (psi)	$A_{s\,(max)}$ (in.²)
3,000	40,000	$0.0478 h_f \left\{ b + b_w \left[\frac{0.582}{h_f}(d) - 1 \right] \right\}$
	60,000	$0.0319 h_f \left\{ b + b_w \left[\frac{0.503}{h_f}(d) - 1 \right] \right\}$
4,000	40,000	$0.0638 h_f \left\{ b + b_w \left[\frac{0.582}{h_f}(d) - 1 \right] \right\}$
	60,000	$0.0425 h_f \left\{ b + b_w \left[\frac{0.503}{h_f}(d) - 1 \right] \right\}$

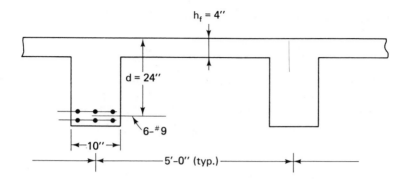

FIGURE 3-6 Sketch for Example 3-2

EXAMPLE 3-2

For the T-beam shown in Fig. 3-6, determine the dependable resisting moment strength M_R if $f'_c = 3{,}000$ psi and $f_y = 40{,}000$ psi. The beam span length is 24 ft.

1. Find the effective flange width:

$$\tfrac{1}{4} \text{ span length } = \frac{24(12)}{4} = 72 \text{ in.}$$

$$b_w + 16h_f = 10 + 16(4) = 74 \text{ in.}$$

$$\text{Beam spacing} = 60 \text{ in.}$$

$$\text{Use } b = 60 \text{ in.}$$

2. Assume that the steel yields and find N_T.

$$N_T = A_s \, f_y = 6.0(40{,}000) = 240{,}000 \text{ lb}$$

3. The flange itself is capable of furnishing a compression force of

$$N_C = (0.85f'_c)bh_f = 0.85(3{,}000)(60.0)(4) = 612{,}000 \text{ lb}$$

4. Since $612{,}000 > 240{,}000$, the flange furnishes sufficient compression area and the stress block lies entirely in the flange. Therefore, analyze the T-beam as a rectangular T-beam of width $b = 60$ in. (Note that this beam, although a rectangular T-beam for determination of M_R, will probably be a true T-beam at the balanced condition when maximum A_s is checked.)

5. Check $\rho_{\min}$.

$$\rho_{\min} = \frac{200}{f_y} = \frac{200}{40{,}000} = 0.005$$

$$\text{Actual } \rho = \frac{A_s}{b_w d} = \frac{6.0}{10(24)} = 0.025 > 0.005 \qquad \text{(OK)}$$

71

6. The actual steel ratio (use this to determine $\overline{k}$) is

$$\rho = \frac{A_s}{bd} = \frac{6.0}{60(24)} = 0.0042$$

One should be careful, at this point, not to confuse the actual steel ratio to be utilized for moment strength with the actual steel ratio (from step 5) used for comparison with ρ_{min}. The two are calculated in *different manners*.

7. For $\rho = 0.0042$, go to Table A-6 and find the required $\overline{k}$.

$$\text{Required } \overline{k} = 0.1625 \text{ ksi}$$

8.
$$M_R = \phi b d^2 \overline{k}$$

$$= 0.9(60)(24)^2\left(\frac{0.1625}{12}\right) = 421.2 \text{ ft-kips}$$

9. Check the beam ductility by comparing $A_{s(max)}$ with actual A_s.

$$A_{s(max)} = 0.0478 h_f\left[b + b_w\left(\frac{0.582d}{h_f} - 1\right)\right]$$

$$= 0.0478(4)\left\{60 + 10\left[\frac{0.582(24)}{4} - 1\right]\right\}$$

$$= 16.24 \text{ in.}^2$$

The actual $A_s = 6.00$ in.². Since $16.24 > 6.00$, the beam is ductile and the steel yields at the ultimate moment, as assumed.

3-4 ANALYSIS OF BEAMS HAVING IRREGULAR CROSS SECTIONS

Beams having other than rectangular and T-shaped cross sections are common, particularly in structures utilizing precast elements. The approach for the analysis of such beams is to utilize the internal couple in the normal way, taking into account any variation in the shape of the compressive stress block. The method is similar to that used for true T-beam analysis.

EXAMPLE 3–3

Find the dependable resisting moment strength M_R for the beam shown in Fig. 3-7. (The ledges in the beam cross section will possibly be used for support of precast slabs.) $f_y = 40,000$ psi (A615 grade 40) and $f'_c = 3,000$ psi.

1. The effective flange width may be considered to be 7 in.

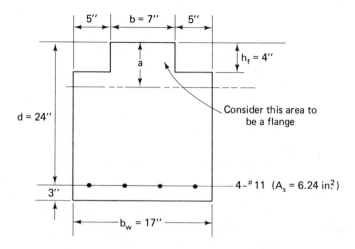

FIGURE 3-7 Sketch for Example 3–3

2. Assume that the steel yields and find N_T.

$$N_T = A_s \, f_y = 6.24(40,000) = 249,600 \text{ lb}$$

3. Determine the amount of compression that the 7 in. × 4 in. area (above the ledges) is capable of furnishing.

$$N_C = (0.85 f'_c) h_f b$$
$$= 0.85(3,000)(4)(7) = 71,400 \text{ lb}$$

4. Since 249,600 > 71,400, the compressive stress block must extend below the ledges to provide the remaining compression.

$$249,600 - 71,400 = 178,200 \text{ lb}$$

5. The remaining compression will be furnished by additional beam area below the ledges. Referring to Fig. 3-7,

$$a = \frac{N_T - N_C}{(0.85 f'_c) b_w} + h_f$$
$$= \frac{178,200}{0.85(3,000)(17)} + 4$$
$$= 8.11 \text{ in. from top of beam}$$

6. Check ρ_{min}.

$$\rho_{min} = \frac{200}{f_y} = \frac{200}{40,000} = 0.005$$

$$\text{Actual } \rho = \frac{A_s}{b_w d} = \frac{6.24}{17(24)} = 0.0153 > 0.005 \qquad \text{(OK)}$$

73

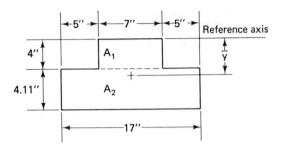

FIGURE 3-8 Compressive area

7. Find the location of N_C at the centroid of the compression area. With reference to Fig. 3-8, the centroid is located at a distance $\bar{y}$ down from a reference axis at the top of the cross section.

$$\bar{y} = \frac{\Sigma(Ay)}{\Sigma A}$$

$$A_1 = 7(4) = 28 \text{ in.}^2 \qquad A_2 = 4.11(17) = 69.87 \text{ in.}^2$$

$$\bar{y} = \frac{28(2) + 69.87\,(4 + 4.11/2)}{28 + 69.87} = 4.89 \text{ in.}$$

from which the lever arm Z may be calculated.

$$Z = d - \bar{y} = 24 - 4.89$$

$$= 19.11 \text{ in.}$$

8. Compute the ideal resisting moment (M_n) and the dependable resisting moment M_R.

$$M_n = N_T\,Z = \frac{249{,}600(19.11)}{12{,}000} = 397.5 \text{ ft-kips}$$

$$M_R = \phi M_n = 0.9(397.5) = 357.7 \text{ ft-kips}$$

9. Check the Code requirement for $A_{s(\text{max})}$ (see Table 3-1).

$$A_{s(\text{max})} = 0.0478\,h_f\left\{b + b_w\left[\frac{0.582(d)}{h_f} - 1\right]\right\}$$

$$= 0.0478(4)\left\{7 + 17\left[\frac{0.582(24)}{4} - 1\right]\right\}$$

$$= 9.44 \text{ in.}^2$$

$$9.44 \text{ in.}^2 > 6.24 \text{ in.}^2 \qquad\qquad (\text{OK})$$

It should be noted in step 9 that the equation from Table 3–1 applies because the area above the ledges behaves as does the flange of a T-beam.

3–5 T-BEAM DESIGN (FOR MOMENT)

The design of the T-sections involves the dimensions of the flange and web and the area of the tension steel, a total of five unknowns. In the normal progression of a design, the flange thickness is determined by the design of the slab and the web size is determined by the shear and moment requirements at the end supports of a beam in continuous construction. Practical considerations, such as column sizes and forming, may also dictate web width. Therefore, when the T-section is designed for positive moment, most of the five unknowns have been previously determined.

The ACI Code, as indicated previously, dictates permissible effective flange width (b). The flange itself generally provides more than sufficient compression area; therefore, the stress block usually lies completely in the flange. Thus most T-beams are only wide rectangular beams with respect to flexural behavior.

The recommended method for design of T-beams will depend on whether the T-beam behaves as a rectangular T-beam or as a true T-beam. The first step will be to answer this question. If the T-beam is determined to be a rectangular T-beam, the design procedure is the same as for the tensile reinforced rectangular beam where the size of the cross section is known (see Section 2-15). If the T-beam is determined to be a true T-beam, an approximate design, followed by analysis, may be used.

EXAMPLE 3–4

Design the T-beam for the floor system shown in Fig. 3-9. The floor has a 4-in. slab supported by 22-ft-span-length beams cast monolithically with the slab. Beams are 8 ft 0 in. on center and have a web width of 12 in. and a total depth = 22 in.; f'_c = 3,000 psi and f_y = 60,000 psi (A615 grade 60). Service load moments are M_{DL} = 62 ft-kips (includes the weight of the floor system) and M_{LL} = 120 ft-kips.

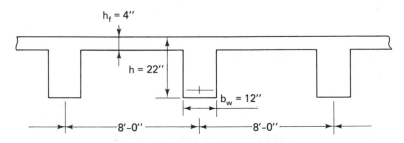

FIGURE 3-9 Typical floor section—Example 3–4

1. Establish the design moment.

$$M_u = 1.4\,M_{DL} + 1.7\,M_{LL}$$
$$= 1.4(62) + 1.7(120)$$
$$= 291 \text{ ft-kips}$$

2. Assume an effective depth $d = h - 3$ in.

$$d = 22 - 3 = 19 \text{ in.}$$

3. Effective flange width:

$$\frac{1}{4} \text{ span length} = 0.25(22)(12) = 66 \text{ in.}$$
$$b_w + 16h_f = 12 + 16(4) = 76 \text{ in.}$$
$$\text{Beam spacing} = 96 \text{ in.}$$

Use an effective flange width $b = 66$ in.

4. Determine whether the beam behaves as a true T-beam or as a rectangular beam by computing the resisting moment (M_R) with the complete effective flange assumed to be in compression. This assumes that the bottom of the compressive stress block coincides with bottom of flange as shown in Fig. 3-10.

$$M_R = \phi(0.85f_c')bh_f\left(d - \frac{h_f}{2}\right)$$
$$= 0.9(0.85)(3)(66)\frac{4(19 - 4/2)}{12} = 858 \text{ ft-kips}$$

5. Since $858 > 291$, the total effective flange need not be completely utilized in compression and the T-beam behaves as a wide rectangular beam with a width b of 66 in.

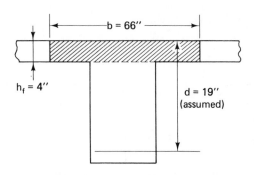

FIGURE 3-10   T-beam for Example 3-4

6. Design as a rectangular beam with b and d as known values (see Section 2–15).

$$\text{Required } \bar{k} = \frac{M_u}{\phi bd^2} = \frac{291(12)}{0.9(66)(19)^2} = 0.1628 \text{ ksi}$$

7. From Table A-7, select the required steel ratio to provide a $\bar{k}$ of 0.1628 ksi.

$$\text{Required } \rho = 0.0028$$

(Recall that for T-beams the steel ratio for moment strength may fall below ρ_{min} and still be acceptable.)

8. Calculate the required steel area.

$$\text{Required } A_s = \rho bd$$
$$= 0.0028(66)(19) = 3.51 \text{ in.}^2$$

9. Select the steel bars. Use 3 No. 10 bars ($A_s = 3.81 \text{ in.}^2$).

$$\text{Minimum } b_w = 10.5 \text{ in.} \quad \text{(OK)}$$

Check the actual d.

$$d = 22 - 1.5 - 0.38 - \frac{1.27}{2} = 19.49 \text{ in.}$$

$$19.49 \text{ in.} > 19 \text{ in.} \quad \text{(OK)}$$

10. Check the ρ_{min} and $A_{s\,(max)}$.

$$\rho_{min} = \frac{200}{f_y} = \frac{200}{60,000} = 0.0033$$

$$\text{Actual } \rho = \frac{A_s}{b_w d} = \frac{3.81}{12(19.49)} = 0.0163 > 0.0033 \quad \text{(OK)}$$

11.
$$A_{s\,(max)} = 0.0319\, h_f \left\{ b + b_w \left[\frac{0.503\,d}{h_f} - 1 \right] \right\}$$
$$= 0.0319(4) \left\{ 66 + 12 \left[\frac{0.503\,(19.49)}{4} - 1 \right] \right\}$$
$$= 10.64 \text{ in.}^2 > 3.81 \text{ in.}^2 \quad \text{(OK)}$$

12. Sketch the design (see Fig. 3–11).

EXAMPLE 3–5

Design a T-beam having a cross section as shown in Fig. 3–12. Assume that the effective flange width given is acceptable. The T-beam will carry a total design moment M_u of 340 ft-kips. $f'_c = 3,000$ psi and $f_y = 60,000$ psi. Use 1½-in. cover and No. 3 stirrups.

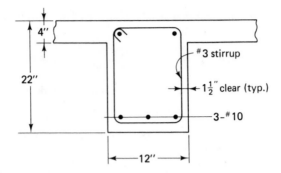

FIGURE 3-11 Design sketch for Example 3–4

1. The design moment $M_u = 340$ ft-kips (given).

2. Assume an effective depth

$$d = 22 - 3 = 19 \text{ in.}$$

3. The effective flange width = 27 in. (given).

4. Determine M_R assuming the flange to be completely in compression.

$$M_R = \phi(0.85f'_c)bh_f\left(d - \frac{h_f}{2}\right)$$

$$= 0.9(0.85)(3)\frac{(27)(3.5)(19 - 3.5/2)}{12} = 312 \text{ ft-kips}$$

5. $M_R < M_u$; therefore, the beam must behave as a true T-beam.

6. Find the approximate lever-arm distance for the internal couple:

$$Z = d - \frac{h_f}{2} = 19 - \frac{3.5}{2} = 17.25 \text{ in.}$$

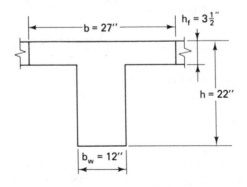

FIGURE 3-12 Sketch for Example 3–5

7. Find the approximate required A_s.

$$A_s = \frac{M_u}{\phi f_y Z} = \frac{340(12)}{0.9(60)(17.25)} = 4.38 \text{ in.}^2$$

8. Select the bars. Try 3 No. 11 bars ($A_s = 4.68$ in.²).

$$\text{Minimum } b_w = 11 \text{ in.} \qquad \text{(OK)}$$

9. Determine the actual effective depth.

$$d = 22 - 1.5 - 0.38 - \frac{1.41}{2} = 19.42 \text{ in.}$$

Use 19.42 in.

Check the beam capacity with an analysis for M_R:

1. The effective flange width has been established as 27 in.

2. The tensile force (assuming steel yields)

$$N_T = A_s f_y = 4.68(60) = 281 \text{ kips}$$

3. The flange compression (assuming full effective flange area is in compression)

$$N_C = (0.85 f_c') b\, h_f$$
$$= 0.85(3)(27)(3.5) = 241 \text{ kips}$$

4. Since $281 > 241$, the compressive stress block must extend below the flange far enough to pick up the remaining compression of

$$281 - 241 = 40 \text{ kips}$$

Hence the analysis is for a true T-beam, as expected.

5. The remaining compression will be obtained by additional web area. Solving for a, the total depth of stress block, as shown in Fig. 3–13:

$$a = \frac{N_T - N_C}{(0.85 f_c') b_w} + h_f$$
$$= \frac{40}{0.85(3)(12)} + 3.5 = 4.81 \text{ in.}$$

6. Check ρ_{min}.

$$\rho_{min} = \frac{200}{f_y} = \frac{200}{60,000} = 0.0033$$

$$\text{Actual } \rho = \frac{A_s}{b_w d} = \frac{4.68}{12(19.42)} = 0.0197$$

$$0.0197 > 0.0033 \qquad \text{(OK)}$$

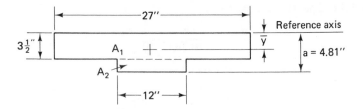

FIGURE 3-13   Compressive area—Example 3-5

7. Locate the centroid of the compressive area from the top of the cross section.

$$\bar{y} = \frac{\Sigma(Ay)}{\Sigma A}$$

$$A_1 = 27(3.5) = 94.5 \text{ in.}^2 \qquad A_2 = 12(1.31) = 15.72 \text{ in.}^2$$

$$\bar{y} = \frac{94.5(1.75) + 15.72(4.15)}{94.5 + 15.72} = 2.09 \text{ in.}$$

from which

$$Z = d - \bar{y} = 19.42 - 2.09$$

$$= 17.33 \text{ in.}$$

8. Compute the resisting moment M_R.

$$M_R = \phi N_T Z = \frac{0.9(281)(17.33)}{12} = 365.2 \text{ ft-kips}$$

M_u was previously determined to be 340 ft-kips; therefore, the design *for flexure* is adequate, since $M_R > M_u$.

9. Check $A_{s(\text{max})}$.

$$A_{s(\text{max})} = 0.0319 h_f \left[b + b_w \left(\frac{0.503\,d}{h_f} - 1 \right) \right]$$

$$= 0.0319(3.5) \left\{ 27 + 12 \left[\frac{0.503(19.42)}{3.5} - 1 \right] \right\} = 5.41 \text{ in.}^2$$

$$5.41 > 4.68 \qquad\qquad\qquad\qquad\qquad (\text{OK})$$

10. Sketch the design (see Fig. 3-14).

It should be noted at this point that the approximate method above will work whether the T-beam is a true T-beam or rectangular T-beam. (In fact, the use of an approximate Z to determine an approximate A_s can be used for the design of any beam because the approximate design is to be checked by an analysis, anyway.) However, the method works

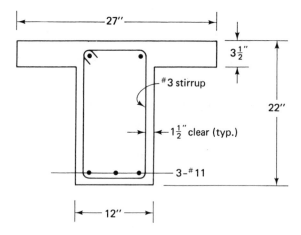

FIGURE 3-14 Design sketch for Example 3-5

particularly well for T-beams, in which, because of their basic shape, the centroid of the compressive stress block will never be very far from the middepth of the flange. An alternative design method for the true T-shape would be to design steel for the flange couple and stem couple individually; the total A_s required would then be the sum of the required steel areas for the two individual couples. This is similar to the approach used for the design of beams with compression reinforcement.

3-6 SUMMARY OF PROCEDURE FOR ANALYSIS OF T-BEAMS (FOR FLEXURE)

1. Establish the effective flange width based on ACI criteria.

2. Assume that the tensile steel yields and compute the total tension.

$$N_T = A_s f_y$$

3. Compute the magnitude of the compression that the flange itself is capable of furnishing.

$$N_C = 0.85 f'_c b h_f$$

4. If $N_T > N_C$, the beam will behave as a true T-beam and the remaining compression, which equals $N_T - N_C$, will be furnished by additional web area. If $N_T < N_C$, the beam will behave as a rectangular beam of width b.

True T-beam

5. Locate the bottom of the compressive stress block in the beam web.

$$a = \frac{N_T - N_C}{0.85 f'_c b_w} + h_f$$

6. Check ρ_{min}.

$$\rho_{min} = \frac{200}{f_y} \qquad \text{Actual } \rho = \frac{A_s}{b_w d}$$

The actual ρ must be larger than ρ_{min}.

7. Locate the centroid of total compressive area using the relationship

$$\bar{y} = \frac{\Sigma(Ay)}{\Sigma A}$$

from which

$$Z = d - \bar{y}$$

8. Compute the resisting moment M_R.

$$M_R = \phi N_C Z \quad \text{or} \quad \phi N_T Z$$

9. Check the ACI Code ductility requirement using the proper expression for $A_{s(max)}$ from Table 3–1. $A_{s(max)}$ must be larger than actual A_s.*

Rectangular T-beam

5. Check ρ_{min}.

$$\rho_{min} = \frac{200}{f_y} \qquad \text{Actual } \rho = \frac{A_s}{b_w d}$$

The actual ρ must be larger than ρ_{min}.

6. Compute the actual steel ratio in order to determine $\bar{k}$.

$$\rho = \frac{A_s}{bd}$$

7. Consult the proper table in Appendix A and find the required $\bar{k}$ for the ρ value from step 6.

8. Compute the resisting moment M_R.

$$M_R = \phi bd^2 \bar{k}$$

9. Check the ACI Code ductility requirement using the proper expression for $A_{s(max)}$ from Table 3–1. $A_{s(max)}$ must be larger than the actual A_s.*

A flow diagram of this procedure is presented in Appendix B.

---

*When checking the maximum steel limitation (step 9), if actual $A_s > A_{s(max)}$, the resisting moment M_R must be recalculated using $A_{s(max)}$, which may be considered an "effective" A_s.

3-7 SUMMARY OF PROCEDURE FOR DESIGN OF T-BEAMS (FOR FLEXURE)

1. Compute the design moment M_u.

3. Establish the effective flange width based on ACI criteria.

4. Compute the resisting moment M_R assuming that the total effective flange is in compression.

$$M_R = \phi(0.85f'_c)bh_f\left(d - \frac{h_f}{2}\right)$$

5. If $M_R > M_u$, the beam will behave as a rectangular T-beam of width b. If $M_R < M_u$, the beam will behave as a true T-beam.

Rectangular T-beam

6. Design as a rectangular beam with b and d as known values. Compute the required $\bar{k}$.

$$\text{Required } \bar{k} = \frac{M_u}{\phi bd^2}$$

7. From the tables in Appendix A, determine the required ρ for the required $\bar{k}$ of step 6.

8. Compute the required A_s.

$$\text{Required } A_s = \rho bd$$

9. Select bars and check the beam width. Check the actual d and compare with the assumed d. If the actual d is slightly in excess of the assumed d, the design will be slightly conservative (on the safe side). If the actual d is less than the assumed d, the design may be on the nonconservative side (depending on steel provided) and should be more closely investigated for possible revision.

10. Check ρ_{min}.

$$\rho_{min} = \frac{200}{f_y} \qquad \text{Actual } \rho = \frac{A_s}{b_w d}$$

The actual ρ must be larger than ρ_{min}. If not, the beam is not acceptable.

11. Check the ACI ductility requirement using proper expressions for $A_{s(max)}$ from Table 3-1. $A_{s(max)}$ must be larger than actual A_s.

12. Sketch the design.

True T-beam

6. Assume that $Z = d - h_f/2$.

7. Compute the required A_s based on the assumption of step 6.

$$A_s = \frac{M_u}{\phi f_y Z}$$

8. Select bars and check the beam width.

9. Determine the actual effective depth d and *analyze* the beam. (See the procedure for analysis of T-beam in Section 3–6.)

10. Sketch the design.

A flow diagram of this procedure is presented in Appendix B.

3–8 DOUBLY REINFORCED BEAMS—INTRODUCTION

As indicated in Chapter 2, the moment strength of a rectangular reinforced concrete beam reinforced with tensile steel only may be determined by the expression $M_R = \phi bd^2 \overline{k}$, where $\overline{k}$ is a function of the steel ratio ρ.

The maximum legal limit of ρ in a given cross section (with tensile steel only) has been established by the ACI Code as $\rho_{max} = 0.75\rho_b$. Therefore, for a given cross section of given material strengths, the maximum moment strength or resistance may be calculated using the $\overline{k}$ that corresponds to ρ_{max}.

Occasionally, practical and architectural considerations may dictate and limit beam sizes whereby it becomes necessary to develop more moment strength from a given cross section. When this situation occurs, the ACI Code, Section 10.3.4, permits the addition of tensile steel over and above the code maximum provided that compression steel is also added in the compression zone of the cross section. The result constitutes a combined tensile and compressive reinforced beam commonly called a *doubly reinforced beam*.

Where beams span more than two supports (continuous construction), practical considerations are sometimes the reason for the existence of main steel in compression zones. In Fig. 3–15, positive moment exists at *A* and *C*; therefore, the main tensile reinforcement would be placed in the bottom of the beam. However, at *B*, a negative moment exists and the bottom of the beam is in compression. The tensile reinforcement must be placed near the top of the beam. It is general practice that at least some of the tension steel in each of the cases above will be extended

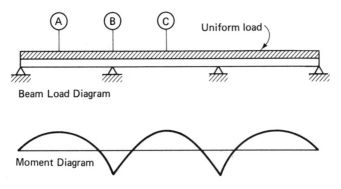

Beam Load Diagram

Moment Diagram

FIGURE 3-15 Continuous beam

the length of the beam and will pass through compression zones. In this case, the compression steel may sometimes be utilized for additional strength. In Chapter 7, it will also be seen that compression reinforcement will aid significantly in reducing long-term deflections. In fact, the use of compression steel to increase the bending strength of a reinforced concrete beam is an inefficient way to utilize steel. More commonly, deflection control will be the reason for the presence of compression steel.

3-9 DOUBLY REINFORCED BEAM ANALYSIS FOR FLEXURE (CONDITION I)

The basic assumptions for the analysis of doubly reinforced beams are similar to those for tensile reinforced beams. One additional significant assumption is that the *compression steel stress* (f'_s) is a function of the strain at the level of the centroid of the compression steel. As previously discussed, the steel will behave elastically up to the point where the strain exceeds the yield strain ϵ_y. In other words, as a limit, $f'_s = f_y$ when the compression steel strain $\epsilon'_s \geq \epsilon_y$. If $\epsilon'_s < \epsilon_y$, the compression steel stress $f'_s = \epsilon'_s E_s$, where E_s is the modulus of elasticity of the steel.

With two different materials, concrete and steel, resisting the compressive force N_C, the total compression will now consist of two forces: N_{C1}, the compression resisted by the concrete and N_{C2}, the compression resisted by the compressive steel. For analysis, the total resisting moment of the beam will be assumed to consist of two parts or two internal couples: the part due to the resistance of the compressive concrete and tensile steel, and the part due to the compressive steel and additional tensile steel. The two internal couples are illustrated in Fig. 3–16.

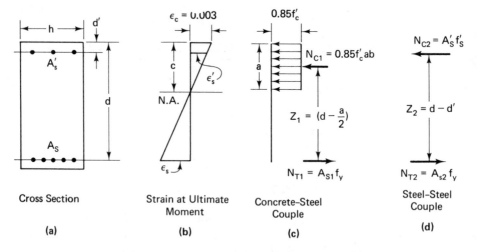

FIGURE 3-16 Doubly reinforced beam analysis

Notation for doubly reinforced beams is as follows:

A'_s = total compression steel cross-sectional area

d = effective depth of tension steel

d' = depth to centroid of compression steel from compression face of beam

A_{s1} = amount of tension steel utilized by the concrete–steel couple

A_{s2} = amount of tension steel utilized by the steel–steel couple

A_s = total tension steel cross-sectional area ($A_s = A_{s1} + A_{s2}$)

M_{n1} = ideal moment strength of the concrete–steel couple

M_{n2} = ideal moment strength of the steel–steel couple

M_n = ideal moment strength of the (doubly reinforced) beam

ϵ_s = unit strain at the centroid of the tension steel

ϵ'_s = unit strain at the centroid of the compression steel

The total moment strength may be developed as the sum of the two internal couples, neglecting the concrete that is displaced by the compression steel.

The strength of the steel–steel couple is evaluated as follows:

$$M_{n2} = N_{T2} Z_2$$

Assuming that $f_s = f_y$ (tensile steel yields),

$$M_{n2} = A_{s2} f_y(d - d')$$

Also, since $\Sigma H_F = 0$ and $N_{C2} = N_{T2}$,

$$A'_s f'_s = A_{s2} f_y$$

If we assume that the compression steel yields and that $f'_s = f_y$, then

$$A'_s f_y = A_{s2} f_y$$

from which

$$A'_s = A_{s2}$$

Therefore,

$$M_{n2} = A'_s f_y(d - d')$$

The strength of the concrete–steel couple is evaluated as follows:

$$M_{n1} = N_{T1} Z_1$$

Assuming that the tensile steel yields and $f_s = f_y$,

$$M_{n1} = A_{s1} f_y\left(d - \frac{a}{2}\right)$$

Also, since $A_s = A_{s1} + A_{s2}$, then

$$A_{s1} = A_s - A_{s2}$$

and since $A_{s2} = A'_s$, then

$$A_{s1} = A_s - A'_s$$

Therefore,

$$M_{n1} = (A_s - A'_s)f_y\left(d - \frac{a}{2}\right)$$

Summing the two couples, we arrive at the theoretical ideal moment strength of a doubly reinforced beam:

$$M_n = M_{n1} + M_{n2}$$
$$= (A_s - A'_s)f_y(d - \frac{a}{2}) + A'_s f_y(d - d')$$

M_R may be found by applying the strength reduction factor to M_n:

$$M_R = \phi M_n$$

The foregoing expressions are based on the assumption that both tension and compression steels yield prior to the concrete strain reaching 0.003. This may be checked by determining the strains that exist at ultimate moment, which depend on the location of the neutral axis. The neutral axis may be located, as previously, by the depth of the compressive stress block and the relationship $a = \beta_1 c$.

$$N_T = N_{C1} + N_{C2}$$

$$A_s f_y = (0.85 f'_c) ab + A'_s f_y$$

$$a = \frac{(A_s - A'_s) f_y}{(0.85 f'_c) b}$$

which may also be expressed as

$$a = \frac{A_{s1} f_y}{(0.85 f'_c) b}$$

With the calculation of the a distance, the neutral axis location c may be determined and the assumptions checked.

The case where both tensile and compressive steels yield prior to the concrete strain reaching 0.003 will be categorized as *condition I* (see Example 3–6). The case where the tensile steel yields but the compressive steel does not yield prior to the concrete strain reaching 0.003 is categorized as *condition II* (see Example 3–7).

EXAMPLE 3–6

Compute the dependable resisting moment strength M_R for a beam having a cross section as shown in Fig. 3-17. Use $f'_c = 3,000$ psi and $f_y = 40,000$ psi.

1. Assume that all the steel yields.

$$f'_s = f_y \quad \text{and} \quad f_s = f_y$$

Therefore, $A_{s2} = A'_s$ [reference, Fig. 3-16(d)].

2.
$$A_s = A_{s1} + A_{s2}$$
$$A_{s1} = A_s - A'_s$$
$$= 7.62 - 3.12 = 4.50 \text{ in.}^2$$

From the concrete–steel couple, the stress block depth may be found.

$$a = \frac{A_{s1} f_y}{(0.85 f'_c) b} = \frac{4.50(40)}{0.85(3.0)(11)} = 6.42 \text{ in.}$$

3. Assuming that the same relationship ($a = \beta_1 c$) exists between the depth of the stress block and the beam's neutral axis as existed in singly reinforced

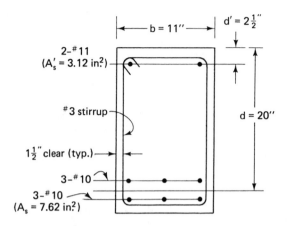

FIGURE 3-17 Sketch for Example 3-6

beams, the neutral axis may now be located for purposes of checking the steel strains. From Fig. 3-16 at the ultimate moment,

$$c = \frac{a}{\beta_1} = \frac{a}{0.85} = \frac{6.42}{0.85} = 7.55 \text{ in.}$$

4. Check the strains to determine whether the assumptions are valid and that both steels yield before concrete crushes; see Fig. 3-16(b). The strains calculated exist at the ultimate moment.

$$\epsilon'_s = \frac{c - d'}{c}(0.003)$$

$$= \frac{7.55 - 2.5}{7.55}(0.003) = 0.0020$$

$$\epsilon_s = \frac{d - c}{c}(0.003)$$

$$= \frac{12.45}{7.55}(0.003) = 0.0049$$

$$\epsilon_y = 0.00138 \qquad \text{(from Table A-1)}$$

Since ϵ'_s and ϵ_s both exceed ϵ_y, the compression and tensile steel have reached their yield stress before the concrete reaches a strain of 0.003. Therefore, the assumptions concerning the steel stresses are OK ($f'_s = f_y$ and $f_s = f_y$).

5. From the concrete–steel couple:

$$M_{n1} = A_{s1}f_y\left(d - \frac{a}{2}\right)$$

$$= 4.50\,(40)\left(20 - \frac{6.42}{2}\right) = 3022 \text{ in.-kips}$$

$$\frac{3,022}{12} = 252 \text{ ft-kips}$$

From the steel–steel couple:

$$M_{n2} = A_s'\, f_y(d - d') = 3.12\,(40)\,(20 - 2.5)$$

$$= 2,184 \text{ in.-kips}$$

$$\frac{2,184}{12} = 182 \text{ ft-kips}$$

$$M_n = M_{n1} + M_{n2}$$

$$= 252 + 182 = 434 \text{ ft-kips}$$

6.
$$M_R = \phi M_n$$

$$= 0.9\,(434)$$

$$= 391 \text{ ft-kips}$$

7. The fact that both the compressive steel and tensile steel yield prior to the concrete reaching a compressive strain of 0.003 implies a theoretical ductile beam. However, despite this ductile behavior, the beam may not satisfy the ACI Code ductility requirement if the actual ρ is in excess of $0.75\rho_b$; therefore, a Code ductility check should be made.

The ACI Code limitation on ρ (section 10.3.3) applies to doubly reinforced beams as well as to singly reinforced beams. Expressions for $A_{s\,(\max)}$ for doubly reinforced beams may be developed in a manner similar to that used for T-beams. This development and the resulting expressions will be found in Appendix B. The intent of the Code to ensure a ductile type beam is satisfied if the concrete–steel couple has ρ less than $0.75\rho_b$ and is, therefore, ductile.

This will usually be satisfactory since the addition of the steel couple comprised of A_s' and A_{s2} will not, of itself, lead to a brittle failure, since this is a steel couple and has toughness typical of steel. However, in cases of slabs and shallow beams where f_s' may be less than f_y, simply ensuring that the concrete–steel couple (with $A_{s1} = A_s - A_s'$) is ductile will not necessarily ensure that the doubly reinforced beam is ductile. In these cases the expressions of Appendix B should be used.

Checking the concrete–steel couple, we obtain

$$\rho = \frac{A_{s1}}{bd} = \frac{4.50}{11(20)} = 0.0205$$

$$\rho_{max} = 0.0278$$

Since $0.0205 < 0.0278$, the Code criterion for ductile behavior is ensured.

3–10 DOUBLY REINFORCED BEAM ANALYSIS FOR FLEXURE (CONDITION II)

As has been pointed out, usually the compression steel (A_s') will reach its yield stress before the concrete reaches a strain of 0.003. However, particularly in shallow beams reinforced with the higher-strength steels, this is not necessarily true. Referring to Fig. 3-16(b), if the neutral axis is located relatively high in the cross section, it is possible that $\epsilon_s' < \epsilon_y$ at ultimate moment. The magnitude of ϵ_s' (and therefore f_s') will depend on the actual location of the neutral axis. The depth of the compressive stress block a also depends on c, since $a = \beta_1 c$.

The total compressive force must be equal to the total tensile force $A_s f_y$ and an equilibrium equation can be written to solve for the exact required value of c. This turns out to be a quadratic equation.

This situation and its solution may be observed in the following example.

EXAMPLE 3–7

Compute the dependable resisting moment strength M_R for a beam having a cross section as shown in Fig. 3-18. Use $f_c'= 5{,}000$ psi and $f_y= 60{,}000$ psi.

1. Assume that all the steel yields. This will result in

$$A_{s2} = A_s'$$

2. With reference to Fig. 3–19(b),

$$a = \frac{(A_s - A_s')f_y}{(0.85f_c')b}$$

$$= \frac{3.10(60)}{0.85(5)(11)}$$

$$= 3.98 \text{ in.}$$

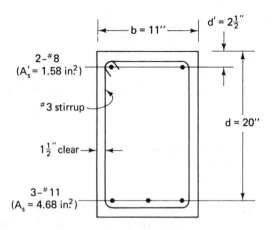

FIGURE 3-18   Sketch for Example 3-7

3. Locate the neutral axis.

$$a = \beta_1 c \qquad \beta_1 = 0.80 \qquad \text{(reference, ACI Code, Section 10.2.7c)}$$

$$c = \frac{a}{\beta_1} = \frac{3.98}{0.80} = 4.98 \text{ in.}$$

4. By similar triangles of Fig. 3-19(a), check the steel strains:
   Compressive steel:

$$\epsilon'_s = \frac{0.003\,(c - d')}{c}$$

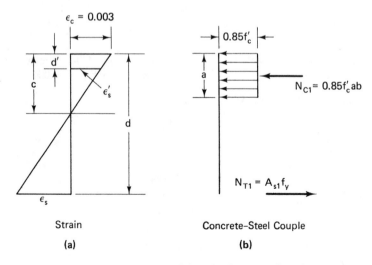

FIGURE 3-19   Compatibility check—Example 3-7

92

$$\epsilon'_s = \frac{2.48}{4.98} \,(0.003) = 0.0015$$

Tensile steel:

$$\epsilon_s = \frac{0.003\,(d - c)}{c}$$

$$\epsilon_s = \frac{20 - 4.98}{4.98}(0.003) = 0.00905$$

For grade 60 steel, $\epsilon_y = 0.00207$ (from Table A–1). Since $\epsilon_s > \epsilon_y > \epsilon'_s$, the tensile steel *has* yielded and the compression steel *has not* yielded. Therefore, the assumptions of step 1 are incorrect.

5. With the original assumptions incorrect, a solution for the location of the neutral axis must be established. With reference to Fig. 3-16, c will be determined by using the condition that horizontal equilibrium exists. That is, $\Sigma H_F = 0$.

$$N_T = N_{C1} + N_{C2}$$

$$A_s f_y = (0.85\, f'_c)\, ba + f'_s A'_s$$

However,

$$a = \beta_1 c$$

and

$$f'_s = \epsilon'_s E_s = \left[\frac{c - d'}{c}\,(0.003) \right] E_s$$

Then, by substitution,

$$A_s f_y = (0.85 f'_c)\, b\beta_1 c + \left[\frac{c - d'}{c}\,(0.003) \right] E_s A'_s$$

Multiplying by c and expanding, we obtain

$$A_s f_y c = (0.85 f'_c)\, b\beta_1 c^2 + c(0.003) E_s A'_s - d'\,(0.003) E_s A'_s$$

Rearranging yields

$$(0.85 f'_c b\beta_1)\, c^2 + (0.003 E_s A'_s - A_s f_y)\, c - d'\,(0.003) E_s A'_s = 0$$

With $E_s = 29{,}000$ ksi, the expression becomes

$$(0.85 f'_c b\beta_1)\, c^2 + (87 A'_s - A_s f_y)\, c - 87\, d' A'_s = 0$$

$$\text{where } A_s = 4.68 \text{ in.}^2$$

$$f_y = 60 \text{ ksi}$$

$$f'_c = 5 \text{ ksi}$$

$$b = 11 \text{ in.}$$

$$\beta_1 = 0.80$$

$$A_s' = 1.58 \text{ in.}^2$$

$$d' = 2.5 \text{ in.}$$

Substitution yields

$$\left[0.85\,(5)(11)(0.80) \right] c^2 + \left[87\,(1.58) - 4.68\,(60) \right] c - 87\,(2.5)(1.58) = 0$$

$$37.4\,c^2 - 143.34\,c - 343.65 = 0$$

$$c^2 - 3.83\,c - 9.19 = 0$$

This may be solved using the usual formula for the roots of a quadratic equation:

$$\frac{-b \pm \sqrt{b^2 - 4ac}}{2a}$$

where the coefficients are

$$a = 1.0$$

$$b = 3.83$$

$$c = 9.19$$

Or, the square may be completed as follows:

$$c^2 - 3.83c = 9.19$$

$$c^2 - 3.83c + \left(\frac{-3.83}{2} \right)^2 = 9.19 + \left(\frac{-3.83}{2} \right)^2$$

$$c^2 - 3.83c + 3.67 = 9.19 + 3.67 = 12.86$$

$$(c - 1.92)^2 = 12.86$$

$$c - 1.92 = \sqrt{12.86} = 3.59$$

$$c = 5.51 \text{ in.}$$

The solution of the quadratic equation for c may be simplified as follows:

$$c = \pm \sqrt{Q + R^2} - R$$

$$\text{where } R = \frac{87A_s' - A_s f_y}{1.7 f_c' b \beta_1}$$

$$Q = \frac{87 d' A_s'}{0.85 f_c' b \beta_1}$$

Note that basic units are kips and inches, so that the value of f_y, for instance, must be in ksi, not in psi.

6. With this value of c, all the remaining unknowns may be found:

$$f'_s = \frac{c - d'}{c}(87) = \frac{5.51 - 2.50}{5.51}(87)$$

$$= 47.5 \text{ ksi} < 60 \text{ ksi} \quad \text{(as expected)}$$

7. $$a = \beta_1 c = 0.80(5.51) = 4.41 \text{ in.}$$

8. $$N_{C1} = (0.85 f'_c)\, ab = 0.85(5)(4.41)(11.0) = 206.2 \text{ kips}$$

$$N_{C2} = A'_s f'_s = 47.5(1.58) = 75.1 \text{ kips}$$

$$N_C = 281.3 \text{ kips}$$

$$(\text{check}) \; N_T = A_s f_y = 4.68(60) = 281 \text{ kips}$$

$$N_T \simeq N_C \qquad\qquad \text{(OK)}$$

9. $$M_{n1} = N_{C1} Z_1 = N_{C1}\left(d - \frac{a}{2}\right) = 206.2\left(20 - \frac{4.41}{2}\right) = 3{,}670 \text{ in.-kips}$$

$$M_{n2} = N_{C2} Z_2 = N_{C2}(d - d') = 75.1(20 - 2.5) = 1{,}314$$

$$M_n = 4{,}984 \text{ in.-kips} = 415.3 \text{ ft-kips}$$

10. $$M_R = \phi M_n$$

$$= 0.9(415.3)$$

$$= 373.8 \text{ ft-kips}$$

11. Ductility will be checked based on ACI Code requirements. Since $f'_s < f_y$, the amount of tension steel required for the steel–steel couple (A_{s2}) will be less than A'_s. Find the new *effective* A_{s1}.

$$N_{T2} = N_{C2}$$

$$A_{s2} f_y = A'_s f'_s$$

$$A_{s2} = \frac{A'_s f'_s}{f_y} = \frac{1.58(47.5)}{60} = 1.25 \text{ in.}^2$$

Then

$$A_{s1} = A_s - A_{s2} = 4.68 - 1.25 = 3.43 \text{ in.}^2$$

For the concrete–steel couple:

$$\rho = \frac{A_{s1}}{bd} = \frac{3.43}{11(20)} = 0.0156$$

$$0.75 \rho_b = 0.0252 \qquad \text{(Table A–4)}$$

$$0.0156 < 0.0252 \qquad\qquad \text{(OK)}$$

3-11 SUMMARY OF PROCEDURE FOR ANALYSIS OF DOUBLY REINFORCED BEAMS (FOR FLEXURE)

1. Assume that all the steel yields, $f_s = f'_s = f_y$. Therefore,

$$A_{s2} = A'_s$$

2. Using the concrete steel couple and $A_{s1} = A_s - A'_s$, compute the depth of the compression stress block.

$$a = \frac{A_{s1} f_y}{(0.85 f'_c)\, b} = \frac{(A_s - A'_s) f_y}{(0.85 f'_c) b}$$

3. Compute the location of the neutral axis.

$$c = \frac{a}{\beta_1}$$

4. Using the strain diagram, check the strains in both the compressive and tensile steels to determine whether the assumption in step 1 is valid.

$$\epsilon'_s = \frac{0.003\,(c - d')}{c}$$

$$\epsilon_s = \frac{0.003\,(d - c)}{c}$$

Assuming that $\epsilon_s \geqslant \epsilon_y$, that is, that the tensile steel has yielded, the two following conditions may exist:

a. Condition I: $\epsilon'_s \geqslant \epsilon_y$. This indicates that the assumption of step 1 is correct and the compression steel has yielded.

b. Condition II: $\epsilon'_s < \epsilon_y$. This indicates that the assumption of step 1 is incorrect and the compression steel has not yielded.

Note that it is possible for two other conditions to exist. These would occur if $\epsilon_s < \epsilon_y$, with the resulting tensile steel stress being at less than the yield stress. This situation would be rare but could occur in the analysis of *existing* overreinforced, doubly reinforced beams and slabs. Conditions I and II assume that $\epsilon_s \geqslant \epsilon_y$.

Condition I

5. If ϵ'_s and ϵ_s both exceed ϵ_y, compute the theoretical moment capacities M_{n1} and M_{n2}.

For a steel-steel couple:

$$M_{n2} = A'_s f_y (d - d')$$

For a concrete-steel couple

$$M_{n1} = A_{s1}f_y\left(d - \frac{a}{2}\right)$$

and

$$M_n = M_{n1} + M_{n2}$$

6. $M_R = \phi M_n$

7. Check the ductility requirement. Ensure that the actual ρ for the concrete-steel couple does not exceed $0.75\rho_b$ (Table A–4) or ensure that the tension steel area is not in excess of $A_{s(max)}$ as determined by expressions found in Appendix B.

$$\text{Actual } \rho = \frac{A_{s1}}{bd}$$

Condition II

5. If ϵ'_s is less than ϵ_y and $\epsilon_s \geqslant \epsilon_y$, compute c using the following formula:

$$(0.85f'_c b\beta_1)\, c^2 + (87A'_s - A_s f_y)\, c - 87d'A'_s = 0$$

and solve the quadratic equation for c, or use the simplified formula approach from Example 3–7, step 5. Note that the basic units are kips and inches.

6. Compute the compressive steel stress (to be less than f_y).

$$f'_s = \frac{c - d'}{c}\ (87)$$

7. Solve for a using

$$a = \beta_1 c$$

8. Compute the compressive forces.

$$N_{C1} = (0.85f'_c)\, ba$$

$$N_{C2} = A'_s f'_s$$

Check these by computing the tensile force:

$$N_T = A_s f_y$$

N_T should equal $N_{C1} + N_{C2}$.

9. Compute the ideal resisting moment strengths of the individual couples.

$$M_{n1} = N_{C1}\left(d - \frac{a}{2}\right)$$

and

$$M_{n2} = N_{C2}(d - d')$$

$$M_n = M_{n1} + M_{n2}$$

10. $M_R = \phi M_n$

11. Check the ductility requirements. Ensure that the actual ρ for the concrete–steel couple does not exceed $0.75\rho_b$ (Table A–4). A_{s1} is computed based on the fact that the compressive steel stress does not reach f_y.

$$A_{s1} = A_s - \frac{A_s' f_s'}{f_y}$$

and

$$\text{Actual } \rho = \frac{A_{s1}}{bd}$$

Or, the ductility requirements may be checked by comparing A_s with $A_{s(\max)}$ as determined by expressions found in Appendix B, in which the actual balanced condition is investigated.

3-12 DOUBLY REINFORCED BEAM DESIGN FOR FLEXURE

If a check shows that a singly reinforced rectangular section is inadequate and the size of the beam cannot be increased, a doubly reinforced section may be designed using a procedure which basically consists of the separate design of the two component couples such that their summation will result in a beam of the required strength.

EXAMPLE 3–8

Design a rectangular reinforced concrete beam to carry service moments of $M_{DL} = 170$ ft-kips and $M_{LL} = 270$ ft-kips. Physical limitations require that $b = 14$ in. and $h = 30$ in. If compression steel is needed, $d' = 3$ in. $f_c' = 3{,}000$ psi and $f_y = 40{,}000$ psi. The beam cross section is shown in Fig. 3–20.

1. Assume that $d = h - 4 = 26$ in.(because of the probability of two rows of steel).

2. Calculate the design moment.

$$M_u = 1.4 M_{DL} + 1.7 M_{LL}$$

$$= 1.4(170) + 1.7(270)$$

$$= 238 + 459 = 697 \text{ ft-kips}$$

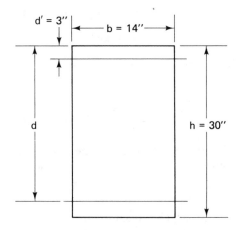

FIGURE 3-20 Sketch for Example 3–8

3. Determine if a singly reinforced beam will work. From Table A–4,

$$\text{Maximum } \bar{k} = 0.8695 \text{ ksi}$$

$$\text{Maximum possible } M_R = \phi bd^2 \bar{k}$$

$$= \frac{0.9(14)(26)^2(0.8695)}{12} = 617 \text{ ft-kips}$$

4. 617 ft-kips < 697 ft-kips. Therefore, a doubly reinforced beam is required.

5. Provide a concrete–steel couple having a steel ratio of about 90% of ρ_{max}.

$$\rho = 0.90(0.0278) = 0.0250$$

From Table A–6,

$$\bar{k} = 0.8039 \text{ ksi}$$

6. The dependable strength or capacity of a concrete–steel couple will then be

$$M_{R1} = \phi bd^2 \bar{k} = \frac{0.9(14)(26)^2(0.8039)}{12}$$

$$= 571 \text{ ft-kips}$$

The required steel area for the concrete–steel couple may now be determined.

$$\text{Required } A_{s1} = \rho bd = 0.0250(14)(26)$$

$$= 9.1 \text{ in.}^2$$

7. The steel–steel couple must be proportioned to have moment strength equal to the balance of the design moment.

$$\text{Required } M_{R2} = M_u - M_{R1} = 697 - 571$$
$$= 126 \text{ ft-kips}$$

8. Considering the steel–steel couple, we have

$$M_{R2} = \phi \, N_{C2}(d - d')$$

$$N_{C2} = \frac{M_{R2}}{\phi(d - d')} = \frac{126(12)}{0.9(26 - 3)} = 73.0 \text{ kips}$$

9. Since $N_{C2} = A'_s f'_s$, compute f'_s by using the neutral-axis location of the concrete–steel couple and checking the strain ϵ'_s in the compression steel (see Fig. 3–21).

$$a = \frac{A_{s1} f_y}{(0.85 f'_c) b} = \frac{9.1(40)}{0.85(3)(14)} = 10.2 \text{ in.}$$

$$c = \frac{a}{\beta_1} = \frac{10.2}{0.85} = 12.0 \text{ in.}$$

$$\epsilon'_s = \frac{c - d'}{c}(0.003) = \frac{9}{12}(0.003) = 0.00225$$

$$\epsilon_y = 0.00138 \quad \text{(from Table A–1)}$$

Since $\epsilon_y < \epsilon'_s$, the compression steel will yield before the concrete strain reaches 0.003, and $f'_s = f_y$ as a limit.

10. Determine the required compression steel.

$$\text{Required } A'_s = \frac{N_{C2}}{f'_s} = \frac{N_{C2}}{f_y} = \frac{73}{40} = 1.83 \text{ in.}^2$$

11. Determine the required tension steel for the steel–steel couple. Since $f'_s = f_y$,

$$A_{s2} = A'_s = 1.83 \text{ in.}^2$$

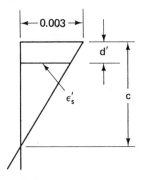

FIGURE 3-21   Concrete strain diagram for Example 3–8

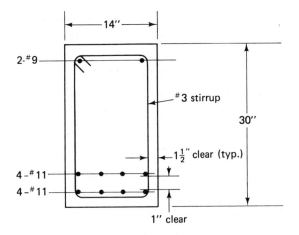

FIGURE 3-22 Design sketch for Example 3-8

12. Determine the *total* tension steel required.

$$A_s = A_{s1} + A_{s2} = 9.1 \text{ in.}^2 + 1.83 \text{ in.}^2 = 10.93 \text{ in.}^2$$

13. Select the compression steel (1.83 in.² required). Use 2 No. 9 bars ($A'_s = 2.00$ in.²).

14. Select the tension steel (10.93 in.² required). Use 8 No. 11 bars ($A_s = 12.5$ in.²). Place these in two layers (four bars per layer).

Required beam width = 14 in.　　　　　　(OK)

15. Assume No. 3 stirrups and determine the actual depth to the centroid of the bar group by considering the total depth (30 in.), required cover (1½ in.), stirrup size (No. 3), tension bar size (No. 11), and required 1-in. mimimum clear space between layers.

Actual $d = 30 - 1.5 - 0.38 - 1.41 - 0.5 = 26.2$ in.

The assumed d was 26 in.

$26.2 > 26$　　　　　　(OK)

16. Figure 3–22 is a sketch of the design.

3-13 SUMMARY OF PROCEDURE FOR DESIGN OF DOUBLY REINFORCED BEAMS (FOR FLEXURE)

The size of the beam cross section is fixed.

1. Assume that $d = h - 4$ in.

2. Establish the total design moment M_u.

3. Check to see if doubly reinforced beam is necessary. From tables in Appendix A, obtain maximum $\overline{k}$ and compute maximum M_R for a singly reinforced beam.

$$\text{Maximum } M_R = \phi bd^2\overline{k}$$

4. If $M_R < M_u$, design the beam as a compressive reinforced beam. If $M_R \geqslant M_u$, the beam can be designed as a beam reinforced with tension steel only.

For a compressive reinforced beam

5. Provide a concrete–steel couple having the steel ratio

$$\rho = 0.90(\rho_{max}) = 0.90(0.75\rho_b)$$

With this ρ, enter the appropriate table and determine $\overline{k}$.

6. Determine the moment capacity of the concrete–steel couple.

$$M_{R1} = \phi bd^2\overline{k}$$

Find the steel required for the concrete–steel couple.

$$\text{Required } A_{s1} = \rho bd$$

7. Find the remaining moment that must be resisted by the steel–steel couple.

$$\text{Required } M_{R2} = M_u - M_{R1}$$

8. Considering the steel–steel couple, find the required compressive force in the steel (assume that $d' = 3$ in.).

$$N_{C2} = \frac{M_{R2}}{\phi(d - d')}$$

9. Since $N_{C2} = A_s'f_s'$ compute f_s' so that A_s' may eventually be determined. This can be accomplished by using the neutral-axis location of the concrete–steel couple and checking the strain ϵ_s' in the compression steel with ϵ_y from Table A–1.

$$a = \frac{A_{s1}f_y}{(0.85f_c')b}$$

$$c = \frac{a}{\beta_1}$$

$$\epsilon_s' = \frac{0.003(c - d')}{c}$$

If $\epsilon_s' \geqslant \epsilon_y$, the compressive steel has yielded at ultimate moment and $f_s' = f_y$. If $\epsilon_s' < \epsilon_y$, then calculate $f_s' = \epsilon_s'E_s$ and use this stress in the following steps.

10. Since $N_{C2} = A_s'f_s'$,

$$\text{Required } A_s' = \frac{N_{C2}}{f_s'}$$

11. Determine required A_{s2}

$$A_{s2} = \frac{f'_s A'_s}{f_y}$$

12. Find the total tension steel required.

$$A_s = A_{s1} + A_{s2}$$

13. Select the compressive steel (A'_s).

14. Select the tensile steel (A_s). Check the required beam width. Preferably, place the bars in one layer.

15. Check the actual d and compare with the assumed d. If the actual d is slightly in excess of the assumed d, the design will be slightly conservative (on the safe side). If the actual d is less than the assumed d, the design may be on the unconservative side and an analysis and possible revision should be considered.

16. Sketch the design.

3-14 ADDITIONAL CODE REQUIREMENTS FOR DOUBLY REINFORCED BEAMS

The compression steel in beams, whether it is in place to increase flexural strength or to control deflections, will act similar to all typical compression members in that it will tend to buckle as shown in Fig. 3–23. Should this buckling occur, it will naturally be accompanied by spalling of the concrete cover. To help guard against this type of failure, the ACI Code requires that the compression bars be tied into the beam in a manner similar to that used for reinforced concrete columns (to be discussed in Chapter 9). Compression reinforcement in beams or girders must be enclosed by ties or stirrups. The size of the ties or stirrups is to be at least No. 3 for No. 10 longitudinal bars or smaller and No. 4 for No. 11 longitudinal bars or larger. The spacing of the ties or stirrups is not to exceed the smaller of 16 longitudinal bar diameters, 48 tie (or stirrup) bar diameters, or the least dimension of the beam. Alternatively, welded wire fabric of equivalent area may be used. The ties or stirrups are to be

FIGURE 3-23 Possible failure mode for compression steel

used throughout the area where compression reinforcement is required (see the ACI Code, Section 7.11).

PROBLEMS

In the following problems, unless otherwise noted, assume No. 3 stirrups, 1½ in. cover for beams, and 1 in. clear between layers of bars.

3–1. The simple span T-beam shown is part of a floor system of span length 20 ft-0 in. and beam spacing 45 inches on centers. $f'_c = 3,000$ psi and $f_y = 40,000$ psi.

    a. Find the dependable moment capacity M_R and check the steel to ensure that it is within allowable limits.

    b. How much steel would be required in this beam cross section to make the compressive stress block just completely cover the flange?

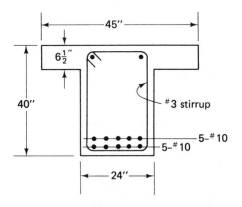

FIGURE P3-1

3–2. Find the dependable moment capacity M_R for the T-beam in the floor system shown. The beam span is 31 ft 6 in. $f'_c = 4,000$ psi and $f_y = 60,000$ psi. Check the steel.

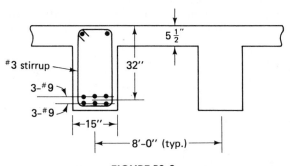

FIGURE P3-2

3-3. The T-beam shown is on a simple span of 30 ft. $f_c'= 3,000$ psi and $f_y = 60,000$ psi. No dead load exists other than the weight of the floor system. Assume that the slab design is adequate.

  a. Find the dependable moment capacity M_R.
  b. Compute the permissible service live load that can be placed on the beam (kips/ft).
  c. Compute the permissible service live load that can be placed on the floor (psf).

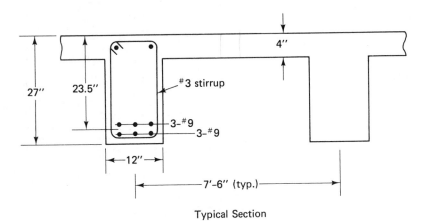

Typical Section

FIGURE P3-3

3-4. The simple span T-beam shown is part of a floor system of span length 20 ft-0 in. and beam spacing of 8 ft-0 in. on centers. Find the dependable moment capacity M_R. $f_c'= 3,000$ psi and $f_y = 60,000$ psi.

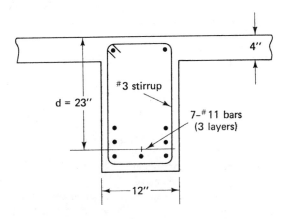

FIGURE P3-4

3–5. Find the dependable moment capacity for the beams of cross section shown. Assume that the physical dimensions are acceptable.

   a. Spandrel beam with a flange on one side only. $f'_c = 4,000$ psi and $f_y = 60,000$ psi.
   b. Box beam. $f'_c = 3,000$ psi and $f_y = 40,000$ psi.

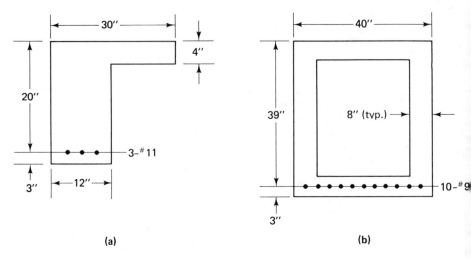

(a)   (b)

FIGURE P3-5

3–6. Design a typical interior tension reinforced T-beam to resist positive moment. A cross section of the floor system is shown. The service loads are 50 psf dead load (this does not include the weight of the beam and slab) and 325 psf live load. The beam is on a simple span of 18 ft. $f'_c = 4,000$ psi and $f_y = 40,000$ psi.

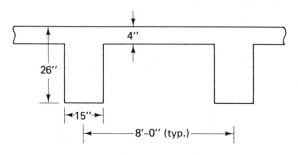

FIGURE P3-6

3–7. A reinforced concrete floor system consists of a 3-in. concrete slab supported by continuous span T-beams of 24-ft spans. The T-beams are spaced 4 ft 8 in. on centers. The web dimensions, determined by negative moment and shear requirements at the supports, are shown. Select the steel required at midspan to resist a total positive design moment M_u of 575 ft-kips (this includes the weight of the floor system). Use $f'_c = 3,000$ psi and $f_y = 60,000$ psi.

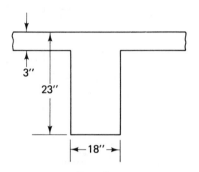

FIGURE P3-7

3–8. Select steel for the beams of cross section shown. The positive service moments are 155 ft-kips live load and 80 ft-kips dead load (this includes beam weight). Use $f_c' = 3,000$ psi and $f_y = 40,000$ psi.

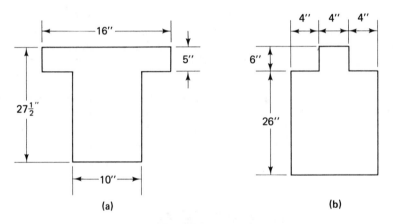

FIGURE P3-8

3–9. Find M_R for the beam of cross section shown (p. 108). Use $f'_c = 4,000$ psi and $f_y = 60,000$ psi.

3–10. The beam of cross section shown is to span 28 ft on simple supports. The uniform load on the beam (in addition to its own weight) will be composed of equal service dead load and live load. $f'_c = 4,000$ psi and $f_y = 60,000$ psi.

   a. Find the service loads that the beam can carry (in addition to its own weight).
   b. Compare the strength (M_R) of the beam as shown with the strength of a beam of similar size reinforced with the maximum allowable area of tension steel only.

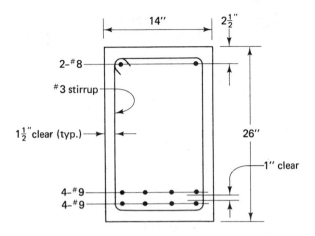

FIGURE P3-9

3–11. Find M_R for the beam cross section shown. $f'_c = 4,000$ psi and $f_y = 40,000$ psi. Bottom tensile steel is 4-No. 11 bars.

3–12. Design a rectangular reinforced concrete beam to resist a total design moment M_u of 780 ft-kips (this includes the moment due to the weight of the beam). The beam size is limited to 15 in. maximum width and 30 in. maximum overall depth. $f'_c = 3,000$ psi and $f_y = 40,000$ psi. If compression steel is required, make $d' = 2\frac{1}{2}$ in.

3–13. Design a rectangular reinforced concrete beam to resist service moments of 136 ft-kips dead load (includes moment due to weight of beam) and 150 ft-kips due to live load. Architectural considerations require that width be limited to 11 in. and overall depth be limited to 23 in. Use $f'_c = 3,000$ psi and $f_y = 40,000$ psi.

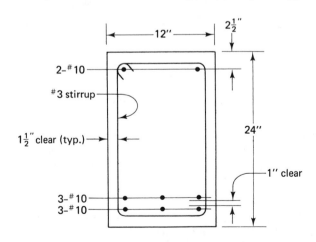

FIGURE P3-10

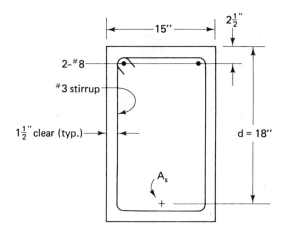

FIGURE P3-11

3–14. Design a rectangular reinforced concrete beam to carry service loads of 1.05 kips/ft dead load (includes beam weight) and 2.47 kips/ft live load. The beam is a simple span and has a span length of 18 ft. The overall dimensions are limited to width of 10 in. and overall depth of 20 in. Use $f'_c = 3,000$ psi and $f_y = 40,000$ psi.

3–15. Redesign the beam of Problem 3–14 for tension reinforcing only and increased width. Keep an overall depth of 20 in.

4

Shear in Beams

4-1 INTRODUCTION

In addition to bending considerations, beams must normally resist shear forces. In reinforced concrete beams, additional steel reinforcing must be added specifically to resist shear if shear is in excess of what concrete itself can resist. It will be seen that in most bending members the critical shear condition is that which results in additional tensile stresses.

The concepts of bending stresses and shearing stresses in homogeneous elastic beams are generally discussed at great length in most strength-of-materials texts. The accepted expressions are

$$f = \frac{Mc}{I} \quad \text{and} \quad v = \frac{VQ}{Ib} \tag{4-1}$$

where f, M, c, and I are as defined in Chapter 1, v is the shear stress, V the external shear, Q the statical moment of area about the neutral axis, and b the width of the cross section.

All points in the length of the beam, where the shear and bending moment are not equal to zero, and at locations other than the extreme fiber or neutral axis are subject to both shearing stresses and bending stresses. The combination of these stresses is of such a nature that maximum normal and shearing stresses at a point in a beam exist on planes that are inclined with respect to the axis of the beam. It can be shown that maximum and minimum normal stresses exist on two perpendicular planes. These planes are commonly called the *principal planes* and the stresses that act on them are called *principal stresses.* The principal stresses in a beam subjected to shear and bending may be calculated using the following formula:

$$f_{\text{pr}} = \frac{f}{2} \pm \sqrt{\frac{f^2}{4} + v^2} \qquad (4\text{--}2)$$

where f_{pr} is the principal stress and f and v are the bending and shear stresses, respectively, calculated from Eqs. (4–1).

The *orientation* of the principal planes may be calculated using the following formula:

$$\tan 2\alpha = \frac{2v}{f} \qquad (4\text{--}3)$$

where α is the angle measured from the horizontal.

The magnitudes of the shearing stresses and bending stresses vary along the length of the beam and with distance from the neutral axis. It follows, then, that the inclination of the principal planes as well as the magnitude of the principal stresses will vary, also. At the neutral axis the principal stresses will occur at a 45° angle. This may be verified by Eq. (4–3), substituting $f = 0$, from which $\tan 2\alpha = \infty$ and $\alpha = 45°$.

In Fig. 4–1, we isolate a small square unit element from the neutral axis of a beam (where $f = 0$). The vertical shear stresses are equal and opposite on the two vertical faces by reason of equilibrium. If these were the only two stresses present, the element would rotate. Therefore, there must exist equal and opposite horizontal shear stresses on the horizontal faces and of the same magnitude as the vertical shear stresses. (The concept of horizontal shear stresses equal in magnitude to the vertical shear stresses at any point in a beam may also be observed in almost any strength-of-materials text.)

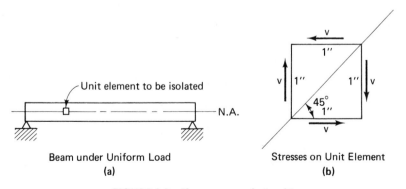

Beam under Uniform Load (a)

Stresses on Unit Element (b)

FIGURE 4-1 Shear stress relationship

If we cut a section A–A through the element at 45°, the stresses combine in such a manner that their effect is as shown in Fig. 4–2.

v = shear stress,
V = shear force,
T = tensile force, and
$0.707\,V$ = component normal to plane A–A.

Since we are dealing with a *unit area*, v (shear stress) = V (shear force). If we take a summation of forces normal to plane A–A ($\Sigma F = 0$),

$$0.707\,V + 0.707\,V = T$$

Since a force = area × stress, the formula above can be expressed as

$$0.707\,v(1) + 0.707\,v(1) = t(1.414)$$

where t = tensile stress

$$1.414v = t(1.414)$$

$$v = t$$

From Eq. (4–2),

$$f_{\mathrm{pr}} = \pm \sqrt{v^2} = \pm v$$

This indicates that there would also be a compressive stress of v on a plane perpendicular to plane A–A.

Hence we can state that there are tensile stresses induced in a beam by shearing forces and that at the neutral axis these are inclined at 45° to the horizontal. As has been shown, this tensile stress is equal in magnitude to the shearing stress and historically has been designated as *diagonal tension*.

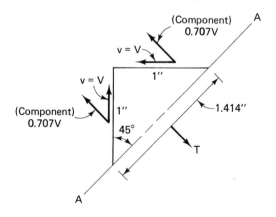

FIGURE 4-2 Diagonal tension due to shear

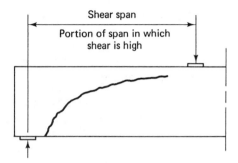

FIGURE 4-3 Typical diagonal tension failure

As previously stated, tensile stresses of various inclinations and magnitudes, resulting from either shear alone or the combined action of shear and bending, exist in all parts of a beam and must be taken into consideration in both analysis and design.

The preceding discussion is a fairly accurate description of what occurs in a plain concrete beam. In the beams with which we are concerned, where the length over which a shear failure could occur (the *shear span*) is in excess of approximately three times the effective depth, the diagonal tension failure would be the mode of failure in shear. Such a failure is shown in Fig. 4–3. For shorter spans the failure mode would actually be some combination of shear, crushing, and/or splitting. For the longer shear spans in plain concrete beams, cracks due to flexural tensile stresses would occur long before cracks due to the diagonal tension. The earlier flexural cracks would initiate the failure and shear would be of little consequence. However, in a concrete beam reinforced for flexure (moment) where tensile strength is furnished by steel, tensile stresses due to flexure and shear will continue to increase with increasing load. The steel placed in the beam to reinforce for moment is not located where the large diagonal tension stresses (due to shear) occur. The problem, then, becomes one of furnishing additional reinforcing steel to resist the diagonal tension stresses.

There has been considerable research over the years in an attempt to establish the exact distribution of the shear stresses over the depth of the beam cross section. Despite extensive studies and ongoing research, the precise shear-failure mechanism is still not fully understood.

Previous codes have recommended design guideliness based on a nominal average shear stress. The 1971 ACI Code expressed this shear stress as

$$v_u = \frac{V_u}{\phi b_w d}$$

where v_u = nominal average total design shear stress (psi)

V_u = total design shear force due to external loads (lb)

ϕ = strength reduction factor = 0.85 for shear

b_w = web width = b for rectangular sections (in.)

d = effective depth as previously defined (in.)

Since we have shown that, at the beam neutral axis, shear stresses and diagonal tension stresses are equal, the justification for this design approach has been that the shear stress is a good *measure* of the diagonal tension stress, although it does not represent the actual diagonal tension stress. Although the general theory of the previous Code has not been changed, the 1977 ACI Code recommends that shear design be based on the shear *force* V_u that exists at a beam cross section. This is a departure from the 1971 ACI Code, which based shear design on shear stress. The shear force may be thought of, in a similar manner, as a measure of the diagonal tension present.

4–2 SHEAR REINFORCEMENT DESIGN PROCEDURE

The basic rationale for the design of the shear reinforcement, or *web reinforcement* as it is usually called in beams, is to provide steel to cross the diagonal tension cracks and subsequently keep them from opening. Visualizing this basic rationale, with reference to Fig. 4–3, it is seen that the web reinforcement may take several forms, such as (1) vertical stirrups, (2) inclined or diagonal stirrups, and (3) the main reinforcement bent at ends to act as inclined stirrups. The most common form of web reinforcement used is the vertical stirrup. The web reinforcement contributes very little to the shear resistance prior to the formation of the inclined cracks but appreciably increases the *ultimate* shear strength of a bending member.

The design of bending members for shear is based on the assumption that the concrete resists part of the shear and any excess over and above what the concrete is capable of resisting has to be resisted by shear reinforcement.

For members that are subject to shear and flexure only, the amount of shear force that the concrete alone, unreinforced for shear, can resist is V_c.

$$V_c = 2\sqrt{f'_c}\ b_w d \qquad\qquad \text{ACI Eq. (11–3)}$$

In the expression for V_c, the terms are as previously defined with units for f'_c in psi, and units for b_w and d in inches. The resulting V_c will be in units of pounds. For rectangular beams, b_w is equivalent to b. The ideal shear strength of the concrete will be reduced to a dependable shear strength by applying a strength reduction factor ϕ of 0.85 (ACI Code, Section 9.3.2).

The design shear force is denoted V_u and results from the application of factored loads. Values of V_u are most conveniently determined using a typical shear force diagram.

Theoretically, no web reinforcement should be required if $V_u \leq \phi V_c$. However, the Code requires that a minimum area of shear reinforcement be provided in all reinforced concrete flexural members except as follows:

1. In slabs and footings.

2. In concrete joist construction as defined by the ACI Code, Section 8.11.

3. In beams with total depth less than 10 in., 2½ times the flange thickness or one-half the width of the web, whichever is greater.

4. Where V_u is less than $\phi V_c/2$.

This provision of the Code is primarily to guard against those cases where an unforeseen overload would cause failure of the member due to shear. Tests have shown the shear failure of a flexural member to be sudden and without warning. In cases where it is not practical to provide shear reinforcement (footings and slabs) and sufficient thickness is provided to resist V_u, the minimum area of web reinforcement is not required. In cases where the minimum area of web reinforcement is required, the amount necessary is determined using the following equation:

$$A_v = 50 \frac{b_w s}{f_y} \qquad \text{ACI Eq. (11–14)}$$

In the preceding equation, and with reference to Fig. 4–4:

A_v = total cross-sectional area of web reinforcement within a distance s; for single-loop stirrups, $A_v = 2A_s$, where A_s is the cross-sectional area of the stirrup bar (in.^{2})

b_w = web width = b for rectangular sections (in.)

s = center-to-center spacing of shear reinforcement in a direction parallel to the longitudinal reinforcement (in.)

f_y = yield strength of web reinforcement steel (psi)

In any span, that portion in which web reinforcement is theoretically necessary can be determined by utilizing the shear diagram. When the applied shear V_u exceeds the capacity of the concrete web ϕV_c, web reinforcement is required. In addition, according to the Code, web reinforcement at least equal to the minimum required must be provided

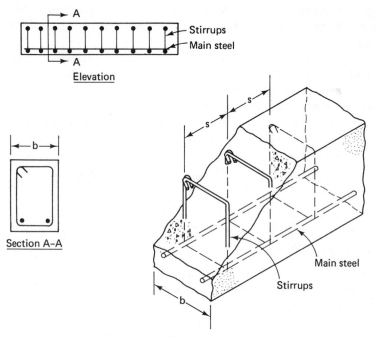

FIGURE 4-4 Isometric section showing stirrups partially exposed

elsewhere in the span, where the applied shear is greater than one-half of ϕV_c. The ACI Code, Section 11.1.1, states that the basis for shear design must be

$$V_u \leq \phi V_n \qquad \text{ACI Eq. (11–1)}$$

where

$$V_n = V_c + V_s \qquad \text{ACI Eq. (11–2)}$$

from which

$$V_u \leq \phi V_c + \phi V_s$$

where V_u, ϕ, and V_c are as previously defined, V_n is the nominal, or ideal, shear strength, and V_s is the nominal shear strength provided by shear reinforcement. V_s for vertical stirrups may be calculated from

$$V_s = \frac{A_v f_y d}{s} \qquad \text{ACI Eq. (11–17)}$$

where all terms are as previously defined. The preceding equation may be derived, with reference to Fig. 4–5, by assuming that the vertical stirrup resists the vertical component of diagonal tension that acts within

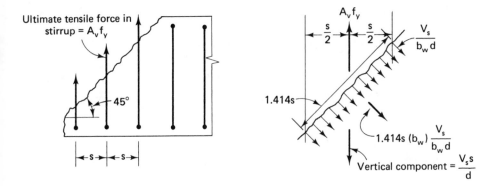

FIGURE 4-5 Determination of stirrup spacing based on required strength

the length $s/2$ on either side of the stirrup. The horizontal component is automatically included in the design of the main longitudinal reinforcement. Fig. 4–5 shows a stirrup intercepting a typical 45° diagonal tension crack.

As a limit, the factored shear force (design shear) V_u will equal the shear strength of the concrete plus the shear strength of the web reinforcement. As previously indicated:

$$V_u \leq (V_c + V_s)\phi$$

Utilizing the *shear stress* concept of the 1971 Code and substituting for V_u yields

$$\text{Shear stress} = \frac{V_u}{\phi b_w d} = \frac{(V_c + V_s)\phi}{\phi b_w d}$$

This may be rewritten as:

$$\text{Shear stress} = \frac{V_c}{b_w d} + \frac{V_s}{b_w d}$$

$V_c/b_w d$ may be considered to be the shear stress capacity of the concrete. $V_s/b_w d$ may be considered to be the excess shear stress, over and above the concrete capacity, to be resisted by the web reinforcement.

The area contributing stress to be resisted by the web reinforcement is $1.414 s b_w$. Therefore, the diagonal force, as depicted in Fig. 4–5, becomes

$$1.414 s b_w \left(\frac{V_s}{b_w d} \right)$$

The vertical component of this force becomes

$$0.707(1.414\,sb_w)\left(\frac{V_s}{b_w d}\right) = sb_w\left(\frac{V_s}{b_w d}\right) = \frac{V_s s}{d}$$

$A_v f_y$ represents the ultimate tensile capacity of the stirrup. Since vertical equilibrium must exist:

$$A_v f_y = \frac{V_s s}{d}$$

from which

$$V_s = \frac{A_v f_y d}{s}$$

In a similar manner, it can be shown that for inclined stirrups at 45°:

$$V_s = \frac{1.414 A_v f_y d}{s} \qquad\qquad \text{ACI Eq. (11–18)}$$

where s is the horizontal center-to-center distance of stirrups parallel to the main longitudinal steel.

It will be more practical if ACI Eqs. (11–17) and (11–18) are rearranged as expressions for spacing, since the stirrup bar size, strength, and beam effective depth are usually predetermined. The design is then for bar spacing. For vertical stirrups,

$$\text{Required } s = \frac{A_v f_y d}{V_s}$$

For 45° stirrups,

$$\text{Required } s = \frac{1.414 A_v f_y d}{V_s}$$

Note that these two equations are for *maximum* spacing of stirrups based on *required strength*. The required nominal shear strength V_s may be determined from the factored applied shear (V_u) diagram and ACI Eqs. (11–1) and 11–2):

$$V_u \leqslant \phi V_c + \phi V_s$$

from which

$$\text{Required } V_s = \frac{V_u - \phi V_c}{\phi}$$

$$= \frac{V_u}{\phi} - V_c$$

A *general procedure* may be adopted with respect to the design of the stirrups:

1. Determine the shear values based on clear span and draw a shear (V_u) diagram.

2. Determine if stirrups are required, and if they are, draw the required V_s diagram.

3. Determine the length of span over which stirrups are required.

4. Select the size of the stirrup bar (use vertical stirrups). See the following notes for reasonable assumptions and required checks.

5. Establish the ACI Code maximum spacing requirements.

6. Determine the spacing requirements based on shear strength to be furnished by web reinforcing.

7. Establish the spacing pattern and show sketches.

Notes on Stirrup Design

1. *Materials and maximum stresses*

   a. To reduce excessive crack widths in beam webs subject to diagonal tension, the ACI Code, Section 11.5.2, limits the design yield strength of shear reinforcement to 60,000 psi.

   b. The value of V_s must not exceed $8\sqrt{f'_c}\,b_w d$ irrespective of amount of web reinforcement (ACI Code, Section 11.5.6.8).

2. *Bar sizes for stirrups*

   a. The most common size stirrup used is a No. 3 bar. Under span and loading conditions where the shear values are relatively large, it may be necessary to use a No. 4 bar. However, it is rare that anything larger than a No. 4 bar stirrup would ever be required. In large beams, multiple stirrup sets are sometimes provided in which a diagonal crack would be crossed by four or more vertical bars at one location of a beam.

   b. It should be noted that when single conventional loop stirrups are used, the web area A_v provided by each stirrup is twice the cross-sectional area of the bar: No. 3 bars: $A_v = 0.22$ in.^{2}; No. 4 bars: $A_v = 0.40$ in.^{2}, since each stirrup crosses a diagonal crack twice.

   c. If possible, never vary the stirrup bar sizes; use the same bar sizes unless all other alternatives are not reasonable. Generally, spacing should be varied and size held constant.

3. *Stirrup spacings*

    a. When stirrups are required, the maximum spacing must not exceed $d/2$ or 24 in., whichever is smaller (ACI Code, Section 11.5.4.1).

        If V_s exceeds $4\sqrt{f'_c}\, b_w\, d$, the maximum spacing must not exceed $d/4$ or 12 in., whichever is smaller (ACI Code, Section 11.5.4.3).

    b. It is usually undesirable to space vertical stirrups closer than 4 in.

    c. It is generally economical and practical to compute the spacing required at several sections and to place stirrups accordingly in groups of varying spacing. Spacing values should be made to not less than ½ in. increments.

    d. The code stipulates that when the support reaction introduces a vertical compression into the end region of a member, the maximum shear will be considered as that at the section a distance d from the face of support (except for brackets, short cantilevers, and special isolated conditions). For stirrup design, the section located a distance d from the face of support will be called the *critical section*. Sections located less than a distance d from the face of the support may be designed for the same V_u as that at the critical section.

        Therefore, stirrup spacing should be constant from the critical section back to face of support based on the spacing requirements *at* the critical section. The first stirrup should be placed a distance $s/2$ ($\pm$) from face of support, where s equals the immediate adjacent required stirrup spacing. For the balance of the span, the spacing is a function of the V_s diagram.

    e. The actual stirrup pattern used in the beam is the designer's choice. The choice will be governed by strength requirements and economy. Many patterns will satisfy the strength requirements. In most cases the shear decreases from the support to the center of the span, indicating that the stirrup spacing could be continually increased from the critical section up to the maximum spacing allowed by the Code. This would create tedious design, detailing, and bar placing operations, but would, nevertheless, result in the least steel used. This is not warranted economically or within the framework of the philosophy outlined in Appendix B. In the usual uniformly loaded beams, no more than two or three different spacings should

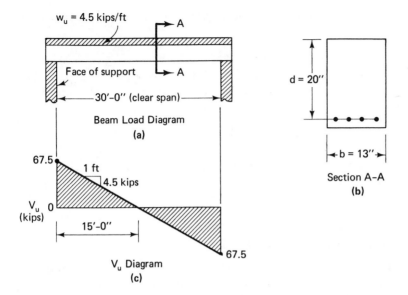

FIGURE 4-6 Sketch for Example 4-1

be used within a pattern. Longer spans and/or concentrated loads may warrant more detailed spacing patterns.

The data in Table A-5 will help to simplify shear calculations.

EXAMPLE 4-1

A simply supported rectangular reinforced concrete beam, 13 in. wide and having an effective depth of 20 in., carries a total design load (w_u) of 4.5 kips/ft on a 30-ft clear span. (The given load includes the weight of the beam.) Design the web reinforcement if f'_c= 3,000 psi and f_y = 40,000 psi (see Fig. 4–6).

1. Draw the shear force (V_u) diagram.

$$\text{Maximum } V_u = \frac{w_u \ell}{2} = \frac{4.5(30)}{2} = 67.5 \text{ kips}$$

2. Draw the required V_s diagram.

$$\text{Required } V_s = \frac{V_u}{\phi} - V_c$$

$$V_c = 2\sqrt{f'_c}\, b_w d = 0.110(13)(20) = 28.6 \text{ kips}$$

The ACI Code, Section 11.5.5.1, requires that stirrups be supplied if $V_u >$ ½ϕV_c.

$$\text{½}\phi V_c = \frac{0.85(28.6)}{2} = 12.16 \text{ kips}$$

Stirrups *are* required since $67.5 > 12.16$. At the face of the support,

$$\text{Required } V_s = \frac{67.5}{0.85} - 28.6 = 50.8 \text{ kips} \quad \text{ok}$$

The total uniform load of 4.5 kips/ft is the slope of the V_u diagram. Since V_s is a function of V_u/ϕ, the slope of the V_s diagram will be the slope of the V_u diagram divided by ϕ.

$$\text{Slope of } V_s = \frac{4.5}{\phi} = \frac{4.5}{0.85} = 5.29 \text{ kips/ft} \quad \text{ok}$$

The point where the V_s diagram passes through zero, measured from face of support, may then be determined.

$$V_s = 0 \text{ at } \frac{50.8}{5.29} = 9.60 \text{ ft} \quad \text{ok}$$

The V_s diagram is shown in Fig. 4–7.

3. Find the length of span over which stirrups are required. Since stirrups must be provided to the point where $V_u = \frac{1}{2}\phi V_c = 12.16$ kips, find where this shear exists on the V_u diagram of Fig. 4–6(c). From the face of the support,

$$\frac{67.5 - 12.16}{4.5} = 12.3 \text{ ft}$$

4. Assume a No. 3 stirrup ($A_v = 0.22$ in.²). Check the spacing required at the critical section where the stirrups will be most closely spaced. At the critical section, with reference to Fig. 4–7, noting that $d = 20$ in.,

$$V_s = 50.8 - \frac{20}{12}(5.29) = 41.98 \text{ kips}$$

$$\text{Required } s = \frac{A_v f_y d}{V_s} = \frac{0.22(40)(20)}{41.98} = 4.19 \text{ in.}$$

· Use 4 in. This is the maximum spacing allowed in the portion of the beam between the face of support and the critical section, which lies a distance d from the face of the support. It is based on the amount of shear strength that must be provided by the shear reinforcing. Had the required spacing in this case turned out to be less than 4 in., a larger bar would be selected for the stirrups.

5. The ACI Code maximum spacing requirements will be determined next. Compare V_s at the critical section with $4\sqrt{f'_c}\, b_w d$:

$$4\sqrt{f'_c}\, b_w d = 0.219(13)(20) = 56.94 \text{ kips}$$

$$V_s = 41.98 \text{ kips at critical section}$$

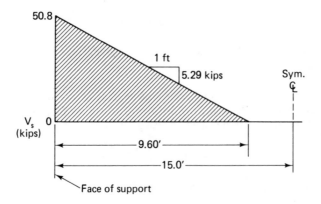

FIGURE 4-7 V_s diagram for Example 4–1

Since $41.98 < 56.94$, maximum spacing will be $d/2$ or 24 in., whichever is smaller.

$$\frac{d}{2} = \frac{20}{2} = 10 \text{ in.}$$

When $V_s > 4\sqrt{f'_c}\,b_w d$, maximum spacing should be $d/4$ or 12 in., whichever is smaller.

A second criterion is based on the code minimum area requirement (ACI Code, Section 11.5.5.3). ACI Eq. (11–14) may be rewritten in the form

$$s_{\max} = \frac{A_v f_y}{50 b_w} = \frac{0.22(40,000)}{50(13)} = 13.5 \text{ in.}$$

where the units of f_y should be carefully noted as psi.

Of the foregoing two maximum spacing criteria, the smaller value will control. Therefore, the 10-in. maximum spacing controls throughout the beam wherever stirrups are required.

6. Determine the spacing requirements based on shear strength to be furnished. We know, at this point, that the spacing required at the critical section is 4.19 in. and that the maximum spacing allowed in this beam is 10 in. where stirrups are required.

To establish a spacing pattern for the rest of the beam, the spacing required should be established at various distances from face of support. This will permit the placing of stirrups in groups, with each group having a different spacing. The number of locations at which to determine the spacing required is a judgment factor and should be a function of the V_s diagram.

To simplify required spacing computations at various locations from the face of support, a table or plot may be utilized along with the following expressions:

123

$$\text{Required } s = \frac{A_v f_y d}{V_s}$$

The value of V_s may be taken from the V_s diagram (Fig. 4–7):

$$V_s = V_{s(max)} - mx$$

where $V_{s(max)}$ is the value of V_s (kips) at the face of support, m the slope of the V_s diagram (kips/ft) and x the distance (ft) from the face of support to the point where the required spacing is being found. Therefore,

$$\text{Required } s = \frac{A_v f_y d}{V_{s(max)} - mx}$$

The required spacing will be arbitrarily found at 2-ft intervals. At 2 ft from the face of support,

$$\text{Required } s = \frac{0.22(40)(20)}{50.8 - 5.29(2)} = 4.37 \text{ in.}$$

Similarly, required spacing may be found at other points along the beam. The results of these calculations are tabulated and plotted in Fig. 4–8. Note that required spacings need not be determined beyond the point where s_{max} (10 in.) has been exceeded.

7. The plot of Fig. 4–8 may be readily used to aid in establishing a final pattern. For example, 5 in. spacing could be started about 3 ft from the face of support

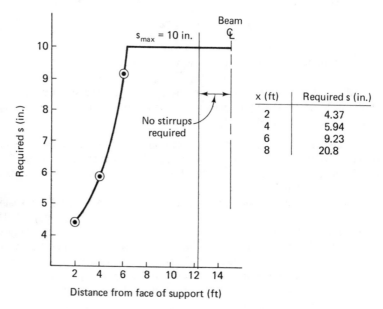

x (ft)	Required s (in.)
2	4.37
4	5.94
6	9.23
8	20.8

FIGURE 4-8 **Stirrup spacing requirements—Example 4–1**

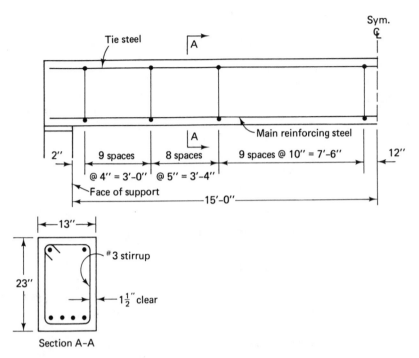

FIGURE 4-9 Design sketch for Example 4–1

and the maximum spacing of 10 in. could be started at about 6.2 ft from the face of support. The first stirrup will be placed away from the face of support a distance approximately equal to one-half the required spacing at the critical section. The design sketches are shown in Fig. 4–9. Note that the stirrups, at maximum spacing, have been run all the way to the center of the beam. This is common practice and is conservative.

EXAMPLE 4–2

A continuous reinforced concrete beam shown in Fig. 4–10 is 15 in. wide and has an effective depth of 31 in. The beam design loads and shear diagram are shown. (The design uniform load includes the weight of the beam.) Design the web reinforcement if $f'_c = 3,000$ psi and $f_y = 40,000$ psi.

1. The shear force (V_u) diagram is shown in Fig. 4–11.

2. Required V_s diagram:

$$V_c = 2\sqrt{f'_c}\, b_w d$$

$$= 0.110(15)(31) = 51.15 \text{ kips}$$

$$\tfrac{1}{2}\phi V_c = \tfrac{1}{2}(0.85)(51.15) = 21.74 \text{ kips}$$

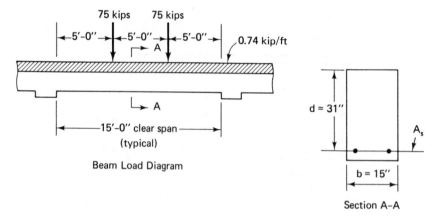

FIGURE 4-10 Sketch for Example 4-2

Stirrups *are* required, since 80.55 > 21.74. Calculate required V_s at the face of the support.

$$\text{Required } V_s = \frac{V_u}{\phi} - V_c = \frac{80.55}{0.85} - 51.15 = 43.61 \text{ kips}$$

Calculate the required V_s at concentrated load.

$$\text{Required } V_s = \frac{76.85}{0.85} - 51.15 = 39.26 \text{ kips}$$

$$\text{Slope of } V_s \text{ diagram} = \frac{0.74}{0.85} = 0.87 \text{ kips/ft}$$

The V_s diagram is shown in Fig. 4-12.

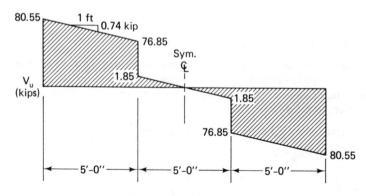

FIGURE 4-11 V_u diagram for Example 4-2

126

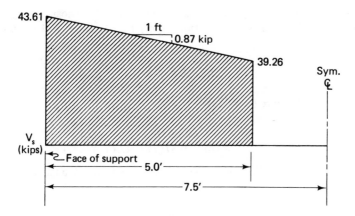

FIGURE 4-12 V$_s$ diagram for Example 4-2

3. Length of span over which stirrups are required:

$$\tfrac{1}{2}\phi V_c = 21.74 \text{ kips}$$

From Fig. 4–11, the point where V_u is equal to 21.74 kips may be determined by inspection to be at the concentrated load, 5 ft 0 in. from the face of the support. No stirrups are required between the two concentrated loads.

4. Assume a No. 3 stirrup ($A_v = 0.22$ in.²). The spacing required at the critical section is

$$V_s = 43.61 - \frac{31}{12}(0.87) = 41.35 \text{ kips}$$

$$\text{Required } s = \frac{A_v f_y d}{V_s} = \frac{0.22(40)(31)}{41.35} = 6.60 \text{ in.}$$

Use 6 in.

5. Determine the ACI Code maximum spacing requirements.

$$4\sqrt{f_c'}\, b_w d = 0.219(15)(31) = 101.8 \text{ kips}$$

Comparing with V_s at the critical section, we obtain

$$41.35 < 101.8$$

Therefore, $s_{max} = d/2$ or 24 in., where $d/2 = 15.5$ in. Also, check

$$s_{max} = \frac{A_v f_y}{50 \, b_w} = \frac{0.22(40,000)}{50(15)} = 11.7 \text{ in.}$$

Of the s_{max} requirements, the latter controls. Use $s_{max} = 11\tfrac{1}{2}$ in.

6. Determine the spacing requirements based on the shear strength to be furnished. The following expressions apply (see Example 4–1 for notation):

$$V_s = V_{s\,(\text{max})} - mx$$

$$= 43.61 - 0.87(x)$$

$$\text{Required } s = \frac{A_v f_y d}{V_s}$$

$$= \frac{0.22(40)(31)}{43.61 - 0.87(x)} = \frac{272.8}{43.61 - 0.87(x)}$$

The results of substitution and calculation are shown tabulated and plotted in Fig. 4–13.

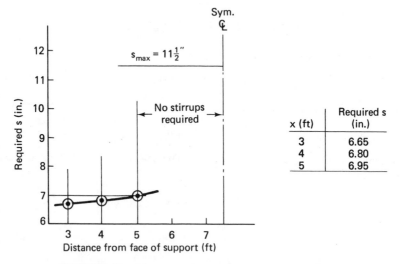

FIGURE 4-13   Stirrup spacing requirements—Example 4-2

7. Since no stirrups are required in the distance between the concentrated loads, it is apparent, with reference to Fig. 4–13, that the maximum spacing of 11½ in. need not be used in that distance. A spacing of 6 in. will be used between the face of support and the concentrated load. The center portion of the beam will be reinforced with stirrups at approximately the maximum spacing as a conservative measure. The design sketches are shown in Fig. 4–14. Note symmetry about beam centerline.

4-3 TORSION—INTRODUCTION

Torsion occurs when a member is loaded so that it twists about its longitudinal axis. This is easily visualized in machines, for instance in the transmission of power through shafts as is found in electric motors,

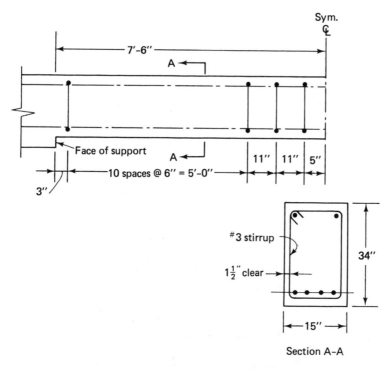

FIGURE 4-14 Design sketches for Example 4-2

automotive drive trains, and so on. Structural members found in buildings may also be subjected to torsional loads. Perhaps the most common situation is that of the spandrel girder in Fig. 4-15.

Beam B-1 frames into the spandrel girder as shown in Fig. 4-15(a). In Fig. 4-15(b), it is seen that some moment is developed at the supports of beam B-1 as a result of their rigidity. The moment developed at the left side of beam B-1 becomes a torsional load on the spandrel girder.

Torsion in noncircular members produces complex behavior in which the member cross section will warp as it twists. Torsional shear will be created in the faces of the member and will tend to produce diagonal tension cracks similar to those produced by flexural shear. However, the torsional shear will be in opposite directions on opposite sides of the member. Since shear and torsion normally occur simultaneously (or interact) within a member, the diagonal tension effects will be additive on one of the faces. When this occurs and when the tensile strength of the concrete is exceeded, a failure at approximately 45° to the horizontal will be initiated. This may be observed in Fig. 4-16. Obviously, steel must be introduced to cross any crack that forms in order to prevent a

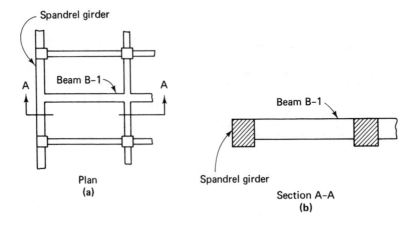

Spandrel girder

Beam B-1

A

A

Beam B-1

Plan
(a)

Spandrel girder

Section A-A
(b)

FIGURE 4-15   Torsion in beams

failure. The torsional failure occurring on a warped section may also be observed by twisting a piece of chalk to fracture.

4-4 DESIGN OF TORSION REINFORCEMENT

An approach to the design of torsion reinforcement is furnished in Chapter 11 of the ACI Code (ACI 318–77). The Code stipulates in Section 11.6.7 that "torsion reinforcement, where required, shall be provided in addition to reinforcement required to resist shear, flexure, and axial forces. The reinforcement required for torsion may be combined with

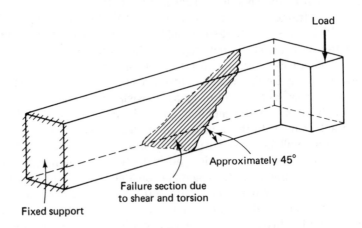

Load

Approximately 45°

Failure section due
to shear and torsion

Fixed support

FIGURE 4-16   Beam failure due to combined shear and torsion

that required for other forces, provided the area furnished is the sum of the individually required areas and the most restrictive requirements for spacing and placement are met.''

The Code requires that a reinforced concrete member be designed for torsion when the factored torsional moment T_u is in excess of a given maximum (Section 11.6.1).

In a manner similar to the design of shear reinforcement, torsional reinforcement must be provided when the factored torsional moment T_u that exists in the member is greater than ϕ times the nominal torsional moment strength T_c provided by the concrete. Assuming that torsion reinforcement is required, it must consist of additional longitudinal steel as well as additional closed web reinforcement (stirrups). The strength provided by the additional torsion reinforcement is denoted T_s. The basic design equations for torsion are

$$T_u \leq \phi T_n$$

where

$$T_n = T_c + T_s$$

Expressions for both T_c and T_s, as well as for the area of additional longitudinal steel required, are all contained in the ACI Code, Section 11.6.

The Code also stipulates (Section 11.6.8.2) that "the spacing of longitudinal bars, not less than #3, distributed around the perimeter of the closed stirrups, shall not exceed 12 inches. At least one longitudinal bar shall be placed in each corner of the stirrups."

Since torsional considerations are not common in the analysis and design of traditional reinforced concrete structures, it is the authors' opinion that rigorous treatment of torsion is beyond the scope of this book. The reader is referred to more theoretical texts for design examples and extensive theoretical discussions.

PROBLEMS

4–1. A simply supported beam carries a total design load w_u of 5.0 kips/ft. The span length is 28 ft center to center of supports and the supports are 12 in. wide. $b = 14$ in., $d = 20$ in., $f_c' = 4,000$ psi, and $f_y = 60,000$ psi. Determine spacings required for No. 3 stirrups and show the pattern with a sketch. (Recall that clear span is used for determining shears.)

4–2. Design stirrups for the beam shown. The supports are 12 in. wide and the loads shown are service loads. The dead load includes the weight of the beam.

$b = 16$ in., $d = 20$ in., $f'_c = 3,000$ psi, and $f_y = 40,000$ psi. Sketch the stirrup pattern.

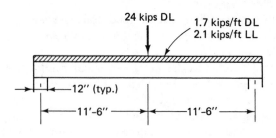

24 kips DL

1.7 kips/ft DL
2.1 kips/ft LL

12" (typ.)

11'-6" 11'-6"

FIGURE P4-2

4–3. Design stirrups for the beam shown. Service loads are 1.5 kips/ft dead load (includes beam weight) and 1.9 kips/ft live load. The supports are 12 in. wide. $b = 13$ in. and $d = 24$ in. for both top and bottom steel. $f'_c = 3,000$ psi and $f_y = 40,000$ psi. Sketch the stirrup arrangement.

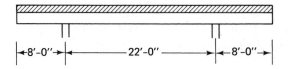

8'-0" 22'-0" 8'-0"

FIGURE P4-3

4–4. Design stirrups for the beam shown. The supports are 12 in. wide and the loads shown are factored design loads. The dead load includes the weight of the beam. $b = 14$ in., $d = 24$ in., $f'_c = 3,000$ psi, and $f_y = 40,000$ psi. Sketch the stirrup pattern.

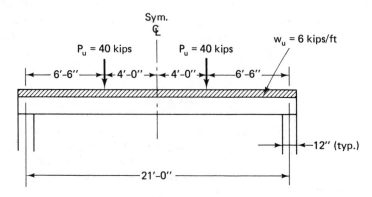

Sym.
$\mathcal{C}$

$P_u = 40$ kips $P_u = 40$ kips $w_u = 6$ kips/ft

6'-6" 4'-0" 4'-0" 6'-6"

12" (typ.)

21'-0"

FIGURE P4-4

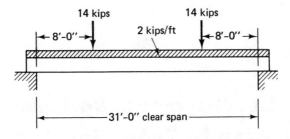

FIGURE P4-5

4-5. Design the rectangular reinforced concrete beam for *moment and shear.* Use only tension steel for flexure. The loads shown are service loads. The uniform load is composed of 1 kip/ft dead load (does not include beam weight) and 1 kip/ft live load. The concentrated loads are dead load. Assume the supports to be 12 in. wide. f'_c = 4,000 psi and f_y = 60,000 psi. Show design sketches, including the stirrup pattern.

5

Development, Splices,
and Simple-Span Bar Cutoffs

5–1 DEVELOPMENT LENGTH—INTRODUCTION

One of the fundamental assumptions of reinforced concrete design is that at the interface of the concrete and the steel bars, perfect bonding exists and that no slippage occurs. Based on this assumption, it follows that some form of bond stress exists at the contact surface between the concrete and the steel bars. In beams this bond stress is caused by the change in bending moment along the length of the beam and the accompanying change in the tensile stress in the bars, and has historically been termed *flexural bond*. The actual distribution of bond stresses along the reinforcing steel is highly complex, due primarily to the presence of concrete cracks. Research has indicated that large local variations in bond stress are caused by flexural and diagonal cracks and very high bond stresses have been measured adjacent to these cracks. These high bond stresses may result in small local slips adjacent to the cracks, with resultant widening of cracks as well as increased deflections. Generally, this will be harmless as long as failure does not propagate all along the bar with resultant complete loss of bond. It is possible, if *end anchorage* is reliable, that the bond can be severed along the entire length of bar, excluding the anchorage, without endangering the carrying capacity of the beam. The resulting behavior would be similar to that of a tied arch.

End anchorage may be considered reliable if the bar is embedded into concrete a prescribed distance known as the *development length*, ℓ_d, of the bar. If in the beam the actual extended length of a bar is equal

90° Bends for Development of #18 Bars—Seabrook Station, New Hampshire.

to or greater than this required development length, no premature bond failure will occur. That is, the predicted strength of the beam will not be controlled by bond but rather by some other factor.

Hence, the main requirement for safety against bond failure is that the length of the bar, from any point of given steel stress f_s (or, as a maximum, f_y) to its nearby free end must be at least equal to its development length. If this requirement is satisfied, the magnitude of the flexural bond stress along the beam is of only secondary importance, since the integrity of the members is assured even in the face of possible minor local failures. However, if the actual available length is inadequate for full development, special anchorages, such as hooks, must be provided to ensure adequate strength.

Current design methods based on both the 1971 and 1977 ACI Codes disregard high localized bond stress (which, prior to the 1971 Code, was always calculated) even though it may result in localized slip between steel and concrete adjacent to the cracks. Instead, attention is directed toward providing adequate length of embedment, past the location at which the bar is fully stressed, which will ensure development of the full strength of the bar.

5-2 DEVELOPMENT LENGTH—TENSION BARS

The ACI Code, Section 12.2, specifies the development length ℓ_d for deformed tension bars and wires as follows:

ℓ_d= basic development length × any applicable modification factors

A. Basic Development Length (B.D.L.)

1. For No. 11 bars and smaller,

$$\text{B.D.L.} = 0.04 A_b \frac{f_y}{\sqrt{f_c'}} \qquad \text{(in.)}$$

but not less than $0.0004\, d_b\, f_y$, where

$$A_b = \text{area of individual bar (in.}^2)$$

$$d_b = \text{nominal diameter of bar (in.)}$$

and f_y and f_c' are in psi.

2. For No. 14 bars,

$$\text{B.D.L.} = 0.085 \frac{f_y}{\sqrt{f_c'}} \qquad \text{(in.)}$$

3. For No. 18 bars,

$$\text{B.D.L.} = 0.11 \frac{f_y}{\sqrt{f_c'}} \qquad \text{(in.)}$$

4. For deformed wire,

$$\text{B.D.L.} = 0.03 \, d_b \frac{f_y}{\sqrt{f_c'}} \qquad \text{(in.)}$$

B. Modification Factors (To Be Used If Applicable)

1. Top bars: 1.4.
2. Bars with $f_y > 60,000$ psi: $2 - 60,000/f_y$.

3. For lightweight concrete when average splitting tensile strength f_{ct} is not specified:

"All-lightweight" concrete: 1.33

"Sand-lightweight" concrete: 1.18

4. Reinforcement spaced laterally at least 6 in. o.c. and at least 3 in. clear from the side face of the member: 0.8.

5. Reinforcement in excess of that required:

$$\frac{A_s \text{ required}}{A_s \text{ provided}}$$

6. Bars enclosed within a spiral which is not less than ¼ in. in diameter and not more than 4 in. in pitch: 0.75

The basic development lengths in part A must be modified by the applicable factors of part B. In no case may ℓ_d be less than 12 in. except in lap splices and the development of web reinforcement. Table A-12 in Appendix A furnishes basic development lengths for various concrete strengths, steel strengths, and bar sizes.

A top bar, for purposes of determining a modification factor, is categorized as a horizontal bar having more than 12 in. of concrete cast below it. This condition lends itself to the formation of entrapped air and moisture on the underside of the bars, resulting in loss of bond and slip at lesser loads. Where bundled bars are used, the ACI Code, Section 12.4, stipulates that calculated development lengths are to be made for individual bars within a bundle and then increased by 20% for three-bar bundles and by 33% for four-bar bundles. Bundled bars consist of a group of not more than four parallel reinforcing bars in contact with each other and assumed to act as a unit. Additional criteria are found in the ACI Code, Section 7.6.6.

5-3 DEVELOPMENT LENGTH—COMPRESSION BARS

For deformed bars in compression, the basic development length is computed as follows:

$$\text{B.D.L.} = 0.02 \, d_b \frac{f_y}{\sqrt{f'_c}}$$

but not less than $0.0003 f_y \, d_b$, or 8 in., according to the ACI Code, Section 12.3. The following modification factors may be applied to the basic development length for compression bars:

1. Reinforcement in excess of that required:

$$\frac{A_s \text{ required}}{A_s \text{ provided}}$$

2. Bars enclosed within a spiral which is not less than ¼ in. in diameter and not more than 4 in. in pitch: 0.75

In summary, a bar must have sufficient anchorage or development length so that the bar can develop its capacity in tension or compression. This is the same as saying that the end of the bar can be no closer to its point of maximum stress than the calculated ℓ_d values.

EXAMPLE 5–1

Calculate the anchorage or development length ℓ_d required for No. 8 top bars (more than 12 in. concrete below bars) in a sand-lightweight concrete member. $f_y = 60,000$ psi and $f'_c = 4,000$ psi.

1. Basic development length (use the largest value):

$$\text{B.D.L.} = 0.04A_b \frac{f_y}{\sqrt{f'_c}}$$

$$= 0.04\,(0.79)\left(\frac{60,000}{\sqrt{4,000}}\right) = 30 \text{ in.}$$

or

$$\text{B.D.L.} = 0.0004d_b f_y$$

$$= 0.0004\,(1.0)\,(60,000) = 24 \text{ in.}$$

Therefore, B.D.L. = 30 in. (Check this with TableA–12.)

2. Modification factors:

   a. Top bars: use 1.4.

   b. Sand-lightweight concrete: use 1.18.

3. Development length required:

$$\ell_d = 30\,(1.4)\,(1.18) = 49.6 \text{ in.}$$

EXAMPLE 5–2

Calculate the anchorage or development length required for the No. 9 bars as shown in the top of a 15-in. slab, (Fig. 5–1). Note that these bars are the tension reinforcement for negative moment in the slab at the supporting beam. $f_y = 60,000$ psi and $f'_c = 4,000$ psi (normal-weight concrete).

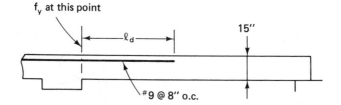

FIGURE 5-1 Sketch for Example 5-2

1. Basic development length (use the largest value):

$$\text{B.D.L.} = 0.04 A_b \frac{f_y}{\sqrt{f_c'}}$$

$$= 0.04(1.0)\frac{60,000}{\sqrt{4,000}} = 37.9 \text{ in.}$$

or

$$\text{B.D.L.} = 0.0004 d_b f_y$$

$$= 0.0004(1.128)(60,000) = 27.1 \text{ in.}$$

Therefore, B.D.L. = 37.9 in. (Check with Table A-12.)

2. Modification factors:

   a. Top bars: use 1.4.

   b. Bar spacing > 6 in. o.c.: use 0.8.

3. Development length required:

$$\ell_d = 37.9(1.4)(0.8) = 42.4 \text{ in.}$$

5-4 ANCHORAGE—HOOKS

In the event that the desired development length in tension cannot be furnished, it is necessary to provide mechanical anchorage at the end of the bars. Although the ACI Code (Section 12.6) allows any mechanical device to serve as anchorage if its adequacy is verified by test, anchorage for main or primary reinforcement is usually accomplished by means of a 90° or 180° hook. The dimensions and bend radii for these hooks have been standardized by the ACI Code. Additionally, 90° and 135° hooks have been standardized for stirrups and tie reinforcement. Hooks in compression bars are ineffective and cannot be used as anchorage. Standard reinforcement hooks are shown in Fig. 5–2. The bend diameters are measured on the inside of the bar.

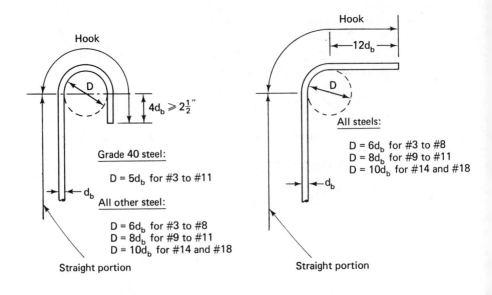

180° Hook

90° Hook

Primary Reinforcement

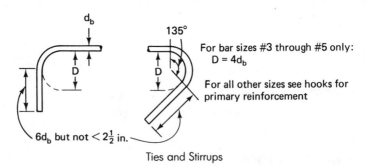

Ties and Stirrups

FIGURE 5-2 ACI standard hooks

According to the ACI Code, Section 12.5, standard hooks in tension may be considered to develop a tensile stress in the steel reinforcement of

$$f_h = \xi\sqrt{f'_c}$$

where values of ξ (Greek lowercase xi) are given in Table 5–1. The values of ξ may be increased by 30% where the bars are enclosed perpendicular to the plane of the hook. Enclosure may consist of external concrete or internal closed ties, spirals, or stirrups.

Anchorage of web reinforcement must be furnished in accordance

TABLE 5-1

TABLE 5-1
ξ Values (ACI Code, Table 12.5.1)

| Bar Number | $f_y = 60$ ksi | | $f_y = 40$ ksi: |
	Top Bars	Other Bars	All Bars
3–5	540	540	360
6	450	540	360
7–9	360	540	360
10	360	480	360
11	360	420	360
14	330	330	330
18	220	220	220

with ACI Code, Section 12.14. Stirrups should be carried as close to the compression and tension surfaces as possible. Close proximity to the compression face is necessary since flexural tension cracks penetrate deeply as ultimate load is approached.

The ACI Code stipulates that ends of single-leg, simple-U, or multiple-U stirrups shall be anchored by one of the following means:

a. A standard hook plus an embedment of $0.5\ell_d$ where the embedment shall be taken as the distance between middepth of member $d/2$ and point of tangency of the hook.

b. Embedment $d/2$ above or below middepth on the compression side for a full development length ℓ_d but not less than 24 bar diameters or 12 inches.

c. For No. 5 bars and smaller, bending around longitudinal reinforcement through at least 135°, plus, for stirrups with $f_y > 40,000$ psi, an embedment of $0.33 \ \ell_d$. The $0.33 \ \ell_d$ embedment shall be taken as the distance between middepth of member $d/2$ and point of tangency of the hook.

The web reinforcement anchorage is illustrated in Fig. 5–3.

In addition, the ACI Code, Section 12.14.5, establishes criteria with respect to lapping of double-U stirrups or ties (without hooks) to form a closed stirrup. Legs shall be considered properly spliced when length of laps are $1.7 \ \ell_d$ as depicted in Fig. 5–3. Each bend of each simple-U stirrup must enclose a longitudinal bar.

EXAMPLE 5–3

Determine the anchorage or development length required for the conditions shown in Fig. 5–4. $f'_c = 3,000$ psi (normal-weight concrete) and $f_y = 40,000$ psi. The No. 8 bars may be categorized as top bars.

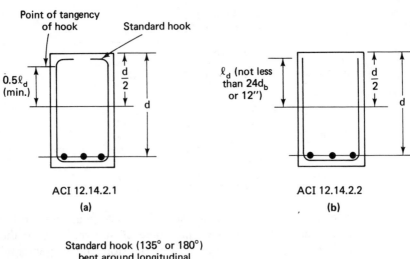

ACI 12.14.2.1

(a)

ACI 12.14.2.2

(b)

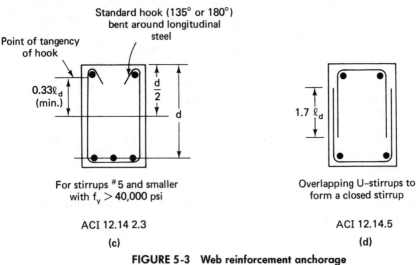

For stirrups #5 and smaller
with $f_y > 40{,}000$ psi

ACI 12.14 2.3

(c)

Overlapping U-stirrups to
form a closed stirrup

ACI 12.14.5

(d)

FIGURE 5-3 Web reinforcement anchorage

a. Anchorage into exterior column

1. *Anchorage using straight bars*: The basic development length is 23 in., from Table A–12. Use a modification factor of 1.4, since the bars are top bars.

$$\ell_d = 23(1.4) = 32.2 \text{ in. required}$$

Since 32.2 in. > 24 in., additional anchorage is required. This may be accomplished through the use of a standard ACI hook, either 90° or 180°.

2. *Anchorage using hooks* (see Fig. 5–5): Let ℓ_e represent a straight length of bar which has resistance to slip in the concrete equal to that of the standard

142

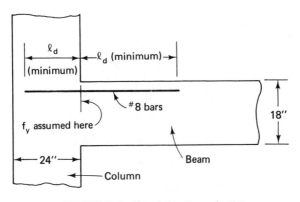

FIGURE 5-4 Sketch for Example 5–3

hook. Since a standard hook is not assumed to develop a bar to f_y, ℓ_e will always be less than ℓ_d. Therefore, there must be provided sufficient straight-length embedment, in addition to ℓ_e, to provide the total required anchorage ℓ_d. This additional straight embedment must have a length of at least $(\ell_d - \ell_e)$.

According to the ACI Code, and as previously mentioned, the standard hook will be considered to develop a tensile stress

$$f_h = \xi\sqrt{f'_c}$$

Also, since the basic development length equals the larger of

$$0.04A_b\frac{f_y}{\sqrt{f'_c}} \quad \text{or} \quad 0.0004d_bf_y$$

f_h may be substituted for f_y and we may compute the straight-length equivalence of the standard hook ℓ_e.

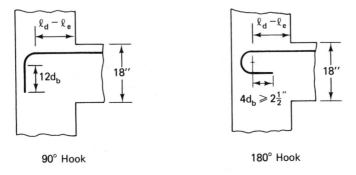

90° Hook   180° Hook

FIGURE 5-5 Sketch for Example 5–3

$$\ell_e = 0.04 A_b \frac{f_h}{\sqrt{f_c'}}$$

or

$$\ell_e = 0.0004 d_b f_h$$

Since $f_h = \xi \sqrt{f_c'}$, substitution will yield

$$\ell_e = 0.04 A_b \xi \quad \text{or} \quad 0.0004 d_b \xi \sqrt{f_c'}$$

For the No. 8 bars that we are considering,

$$\ell_e = 0.04 (0.79)(360) = 11.4 \text{ in.}$$

or

$$\ell_e = 0.0004 (1.0)(360)(\sqrt{3,000}) = 7.9 \text{ in.}$$

The larger of the two values controls and constitutes the straight-length equivalence for the standard hook. Values of ℓ_e may be obtained from Table A–13 in Appendix A. This table has been developed using the foregoing expressions.

If we modify ℓ_e (since bars are top bars), then

$$\ell_e = 11.4(1.4) = 16 \text{ in.}$$

and

$$\ell_d - \ell_e = 32.2 - 16 = 16.2 \text{ in.}$$

Therefore, 16.2 in. is the straight length of top bar required in addition to the hook for adequate anchorage. Notice in Fig. 5–6 that this straight length of bar begins at the point of tangency, the point at which f_h is considered to

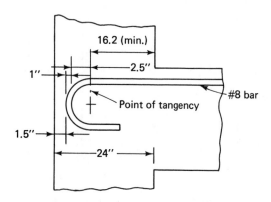

FIGURE 5-6   Sketch for Example 5–3

have been developed by the hook. The total width of the column required is

$$16.2 + 2.5 + 1 + 1.5 = 21.2 \text{ in.} < 24 \text{ in.}$$

Therefore, the total anchorage will fit into the column and the detail is satisfactory.

b. Anchorage into beam

The development length required if bars are straight is 32.2 in., as previously determined. Therefore, the bars must extend at least this distance into the span. The ACI Code has additional requirements for the extension of bars which reinforce for tension due to negative moments. These requirements will be covered in the discussion on continuous construction in Chapter 6.

EXAMPLE 5–4

A reinforced concrete beam of rectangular cross-section is to have an effective depth $d = 20$ in. Assume $f_y = 40,000$ psi, $f'_c = 3,000$ psi, and 1½ in. cover. Check adequacy of anchorage for a No. 3 simple-U stirrup if (a) 90° hooks are used (not bent around longitudinal steel); and (b) no hooks are used.

a. Refer to Fig. 5–3(a). ℓ_d for the No. 3 bar, from Table A–12, is 6 in. This is actually the B.D.L. but modification factors are not applicable for web reinforcement. (Minimum ℓ_d does not apply to hooked web reinforcement.) The minimum anchorage required from middepth to point of tangency of the hook is

$$0.5\ell_d = 0.5(6) = 3 \text{ in.}$$

The actual anchorage available, considering cover and hook geometry is

$$\frac{d}{2} - 1.5 - 3.0d_b$$

$$= \frac{20}{2} - 1.5 - 3.0(\tfrac{3}{8}) = 7.38 \text{ in.}$$

$$7.38 \text{ in.} > 3.0 \text{ in.} \qquad\qquad\qquad\qquad\qquad \text{(OK)}$$

b. Refer to Fig. 5–3(b). ℓ_d for the No. 3 bar is 6 in. The required anchorage is the larger of ℓ_d, 12 in., or $24\,d_b$.

$$24\,d_b = 24(\tfrac{3}{8}) = 9 \text{ in.}$$

Therefore, the 12 in. minimum required anchorage controls. The actual anchorage to middepth of the beam, considering cover, is

$$\frac{d}{2} - 1.5 = \frac{20}{2} - 1.5 = 8.5 \text{ in.}$$

$$8.5 \text{ in.} < 12 \text{ in.} \qquad\qquad\qquad\qquad\qquad \text{(N.G.)}$$

5–5 SPLICES

The need to splice reinforcing steel is a reality due to the limited lengths of steel available. All bars are readily available in lengths up to 60 ft; however, No. 3 and No. 4 bars will tend to bend in handling when longer than 40 ft. Typical stock straight lengths are as follows:

No. 3 bar: 20, 40, 60 ft

No. 4 bar: 30, 40, 60 ft

Nos. 5–18 bars: 60 ft

Splicing may be accomplished by welding, utilizing mechanical connections, or, most commonly for No. 11 bars and smaller, by lapping bars as shown in Fig. 5–7. Lap splices may not be used for bars larger than No. 11 except in footings, as provided in ACI Code, Section 15.8.6. The splice composed of lapped bars is usually more economical than the other types. The lapped bars are commonly tied in contact with each other but may be spaced apart up to one-fifth of the lap length, with an upper limit of 6 in. Splices in regions of maximum moment preferably should be avoided.

5–6 TENSION SPLICES

The required length of lap is based on a *class* into which the splice fits. The required lap length increases with increased stress and increased amount of steel spliced in close proximity and is expressed in terms of the tensile development length ℓ_d for the particular bar. The ACI Code, Section 12.16, for splices in tension, is summarized in Table 5–2. Requirements for welded splices and mechanical connections for tensile reinforcement are contained in the ACI Code, Section 12.16.

In addition, splices in "tension tie members" must be made with a full welded splice or full mechanical connection, and their locations must be staggered a distance at least equal to $1.7\ell_d$, in accordance with the

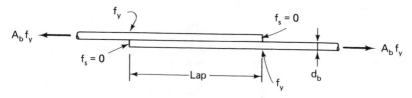

FIGURE 5-7 Stress transfer in tension lap splice

TABLE 5–2
Tension Lap Splices

A_s Provided / A_s Required	Percent of A_s Spliced within Required Lap Length	Splice Class	Minimum Lap Length for Tension Lap Splices*
Equal to or greater than 2 "low stress"	≤ 75	A	1.0 ℓ_d
	> 75	B	1.3 ℓ_d
Less than 2 conservative "high stress"	≤ 50	B	1.3 ℓ_d
	> 50	C	1.7 ℓ_d

*The minimum length of lap should be 12 in., although ℓ_d may be less than 12 in.

ACI Code, Section 12.16.5. Staggering of all tension splices in all types of members is encouraged.

5-7 COMPRESSION SPLICES

The ACI Code, Section 12.17, contains requirements for lap splices for compression bars. For $f'_c = 3,000$ psi or more, the following lap lengths, in multiples of bar diameters d_b, are required:

$$f_y = 40,000 \text{ psi:} \quad 20d_b$$
$$f_y = 60,000 \text{ psi:} \quad 30d_b$$
$$f_y = 75,000 \text{ psi:} \quad 44d_b$$

but not less than ℓ_d for bars in compression. In no case should the lap be less than 12 in. For $f'_c < 3,000$ psi, the length of lap should be increased by one-third. Within ties or spirals of specific makeup (ACI Code, Sections 12.17.2 and 12.17.3) these laps may be reduced to 0.83 or 0.75, respectively, of the foregoing values, but must not be less than 12 in.

Compression splices may also be of the end-bearing type, where bars are cut square, then butted together and held in concentric contact by a suitable device. End-bearing splices must not be used except in members containing closed ties, closed stirrups, or spirals. Welded splices and mechanical connections are also acceptable and are subject to the requirements of the ACI Code, Section 12.17. Special splice requirements for columns are furnished by the ACI Code, Section 12.18.

147

5–8 SIMPLE-SPAN BAR CUTOFFS AND BENDS

The maximum required A_s for a beam is needed only where the moment is maximum. Obviously, this maximum steel may be reduced at points along a bending member where the bending moment is smaller. This is usually accomplished by either stopping or bending the bars in a manner consistent with the theoretical requirements for the strength of the member as well as the requirements of the ACI Code.

Bars can theoretically be stopped or bent in flexural members whenever they are no longer needed to resist moment. However, the ACI Code, Section 12.11.3, requires that each bar be extended for a distance equal to the effective depth of the member or 12 bar diameters, whichever is greater, beyond that point where it is no longer required. This, in effect, prohibits the cutting off of a bar at the theoretical cutoff point but can be interpreted as permitting bars to be bent at the cutoff point. If bars are to be bent, a general practice that has evolved is to commence the bend at a distance equal to one-half the effective depth beyond the theoretical cutoff point. The bent bar should be anchored or made continuous with reinforcement on the opposite face of the member. However, it should be pointed out that the bending of reinforcing bars in slabs and beams to create *truss bars* (see types 3 through 7, Fig. 13–4) has fallen into disfavor over the years because of placement problems and labor involved. It is more common to use straight bars and place in accordance with the strength requirements.

With simple-span flexural members of constant dimensions, we can assume that the required A_s varies directly with the bending moment and that the shape of a *required A_s curve* is identical with that of the moment diagram. The moment diagram (or curve) may then be used as the required A_s curve by merely changing the vertical scale. Steel areas may be used as ordinates, but it is more convenient to use the required number of bars (assuming all bars are of the same diameter). The necessary correlation is established by using the maximum ordinate of the curve as the maximum required A_s in terms of the required number of bars. Because of possible variations in the shape of the moment diagram, either a graphical or a mathematical approach may be more appropriate. A graphical approach requires that the moment diagram be plotted to scale. Example 5–5 lends itself to a mathematical solution.

In determining bar cutoffs it should be remembered that the stopping of bars should be accomplished by utilizing a symmetrical pattern so that the remaining bars will also be in a symmetrical pattern. In addition, the ACI Code, Section 12.12.1, requires that for simple spans at least one-third of the positive moment steel extend into the support a distance of 6 in. In practice, this requirement is generally exceeded. Normally,

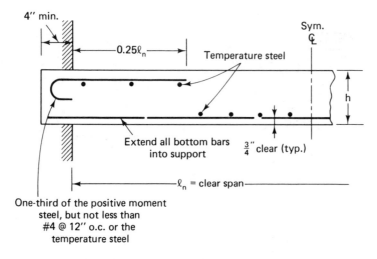

4″ min.

0.25ℓ_n

Temperature steel

Sym.
℄

h

Extend all bottom bars into support

$\frac{3}{4}$″ clear (typ.)

ℓ_n = clear span

One-third of the positive moment steel, but not less than #4 @ 12″ o.c. or the temperature steel

FIGURE 5-8 Recommended bar details–one-way simple-span slabs, tensile reinforced beams similar

recommended bar details for single-span solid concrete slabs indicate that all bottom bars should extend into the support. See Fig. 5–8 for recommended bar details.

Economy sometimes dictates that reinforcing steel should be cut off in a simple span. Example 5–5 shows one approach to this problem.

EXAMPLE 5–5

A simple-span, uniformly loaded beam, shown in Fig. 5–9(a) and (b), requires 6 No. 9 bars for tensile reinforcement. If the effective depth d is 18 in., determine where the bars may be stopped. $f'_c = 3,000$ psi and $f_y = 40,000$ psi. Assume that there is no excess steel (required A_s = furnished $A_s = 6.00$ in.^{2}).

1. Determine where the first two bars may be cut off. The first two bars may be stopped where only four are required. This distance from the centerline of the span is designated x_1 in Fig. 5–9(d).

Since the moment diagram is a second-degree curve (a parabola), offsets to the line tangent to the curve at the point of maximum moment vary as the squares of the distances from the centerline of the span. The solution for the distance x may be formulated as follows:

$$\frac{x_1^2}{\left(\dfrac{\ell}{2}\right)^2} = \frac{y_1}{Y}$$

$$\frac{x_1^2}{10^2} = \frac{2}{6}$$

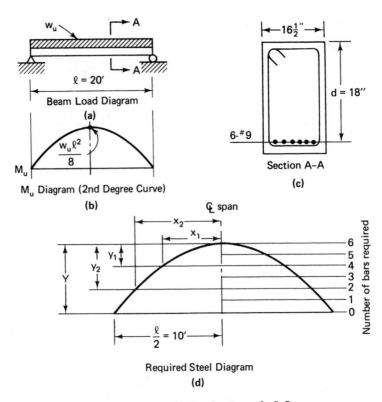

FIGURE 5-9 Sketches for Example 5-5

from which $x_1 = 5.77$ ft. This locates the *theoretical* point where two bars may be terminated. Additionally, bars must be extended past this point a distance d or $12 d_b$, whichever is larger.

$$12 \text{ bar diameters} = 12(1.128)$$

$$= 13.5 \text{ in.}$$

Since $d = 18$ in., the bars should be extended 18 in. Then the minimum distance from the centerline of the span to the cutoff of the first two bars is

$$5.77 + 1.50 = 7.27 \text{ ft}$$

2. Determine where the next two bars may be cut off. The two remaining bars, which should be the corner bars, will continue into the support. With reference to Fig. 5–9(d):

$$\frac{(x_2)^2}{10^2} = \frac{4}{6}$$

$$x_2 = 8.16 \text{ ft}$$

150

Therefore, the minimum distance from centerline of span to the cutoff of the next two bars is

$$8.16 + 1.50 = 9.66 \text{ ft}$$

3. Checking the required development length ℓ_d (see Table A–12 for B.D.L.) for the first two bars yields

$$\ell_d = 29 \text{ in.} = 2.42 \text{ ft}$$

Since this is much less than the distance from the centerline to the actual cutoff point (7.27 ft), the cutoff for the first two bars is satisfactory. Similarly, if we consider the second two bars, for which ℓ_d is also 2.42 ft, the point at which they are stressed to maximum (the theoretical cutoff point for the first two bars), measured from the centerline, is 5.77 ft. The second two bars must extend at least ℓ_d past this point. Again measured from the centerline, the bars must extend at least

$$5.77 + 2.42 = 8.19 \text{ ft}$$

Since this is less than the actual distance to the cutoff point (9.66 ft), the cutoff for the second pair is satisfactory (see Fig. 5–10).

4. The remaining pair of corner bars, representing one-third of the positive-moment steel, must continue into the support. However, since the corner bars are stressed to f_y at a distance of 8.16 ft from the centerline of span (where the second pair of bars may be theoretically cut), the development length of these bars must be satisfied from the end of the bar back to point of maximum stress f_y. The development length ℓ_d required for the corner bars

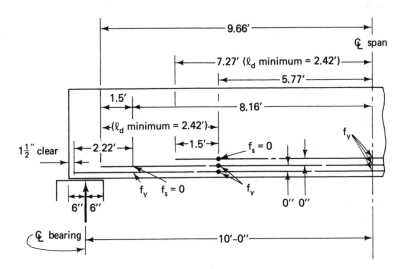

FIGURE 5-10 Sketch for Example 5-5

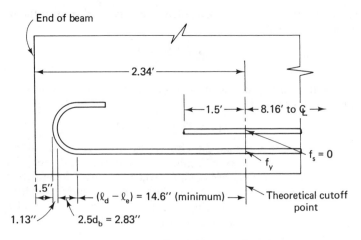

FIGURE 5-11   Sketch for Example 5-5

is 29 in., or 2.42 ft. The straight development length furnished equals 2.22 ft. Since 2.22 ft < 2.42 ft, the development length is inadequate and a hook must be furnished. Use a 180° hook. From Table A–13 for No. 9 bars with f'_c= 3,000 psi and f_y= 40,000 psi,

$$\ell_e = 14.4 \text{ in.}$$

Calculate the straight length required in addition to the hook.

$$\ell_d - \ell_e = 29 - 14.4 = 14.6 \text{ in.}$$

Referring to Fig. 5–11, the required minimum distance from the point where the second pair of bars may be theoretically cut (and where the corner bars are stressed to f_y) to the end of the beam is

$$14.6 + 2.83 + 1.13 + 1.5 = 20.1 \text{ in.}$$

The actual distance furnished is

$$2.34(12) - 1.5 = 26.6 \text{ in.}$$

Therefore, the anchorage is adequate.

A more desirable design would be to run four of the bars into the support, because of the small amount of steel saved by cutting off the second pair of bars.

The bars in the preceding example are terminating in a tension zone. When this occurs the member undergoes a reduction in shear capacity. Therefore, special shear and stirrup calculations must be performed in accordance with the ACI Code, Section 12.11.5, which stipulates that no flexural reinforcement shall be terminated in a tension zone unless *one* of the following conditions is satisfied:

(12.11.5.1) Shear at the cutoff point does not exceed two-thirds of that permitted, including the shear strength of the shear reinforcement provided

$$V_u \leq \frac{2}{3}\phi(V_c + V_s)$$

(12.11.5.2) Stirrup area in excess of that required is provided along each bar terminated over a distance equal to $\frac{3}{4}d$ from point of bar cutoff. The excess stirrup area shall not be less than $60\, b_w s/f_y$. The spacing s shall not exceed $d/(8\beta_b)$ where β_b is the ratio of reinforcement cut off to total area of tension reinforcement at the section.

(12.11.5.3) For No. 11 bars and smaller, the continuing reinforcement provides double the area required for flexure at the cutoff point and shear does not exceed three-quarters of that permitted.

With respect to the preceding conditions, the first is interpreted as a check only. Though the V_s term could be varied, the Code is not specific concerning the length of beam over which this V_s must exist. The third condition involves moving the cutoff point to another location. The second condition is a design method whereby additional stirrups may be introduced if the shear strength is inadequate as determined by the Code, Section 12.11.5.1. This condition will be developed further.

It is convenient to determine the *number* of additional stirrups to be added in the length $\frac{3}{4}d$ along the end of the bar from the cutoff point. The required excess stirrup area is

$$A_v = \frac{60\, b_w s}{f_y}$$

Assuming that A_v is known, the maximum spacing is then

$$s = \frac{A_v f_y}{60\, b_w}$$

Therefore, the number of stirrups N_s to be added in the length $\frac{3}{4}d$ is

$$N_s = \frac{\frac{3}{4}d}{s} + 1 = \frac{\frac{3}{4}d}{\left(\dfrac{A_v f_y}{60\, b_w}\right)} + 1 = \left(\frac{45\, b_w d}{A_v f_y}\right) + 1$$

Also, since maximum $s = d/(8\beta_b)$,

$$N_{s(min)} = \frac{\frac{3}{4}d}{\left(\dfrac{d}{8\beta_b}\right)} + 1 = 6\beta_b + 1$$

The larger resulting N_s controls.

EXAMPLE 5–6

In the beam shown in Fig. 5–12, the location at which the top layer of two No. 9 bars is to be terminated is in a tension zone. At the cutoff point, $V_u = 52$ kips and the No. 4 stirrups are spaced at 11 in. on center. Check shear in accordance with the Code, Section 12.11.5 and redesign the stirrup spacing if

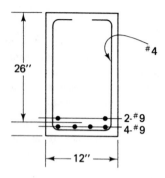

FIGURE 5-12 Sketch for Example 5-6

necessary. Assume $f'_c = 3,000$ psi and $f_y = 40,000$ psi. In accordance with ACI 12.11.5.1, check $V_u \leq \frac{2}{3}\phi(V_c + V_s)$

$$\frac{2}{3}\phi(V_c + V_s) = \frac{2}{3}(0.85)\left(2\sqrt{f'_c}bd + \frac{A_v f_y d}{s}\right)$$

$$= \frac{2}{3}(0.85)\left[2\sqrt{3000}(12)(26) + \frac{(0.40)(40)(26)}{11}\right]$$

$$= 40.9 \text{ kips} < 52 \text{ kips} \qquad \text{(N.G.)}$$

Therefore, add excess stirrups over a length of $\frac{3}{4}d$ along the terminated bars from the cut end in accordance with ACI 12.11.5.2.

$$N_s = \left[\frac{45 b_w d}{A_v f_y} + 1\right] \text{ or } \left[6\beta_b + 1\right]$$

$$\frac{45(12)(26)}{0.40(40,000)} + 1 = 1.88$$

$$6(\tfrac{1}{3}) + 1 = 3$$

Add three stirrups over a length of $\frac{3}{4}d = 19.5$ in. Find the new spacing required in the 19.5 in. length:

$$\frac{19.5}{11} = 1.77 \text{ stirrups at 11 in. spacing}$$

$$+ \underline{3.00} \text{ additional stirrups}$$

$$4.77 \text{ stirrups in 19.5 in. length}$$

New stirrup spacing:

$$\frac{19.5}{4.77} = 4.09 \text{ in.} \qquad \text{Use 4 in.}$$

The stirrup pattern originally designed should now be altered to include the 4 in. spacing along the last 19.5 in. of the cut bars.

It should be noted that the problems associated with terminating bars in a tension zone may be avoided by extending the bars in accordance with ACI Code Section 12.11.5.3 or extending them into the support.

In summary, a general representation of the bar cutoff requirements for positive moment steel in a simple span may be observed in Fig. 5–13. If bars *A* are to be cut off, they must (a) project ℓ_d past the point of maximum positive moment, and (b) project beyond their theoretical cutoff point a distance equal to the effective depth of the member or 12 bar diameters, whichever is greater. The remaining positive moment bars *B* must extend ℓ_d past the theoretical cutoff point of bars *A* and extend at least 6 in. into the support.

5-9 CODE REQUIREMENTS FOR DEVELOPMENT OF POSITIVE MOMENT STEEL AT SIMPLE SUPPORTS

The ACI Code, Section 12.12.3, contains requirements concerning the development of positive moment bars *at simple supports* and at points of inflection. The intent of this Code section is the same as the check on the development of the two bars extending into the support in Example 5–5. The method of Example 5–5 may be used if one is working with the actual moment diagram. The Code approach does not require the use of a moment diagram.

The Code requirement places a restriction on the size of the bar that may be used such that its required ℓ_d does not exceed

$$\frac{M_n}{V_u} + \ell_a$$

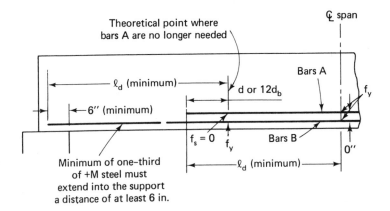

FIGURE 5-13 Bar cutoff requirements for simple spans (positive-moment steel)

where M_n = nominal moment strength $\left[A_s f_y \left(d - \dfrac{a}{2} \right) \right]$ assuming

all reinforcement at the section to be stressed to f_y

V_u = total applied design shear force at the section

ℓ_a = additional embedment length at the support, which is to be taken as the sum of the embedment length beyond the center of the support and the equivalent embedment length of any furnished hook or mechanical anchorage.

The effect of this Code restriction is to require that bars be small enough so that they can become fully developed before the applied moment has increased to the magnitude where they *must* be capable of carrying f_y. M_n/V_u approximates the distance from the section in question to the location where applied moment M_u exists that is equal to M_R (and where f_y must exist in the bars).

Therefore, the distance from the end of the bar to the point where the bar must be fully developed is $(M_n/V_u) + \ell_a$, and bars must be chosen so that their ℓ_d is less than this distance. The Code allows M_n/V_u to be increased by 30% when the ends of the reinforcement are confined by a compressive reaction such as is found in a simply supported beam (a beam supported by a wall).

EXAMPLE 5–7

At the support of a simply supported beam, a cross section exists as is shown in Fig. 5–14. Check the bar diameters in accordance with the ACI Code, Section 12.12.3. Assume a support width of 12 in., 1½ in. cover,

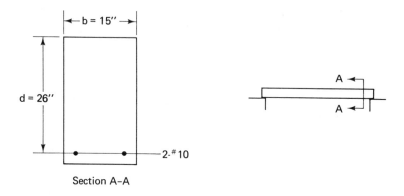

Section A–A

FIGURE 5-14   Sketch for Example 5–7

$f'_c = 4,000$ psi, and $f_y = 60,000$ psi. V_u at the support is 80 kips. Since this beam has its ends confined by a compressive reaction, we will check

$$\ell_d \le 1.3 \left(\frac{M_n}{V_u} \right) + \ell_a$$

From Table A–12, the development length required (B.D.L.) is 48 in.

$$\rho = \frac{A_s}{bd} = \frac{2.54}{15(26)} = 0.0065$$

$$\overline{k} = 0.3676 \text{ ksi}$$

$$M_n = bd^2 \overline{k} = 15(26)^2(0.3676) = 3727.5 \text{ in.-kips}$$

The maximum permissible required ℓ_d is

$$1.3 \frac{M_n}{V_u} + \ell_a = 1.3 \left(\frac{3727.5}{80} \right) + 4.5 = 65.1 \text{ in.}$$

$$65.1 \text{ in.} > 48 \text{ in.} \qquad \text{(O.K.)}$$

The bar diameter is adequate (small enough) and the bar can be developed as required. (If the required ℓ_d were in excess of 65.1 in., the maximum permissible value could be increased by using a hook, thereby increasing ℓ_a. Also, the use of smaller bars would result in a smaller required ℓ_d.)

PROBLEMS

5–1. Determine the development length required for the bars shown. $f'_c = 4,000$ psi and $f_y = 60,000$ psi. Check the anchorage in the column. If it is not satisfactory, design an anchorage using a 180° hook and check its adequacy.

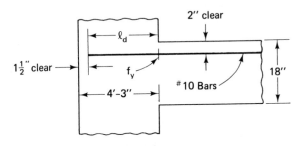

FIGURE P5-1

5–2. Considering the anchorage of the beam bars into the column, determine the largest bar that can be used without a hook. $f'_c = 3,000$ psi and $f_y = 40,000$ psi.

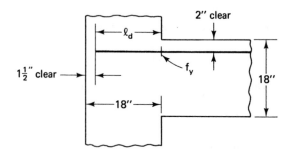

FIGURE P5-2

5-3. A simply supported, uniformly loaded beam carries a total factored design load of 4.8 kips/ft (this includes the beam weight) on a clear span of 34 ft. $f'_c = 3,000$ psi and $f_y = 40,000$ psi. Assume that the supports are 12 in. wide and assume that bars are available in 30 ft. lengths.

    a. Design a rectangular beam (tension steel only).
    b. Determine bar cutoffs.
    c. Locate splices and determine lap length.
    d. Design for shear reinforcement.
    e. Determine if bar cutoffs in the tension zone satisfy the Code, and redesign the stirrups if necessary.

5-4. The beam shown is to carry a total factored uniform load of 10 kips/ft (includes beam weight) and factored concentrated loads of 20 kips. $f'_c = 4,000$ psi and $f_y = 60,000$ psi. Assume that the supports are 12 in. wide.

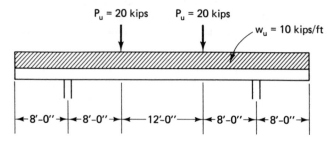

FIGURE P5-4

    a. Design a rectangular beam (tension steel only).
    b. Terminate the bars where possible (use graphical approach).
    c. Design the stirrups. Consider shear at the bar terminations.

6

Continuous Construction
Design Considerations

6–1 INTRODUCTION

A common form of concrete cast-in-place building construction consists of a continuous one-way slab cast monolithically with supporting continuous beams and girders. In this type of system all members contribute in carrying the floor load to the supporting columns (see Fig. 3–1). The slab steel runs through the beams, the beam steel runs through the girders, and the steel from both the beams and girders runs through the columns. The result is that the whole floor system is tied together, forming a highly indeterminate and complex type of rigid structure. The behavior of the members is affected by their rigid connections. Not only will loads applied directly *on* a member produce moment, shear, and a definite deflected shape, but loads applied to *adjacent* members will produce similar effects because of the rigidity of the connections. The shears and moments transmitted through a joint will depend on the relative stiffnesses of all the members framing into that joint. With this type of condition, a precise evaluation of moments and shears resulting from a floor loading is excessively time-consuming and is outside the scope of this text. There are several commercial computer programs available which facilitate these analysis computations.

In an effort to simplify and expedite the design phase, the ACI Code, Section 8.3.3, permits the use of standard moment and shear equations whenever the span and loading conditions satisfy stipulated requirements. This approach applies to continuous nonprestressed one-way slabs, beams, and girders. It is an approximate method and may be used for buildings of the usual type of construction, spans, and story heights. The ACI moment equations result from the product of a coefficient and $w_u\ell_n^2$. The ACI shear equations, similarly, result from the product of a coefficient and $w_u\ell_n$. In these equations, w_u is the factored design uniform load and ℓ_n is the *clear span* for positive moment (and shear) and the average of two adjacent clear spans for negative moment. The application of the equations is limited to the following:

1. For two or more approximately equal spans (with the larger of two adjacent spans not exceeding the shorter by more than 20%).

2. Loads must be uniformly distributed.

3. The maximum allowable ratio of live load to dead load is 3:1 (based on service loads).

These shear and moment equations generally give reasonably conservative values for the stated conditions. If more precision is required, or desired, for economy, or because the stipulated conditions are not satisfied, a more theoretical and precise analysis must be made.

The moment and shear equations are depicted in Fig. 6–1. Their use will be demonstrated later in this chapter.

6–2 CONTINUOUS-SPAN BAR CUTOFFS

In a manner similar to that for simple spans, the area of main reinforcing steel required at any given point is a function of the design moment. As the moment varies along the span, the steel may be modified or reduced in accordance with the theoretical requirements of the member's strength and the requirements of the ACI Code.

Bars can theoretically be stopped or bent in flexural members whenever they are no longer needed to resist moment. A general representation of the bar cutoff requirements for continuous spans (both positive and negative moments) may be observed in Fig. 6–2.

In continuous members the ACI Code, Section 12.12.1, requires that a minimum of one-fourth of the positive-moment steel be extended into the support a distance of at least 6 in. The ACI Code, Section 12.13.3, also requires that at least one-third of the negative-moment steel be extended beyond the extreme position of the point of inflection a distance not less than one-sixteenth of the clear span, the effective depth of the member d, or 12 bar diameters, whichever is greater. If negative-moment bars C [Fig. 6–2(a)] are to be cut off, they must extend at l ast a full development length ℓ_d beyond the face of support. In addition, they must extend a distance equal to the effective depth of the member or 12 bar

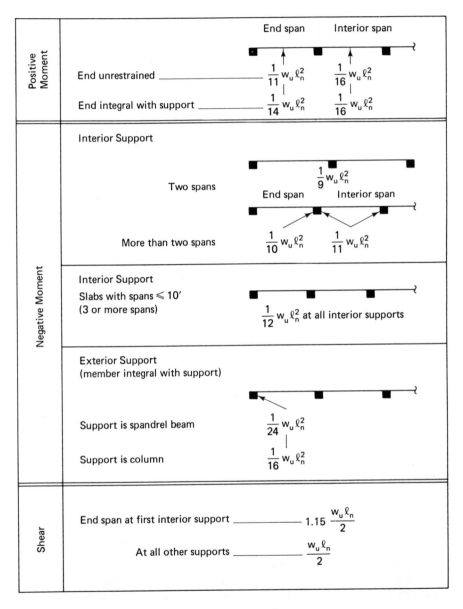

FIGURE 6-1 ACI code coefficients and equations for shear and moment for continuous beams, girders, and one-way slabs

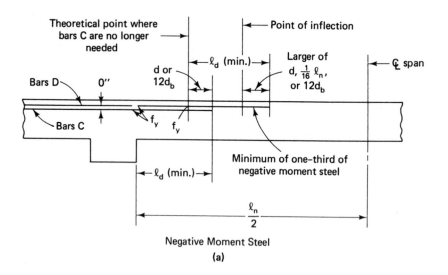

Negative Moment Steel

(a)

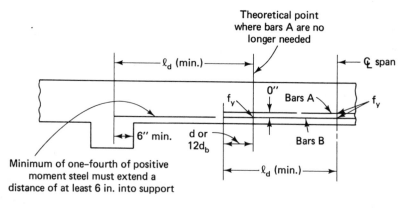

Positive Moment Steel

(b)

FIGURE 6-2 Bar cutoff requirements for continuous slabs

diameters, whichever is larger, beyond the theoretical cutoff point defined by the moment diagram. The remaining negative-moment bars D (minimum of one-third of total negative steel) must extend at least ℓ_d beyond the theoretical point of cutoff of bars C, and, in addition, must extend a distance equal to the effective depth of the member, 12 bar diameters, or one-sixteenth of the clear span, whichever is the greater, past the point of inflection. Where negative moment bars are cut off before reaching the point of inflection, the situation is analogous to the simple beam cutoffs where the reinforcing bars are being terminated in a tension zone. The reader is referred to Example 5–4.

In Fig. 6–2(b), if positive-moment bars *A* are to be cut off, they must project ℓ_d past the point of maximum positive moment as well as a distance equal to the effective depth of the member or 12 bar diameters, whichever is larger, beyond their theoretical cutoff point. Recall that the location of the theoretical cutoff point depends on the amount of steel to be cut and the shape of the applied moment diagram. The remaining positive-moment bars *B* must extend ℓ_d past the theoretical cutoff point of bars *A* and extend at least 6 in. into the support. Comments on terminating bars in a tension zone again apply. Additionally, the size of the positive-moment bars at the point of inflection must meet the requirements of the ACI Code, Section 12.12.3.

Since the determination of cutoff and bend points constitutes a relatively time consuming chore, it has become customary to utilize defined cutoff points which experience has indicated are safe. These defined points may be utilized where the ACI moment coefficients have application, but must be applied with judgment where parameters vary. The recommended bar details and cutoffs for continuous spans may be observed in Fig. 6–3.

EXAMPLE 6–1

The floor system shown in Fig. 6–4 consists of a continuous one-way slab supported by continuous beams. The service loads on the floor are 25 psf dead load (does not include weight of slab) and 200 psf live load. $f'_c = 3,000$ psi and $f_y = 40,000$ psi.

a. Design the continuous one-way floor slab.

b. Design the continuous supporting beam.

The primary difference in this design from previous flexural designs is that, because of continuity, the ACI coefficients and equations will be used to determine design shears and moments.

Part a

Slab thickness

The slab will be designed to satisfy the ACI minimum thickness requirements (ACI Code Section 9.5a) and this thickness will be used to estimate slab weight. With both ends continuous,

$$\text{Minimum } h = \frac{1}{28} \ell_n \left(0.4 + \frac{f_y}{100,000} \right)$$

$$= \frac{1}{28} (11) (12) \left(0.4 + \frac{40,000}{100,000} \right)$$

$$h = 3.77 \text{ in.}$$

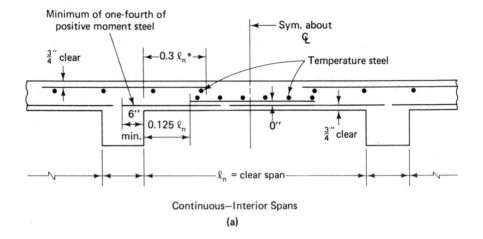

Continuous—Interior Spans

(a)

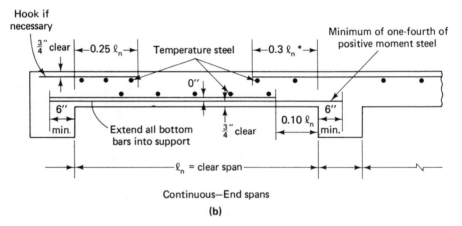

Continuous—End spans

(b)

*If adjacent spans have different span lengths, use the larger of the two.

FIGURE 6-3 Recommended bar details and cutoffs, one-way slabs, tensile reinforced beams similar

With one end continuous,

$$\text{Minimum } h = \frac{1}{24} \ell_n \left(0.4 + \frac{f_y}{100,000} \right)$$

$$= \frac{1}{24} (11)(12)(0.4 + 0.4)$$

$$h = 4.4 \text{ in.}$$

Try a 4½-in.-thick slab. Design a 12-in.-wide segment ($b = 12$ in.).

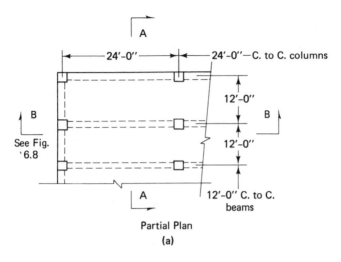

Partial Plan

(a)

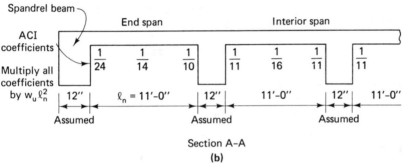

Section A-A

(b)

FIGURE 6-4   Sketches for Example 6–1

Load determination

$$\text{Slab dead load} = \frac{4.5}{12}(150) = 56.3 \text{ psf}$$

$$\text{Total dead load} = 25.0 + 56.3 = 81.3 \text{ psf}$$

$$w_u = 1.4 w_{DL} + 1.7 w_{LL}$$

$$= 1.4(81.3) + 1.7(200)$$

$$= 113.8 + 340.0$$

$$= 453.8 \text{ psf} \quad \text{(design load)}$$

Because we are designing a slab segment that is 12 in. wide, the foregoing loading is the same as 453.8 lb/ft or 0.454 kip/ft.

Determination of moments and shears

Moments are next determined using the ACI moment equations. Refer to Figs. 6–1 and 6–4.

$$+M_u = \frac{1}{14} w_u \ell_n^2 = \frac{1}{14}(0.454)(11)^2 = 3.92 \text{ ft-kips}$$

$$+M_u = \frac{1}{16} w_u \ell_n^2 = \frac{1}{16}(0.454)(11)^2 = 3.43 \text{ ft-kips}$$

$$-M_u = \frac{1}{10} w_u \ell_n^2 = \frac{1}{10}(0.454)(11)^2 = 5.49 \text{ ft-kips}$$

$$-M_u = \frac{1}{11} w_u \ell_n^2 = \frac{1}{11}(0.454)(11)^2 = 4.99 \text{ ft-kips}$$

$$-M_u = \frac{1}{24} w_u \ell_n^2 = \frac{1}{24}(0.454)(11)^2 = 2.29 \text{ ft-kips}$$

Similarly, the shears are determined using the ACI shear equations. In the end span at the face of the first interior support

$$V_u = 1.15 \frac{w_u \ell_n}{2} = 2.87 \text{ kips}$$

while at all other supports

$$V_u = \frac{w_u \ell_n}{2} = 0.454 \left(\frac{11}{2} \right) = 2.50 \text{ kips}$$

Slab design

Using the assumed slab thickness of 4½ in., find the approximate d. Assume No. 5 bars for main steel and ¾-in. cover for the bars in the slab.

$$d = 4.5 - 0.75 - 0.31 = 3.44 \text{ in.}$$

Design of the steel reinforcing

Select the point of maximum moment. This is a negative moment and occurs in the end span at the first interior support and

$$M_u = \frac{w_u \ell_n^2}{10} = 5.49 \text{ ft-kips}$$

$$M_R = \phi b d^2 \overline{k}$$

and, since, for design purposes $M_u = M_R$ as a limit, then

$$\text{required } \overline{k} = \frac{M_u}{\phi b d^2} = \frac{5.49\,(12)}{0.90\,(12)\,(3.44)^2}$$

$$= 0.5155 \text{ ksi}$$

From Table A–6,

$$\rho = 0.0146 < \rho_{max} \qquad \text{(OK)}$$

$$\text{Required } A_s = \rho bd = 0.0146\,(12)\,(3.44)$$

$$A_s = 0.60 \text{ in.}^2$$

Since the steel area required at all other points will be less, the preceding process will be repeated for the other points. The expression

$$\text{Required } \overline{k} = \frac{M_u}{\phi bd^2}$$

can be simplified because all values are constant except M_u:

$$\text{Required } \overline{k} = \frac{M_u(12)}{0.9(12)(3.44)^2} = \frac{M_u}{10.65}$$

where M_u must be in ft-kips. In the usual manner, the required steel ratio ρ and the required steel area A_s may then be determined. The results of these calculations are listed in Table 6–1.

Minimum reinforcement for slabs of constant thickness is that required for shrinkage and temperature reinforcement:

$$\text{Minimum required } A_s = 0.0020bh$$

$$= 0.0020-(12)(4.5) = 0.11 \text{ in.}^2$$

Also, maximum $\rho = 0.0278$ (reference, Table A–4). Therefore, the slab steel requirements for flexure as shown in Table 6–1 are within acceptable limits. The shrinkage and temperature steel may be selected based on the preceding calculation:

$$\text{Use No. 3 bars at 12 in. o.c. } (A_s = 0.11 \text{ in.}^2)$$

Recall that the maximum spacing allowed is the smaller of $5h$ or 18 in. Since $5h = 5(4.5) = 22.5$ in., the 18 in. would control and the spacing is acceptable.

TABLE 6–1
Slab Steel Area Requirements

Location	Moment Equation	$\overline{k}$ (ksi)	Required ρ	A_s (in.²/ft)
End span				
At spandrel	$-\frac{1}{24}\,w_u\ell_n^2$	0.2150	0.0056	0.23
Midspan	$+\frac{1}{14}\,w_u\ell_n^2$	0.3681	0.0100	0.41
Interior spans				
Interior support	$-\frac{1}{11}\,w_u\ell_n^2$	0.4685	0.0131	0.54
Midspan	$+\frac{1}{16}\,w_u\ell_n^2$	0.3221	0.0087	0.36

Check shear strength

Maximum $V_u = 2.87$ kips at the face of the support. A check of shear at the face of the support, rather than at the critical section which is at a distance equal to the effective depth of the member from the face of support, is conservative. The shear strength ϕV_n of the slab, if no shear reinforcing is provided, is the shear strength of the concrete:

$$\phi V_n = \phi V_c = \phi 2 \sqrt{f'_c}\, b_w d$$

$$= 0.85(0.110)(12)(3.44) = 3.86 \text{ kips}$$

$$V_u < \phi V_n$$

Therefore, no shear reinforcing is required.

Select the main steel

Using Table A–3 in Appendix A, establish a pattern in which the number of bar sizes and the number of different spacings are kept to a minimum. The maximum spacing for the main steel is 13.5 in. (the smaller of $3h$ or 18 in.) A work sketch (see Fig. 6–5) is recommended to establish steel pattern and cutoff points. Note, with regard to the steel selection, that No. 4 and No. 5 bars are mixed for positive moment in the end span in an effort to minimize different spacings and still economize on steel provided. At this location No. 5 bars at 9 in. could be used, however maximum spacing would be exceeded if alternate bars were terminated since at cutoff points, the spacings of the bars that remain must also be considered.

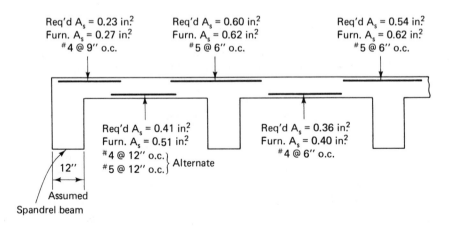

Req'd $A_s = 0.23$ in²
Furn. $A_s = 0.27$ in²
#4 @ 9" o.c.

Req'd $A_s = 0.60$ in²
Furn. $A_s = 0.62$ in²
#5 @ 6" o.c.

Req'd $A_s = 0.54$ in²
Furn. $A_s = 0.62$ in²
#5 @ 6" o.c.

Req'd $A_s = 0.41$ in²
Furn. $A_s = 0.51$ in²
#4 @ 12" o.c.⎫
#5 @ 12" o.c.⎭ Alternate

Req'd $A_s = 0.36$ in²
Furn. $A_s = 0.40$ in²
#4 @ 6" o.c.

12"

Assumed
Spandrel beam

Note: Bar cutoffs and temperature steel may be observed in the design sketch, Fig. 6.7

FIGURE 6-5   Work sketch for Example 6–1

Check anchorage into spandrel beam

This procedure is presented in Chapter 5, Example 5–3. The steel is No. 4 bar at 9 in. o.c.

1. Basic development length = 8.0 in. (from Table A–12).

2. Modification factors to be used are: 1.4 for top bars, 0.8 for bar spacing in excess of 6 in., and for excess steel

$$\frac{\text{Required } A_s}{\text{Provided } A_s} = \frac{0.23}{0.27} = 0.85$$

3. Therefore, the development length required is

$$\text{Required } \ell_d = 8.0(1.4)(0.8)(0.85) = 7.6 \text{ in.}$$

This is less than the Code minimum of 12 in. (ACI Section 12.2.5); therefore, use 12 in. Since the 12 in. length cannot be furnished, a 180° standard hook will be provided.

4. From Table A–13, $\ell_e = 3.9$ in.

5. Using the modification factors of step 2, modify ℓ_e:

$$\ell_e = 3.9(1.4)(0.8)(0.85) = 3.71 \text{ in.}$$

6. The straight embedment length that must be furnished in addition to the hook is

$$\ell_d - \ell_e = 12 - 3.71 = 8.29 \text{ in.}$$

7. Check the total width of beam required (includes cover, hook, and straight embedment as shown in Fig. 6–6):

$$1.5 + 0.5 + 1.25 + 8.29 = 11.54 < 12 \text{ in.} \qquad \text{(OK)}$$

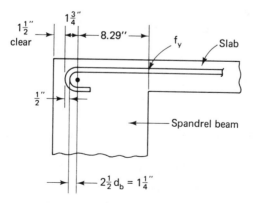

FIGURE 6-6   Hook detail for Example 6–1

Bar cut-off points

For the normal type of construction for which the typical bar cutoff points shown in Fig. 6–3 are used, the cutoff points are located so that all bars terminate in compression zones. Therefore the requirements of the ACI Code Section 12.11.5 need not be checked.

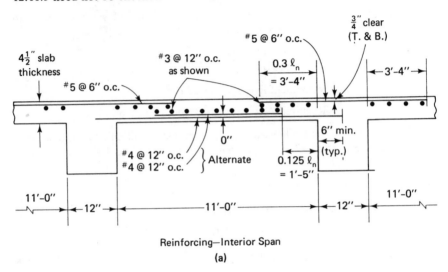

Reinforcing—Interior Span

(a)

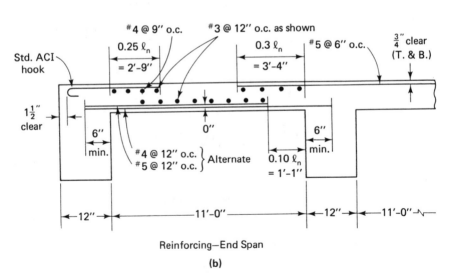

Reinforcing—End Span

(b)

FIGURE 6-7 **Design sketches for Example 6–1**

Design sketches

The final design sketch for the slab is shown in Fig. 6–7. For clarity, the interior and end spans are shown separately.

Part b

Part b of Example 6–1 involves the design of the continuous supporting beam. From Fig. 6–4(a), it is seen that these beams span between columns. The ACI coefficients to be used for moment determination are shown in Fig. 6–8.

Loading

$$\text{Service live load} = 200 \text{ psf} \times 12 = 2,400 \text{ lb/ft}$$

$$\text{Service dead load} = 25 \text{ psf} \times 12 = 300 \text{ lb/ft}$$

$$\text{Weight of slab} = \left(\frac{4.5}{12}\right)(150)(12) = 675 \text{ lb/ft}$$

Assuming a beam width of 12 in. and an overall depth of 30 in. for purposes of member weight estimate (see Fig. 6–9),

$$\text{Weight of beam} = \frac{12(30 - 4.5)}{144}(150) = 318.8 \text{ lb/ft}$$

$$\text{Total service live load} = 2,400 \text{ lb/ft}$$

$$\text{Total service dead load} = 1,293.8 \text{ lb/ft} \qquad \text{say, } 1,294 \text{ lb/ft}$$

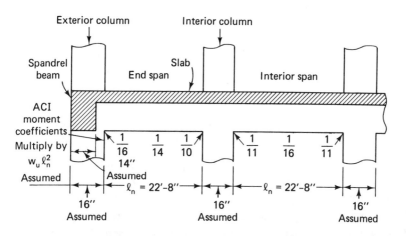

FIGURE 6-8 Section B–B from Fig. 6–4(a)

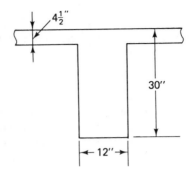

FIGURE 6-9 Sketch for Example 6–1

Design load

$$w_u = 1.4 w_{DL} + 1.7 w_{LL}$$
$$= 1.4(1,294) + 1.7(2,400)$$
$$= 1,812 + 4,080$$
$$= 5,892 \text{ lb/ft} \qquad \text{say, } 5.9 \text{ kips/ft}$$

The loaded beam is depicted in Fig. 6–10.

Design moments and shears

The design moments and shears are calculated utilizing the ACI equations:

$$+ M_u = \frac{1}{14} w_u \ell_n^2 = \frac{1}{14}(5.9)(22.67)^2 = 216.6 \text{ ft-kips}$$

$$+ M_u = \frac{1}{16} w_u \ell_n^2 = \frac{1}{16}(5.9)(22.67)^2 = 189.5 \text{ ft-kips}$$

$$- M_u = \frac{1}{10} w_u \ell_n^2 = \frac{1}{10}(5.9)(22.67)^2 = 303.2 \text{ ft-kips}$$

$$- M_u = \frac{1}{11} w_u \ell_n^2 = \frac{1}{11}(5.9)(22.67)^2 = 275.6 \text{ ft-kips}$$

$$V_u = \frac{w_u \ell_n}{2} = \frac{5.9(22.67)}{2} = 66.9 \text{ kips}$$

$$V_u = 1.15 \frac{w_u \ell_n}{2} = 1.15(5.9)\left(\frac{22.67}{2}\right) = 76.9 \text{ kips}$$

Beam design

Establish concrete dimensions based on the maximum bending moment. This occurs in the end span at the first interior support where the negative moment $M_u = w_u \ell_n^2 / 10$. Since the top of the beam is in tension, the design will be that of a rectangular beam.

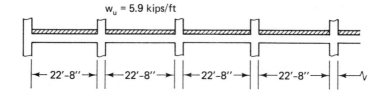

$w_u = 5.9$ kips/ft

|← 22'-8" →| |← 22'-8" →| |← 22'-8" →| |← 22'-8" →| |←—$\vee$

FIGURE 6-10 Beam design load and spans

1. Maximum moment (negative) = 303.2 ft-kips.

2. Assume that $\rho = 0.0135$ (see Table A–4).

3. From Table A–6, $\overline{k} = 0.4828$ ksi.

4. Assume that $b = 12$ in.

$$\text{Required } d = \sqrt{\frac{M_u}{\phi b \overline{k}}} = \sqrt{\frac{303.2(12)}{0.9(12)(0.4828)}}$$

$$= 26.4 \text{ in.}$$

$$d/b \text{ ratio} = \frac{26.4}{12} = 2.2 \qquad \text{which is satisfactory}$$

5. Check the estimated beam weight assuming one layer of No. 11 bars and No. 3 stirrups.

$$\text{Required } h = 26.4 + \frac{1.41}{2} + 0.38 + 1.5 = 28.98 \text{ in.}$$

Use $h = 29$ in. and actual $d = 26.4$ in. Also, check minimum h as in the ACI Code, Table 9.5a.

$$\text{Minimum } h = \frac{1}{18.5}(22.67)(12)\left(0.4 + \frac{40,000}{100,000}\right) = 11.8 \text{ in.} < 29 \text{ in.} \qquad \text{(OK)}$$

Note that the estimated beam weight based on $b = 12$ in. and $h = 30$ in. is slightly conservative, but satisfactory.

6. Design steel reinforcing for points of negative moment as follows:

   a. At the first interior support based on assumed $\rho = 0.0135$ (see steps 1–4):
      Required $A_s = \rho b d$

$$= 0.0135(12)(26.4) = 4.28 \text{ in.}^2$$

   b. At the other interior supports, $-M_u = 275.6$ ft-kips.

$$\text{Required } \overline{k} = \frac{M_u}{\phi b d^2} = \frac{275.6(12)}{0.9(12)(26.4)^2} = 0.4394 \text{ ksi}$$

From Table A–6, $\rho = 0.0121$ and

$$\rho_{\text{min}} < 0.0121 < \rho_{\text{max}}$$

173

Therefore,

$$\text{Required } A_s = \rho bd = 0.0121(12)(26.4) = 3.83 \text{ in.}^2$$

c. At the exterior support—exterior column,
$$- M_u = 189.5 \text{ ft-kips}$$

$$\text{Required } \overline{k} = \frac{M_u}{\phi bd^2} = \frac{189.5(12)}{0.9(12)(26.4)^2} = 0.3021 \text{ ksi}$$

From Table A–6, $\rho = 0.0081$ and

$$\rho_{min} < 0.0081 < \rho_{max}$$

Therefore,

$$\text{Required } A_s = \rho bd = 0.0081(12)(26.4) = 2.57 \text{ in.}^2$$

7. Design steel reinforcing for points of positive moment as follows. At points of positive moment, the top of the beam is in compression; therefore, the design will be that of a T-beam.

   a. End-span positive moment:

   1. Design moment $= 216.6 \text{ ft-kips} = M_u$.

   2. Effective depth $d = 26.4$ in. (see negative-moment design).

   3. Effective flange width:
   $$\tfrac{1}{4} \text{ span length} = 0.25(22.67)(12) = 68 \text{ in.}$$

   $$b_w + 16h_f = 12 + 16(4.5) = 84 \text{ in.}$$

   $$\text{Beam spacing} = 144 \text{ in.}$$

   Use an effective flange width $b = 68$ in. (see Fig. 6–11).

   4. Assuming total flange in compression:
   $$M_R = \phi(0.85f_c')bh_f\left(d - \frac{h_f}{2}\right)$$

   $$= 0.9(0.85)(3)(68)(4.5)\left(\frac{26.4 - 4.5/2}{12}\right) = 1,413 \text{ ft-kips}$$

   5. Since $1,413 > 216.6$ the member behaves as a wide rectangular T-beam with $b = 68$ in. and $d = 26.4$ in.

   6. $$\text{Required } \overline{k} = \frac{M_u}{\phi bd^2}$$

   $$= \frac{216.6(12)}{0.9(68)(26.4)^2} = 0.0609 \text{ ksi}$$

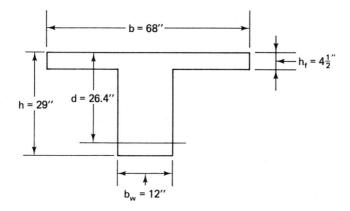

FIGURE 6-11 Beam cross section

7. From Table A–6,

$$\text{Required } \rho = 0.0016$$

8. The required steel area is

$$\text{Required } A_s = \rho b d$$

$$= 0.0016(68)(26.4) = 2.87 \text{ in.}^2$$

9. Use three No. 9 bars ($A_s = 3.0$ in.^{2}).

$$\text{Required } b = 9.5 \text{ in.} \qquad \text{(OK)}$$

10. Check ρ_{min}.

$$\rho_{\text{min.}} = \frac{200}{40,000} = 0.0050$$

$$\text{Actual } \rho = \frac{A_s}{b_w d} = \frac{3.0}{12(26.4)} = 0.0095 > 0.0050 \qquad \text{(OK)}$$

11. Check $A_{s\,(\text{max})}$.

$$A_{s\,(\text{max})} = 0.0478 h_f \left\{ b + b_w \left[\frac{0.582}{h_f}(d) - 1 \right] \right\}$$

$$= 0.0478(4.5) \left\{ 68 + 12 \left[\frac{0.582(26.4)}{4.5} - 1 \right] \right\}$$

$$= 20.9 \text{ in.}^2 > 3.0 \text{ in.}^2 \qquad \text{(OK)}$$

b. Interior span positive moment:

    1. Design moment $= 189.5$ ft-kips $= M_u$

    2. through **5.** See the end-span positive-moment computations. Use an effective flange width $b = 68$ in. and an effective depth $d = 26.4$ in. Also, $M_R = 1413$ ft-kips $> M_u$. Therefore, this member also behaves as a rectangular T-beam.

6.
$$\text{Required } \overline{k} = \frac{M_u}{\phi bd^2}$$

$$= \frac{189.5(12)}{0.9(68)(26.4)^2} = 0.0533 \text{ ksi}$$

7. From Table A–6,

$$\text{Required } \rho = 0.0014$$

8. The required steel area is

$$\text{Required } A_s = \rho bd$$

$$= 0.0014(68)(26.4) = 2.51 \text{ in.}^2$$

9. Use three No. 9 bars ($A_s = 3.0$ in.^{2}).

$$\text{Required } b = 9.5 \text{ in.}$$ (OK)

10. and **11.** The checks for $\rho_{\min}$ and $A_{s\,(\max)}$ are identical to those for the end-span positive moment.

Selection of steel for negative moment

The ACI Code, Section 10.6.6, requires that where flanges are in tension, a part of the main tension reinforcement be distributed over the effective flange width or a width equal to one-tenth of the span, whichever is smaller. The use of smaller bars spread out into part of the flange will also be advantageous where a beam is supported by a spandrel girder or exterior column and the embedment length for the negative moment steel is limited.

$$\frac{\text{Span}}{10} = \frac{22.67(12)}{10} = 27 \text{ in.}$$

$$\text{Effective flange width} = b = 68 \text{ in.}$$

Therefore, distribute the negative-moment bars over a width of 27 in. Figure 6–12 shows suitable bars and patterns to use to satisfy the foregoing Code requirement and furnish the cross-sectional area of steel required for flexure.

The Code also stipulates that if the effective flange width exceeds one-tenth of the span, some longitudinal reinforcement shall be provided in the outer portions of the flange. In this design no additional steel will be furnished. In the authors' opinion this requirement is satisfied by the slab temperature and shrinkage steel (see Figs. 6–7 and 6–19).

Work sketch

A work sketch is developed in Fig. 6–13 which includes the bars previously chosen.

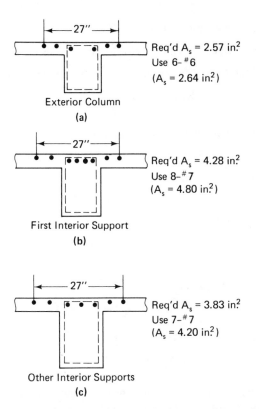

Req'd A_s = 2.57 in.^{2}
Use 6-$^\#$6
(A_s = 2.64 in.^{2})

Exterior Column
(a)

Req'd A_s = 4.28 in.^{2}
Use 8-$^\#$7
(A_s = 4.80 in.^{2})

First Interior Support
(b)

Req'd A_s = 3.83 in.^{2}
Use 7-$^\#$7
(A_s = 4.20 in.^{2})

Other Interior Supports
(c)

FIGURE 6-12   Negative-moment steel for beam of Example 6–1

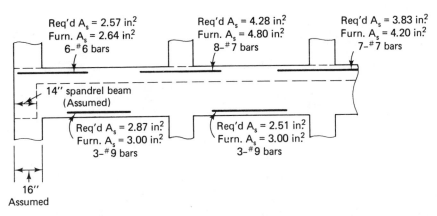

Req'd A_s = 2.57 in.^{2}
Furn. A_s = 2.64 in.^{2}
6-$^\#$6 bars

Req'd A_s = 4.28 in.^{2}
Furn. A_s = 4.80 in.^{2}
8-$^\#$7 bars

Req'd A_s = 3.83 in.^{2}
Furn. A_s = 4.20 in.^{2}
7-$^\#$7 bars

14" spandrel beam
(Assumed)

Req'd A_s = 2.87 in.^{2}
Furn. A_s = 3.00 in.^{2}
3-$^\#$9 bars

Req'd A_s = 2.51 in.^{2}
Furn. A_s = 3.00 in.^{2}
3-$^\#$9 bars

16"
Assumed

FIGURE 6-13   Work sketch for beam design of Example 6–1

Check anchorage into exterior column

1. The basic development length for No. 6 bars = 13 in. (Table A–12).

2. The modification factors to be used are 1.4 for top bars and for excess steel.

$$\frac{\text{Required } A_s}{\text{Provided } A_s} = \frac{2.57}{2.64} = 0.97$$

3. Therefore, the development length required is

$$\ell_d = 13.0\,(1.4)\,(0.97) = 17.7 \text{ in.} > 12 \text{ in.}$$

With 1.5 in. clear, the embedment length available is

$$16 - 1.5 = 14.5 \text{ in.}$$

Since 17.7 in. > 14.5 in., a hook is required. A 90° standard hook will be provided.

4. From Table A–13, $\ell_e = 6.3$ in.

5. Correct ℓ_e for modification factors.

$$\ell_e = 6.3\,(1.4)\,(0.97) = 8.6 \text{ in.}$$

6. The straight embedment length that must be furnished in addition to the hook is

$$\ell_d - \ell_e = 17.7 - 8.6 = 9.1 \text{ in.}$$

7. Check the total width of the column required (includes the cover, hook, and straight embedment as shown in Fig. 6–14):

$$1.5 + 0.75 + 2.25 + 9.1 = 13.6 \text{ in.} < 16 \text{ in.}$$

for the column and 14-in. for the spandrel beam. (OK)

Stirrup design

Established values are $b_w = 12$ in., effective depth $d = 26.4$ in., $f'_c = 3,000$ psi, and $f_y = 40,000$ psi.

1. The shear force diagram may be observed in Fig. 6–15. Since the diagram is

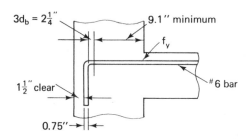

FIGURE 6-14 Anchorage at column

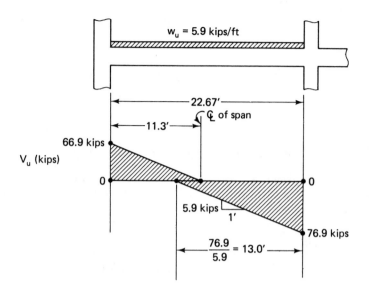

FIGURE 6-15 Stirrup design for Example 6–1

unsymmetrical with respect to the centerline of the span, the stirrup design will be based on the shear in the interior portion of the end span where the maximum values occur. The resulting stirrup pattern will be used throughout the continuous beam.

2. V_s diagram:

$$V_c = (2\sqrt{f'_c})\, b_w\, d$$

$$= 0.110(12)(26.4) = 34.85 \text{ kips}$$

$$\tfrac{1}{2}\phi V_c = \tfrac{1}{2}(0.85)(34.85) = 14.81 \text{ kips}$$

Stirrups are required since $76.9 > 14.81$. Calculate V_s at the face of the support:

$$\text{Required } V_s = \frac{V_u}{\phi} - V_c$$

$$= \frac{76.9}{0.85} - 34.85 = 55.62 \text{ kips}$$

$$\text{Slope of } V_s \text{ diagram} = \frac{5.9}{\phi} = \frac{5.9}{0.85} = 6.94 \text{ kips/ft}$$

The point where the V_s diagram passes through zero, measured from the face of the support:

$$V_s = 0 \quad \text{at} \quad \frac{55.62}{6.94} = 8.01 \text{ ft}$$

The V_s diagram is shown in Fig. 6–16.

179

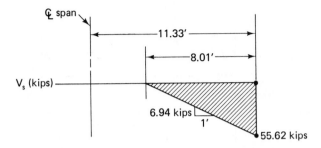

FIGURE 6-16 V_s diagram for Example 6-1

3. Determine the length of the span over which stirrups are required. Stirrups must be provided to the point where $V_u = \frac{1}{2}\phi V_c = 14.81$ kips. Utilizing the V_u diagram of Fig. 6–15 and referencing from the face of the support:

$$\frac{76.9 - 14.81}{5.9} = 10.5 \text{ ft}$$

4. Assume a No. 3 stirrup ($A_v = 0.22$ in.²) and check the spacing at the critical section. With reference to Fig. 6–16,

$$V_s = 55.62 - \left(\frac{26.4}{12}\right)(6.94) = 40.35 \text{ kips}$$

$$\text{Required } s = \frac{A_v f_y d}{V_s} = \frac{0.22(40)(26.4)}{40.35} = 5.76 \text{ in.}$$

Use 5½ in.

5. Determine the ACI Code maximum spacing requirements.

$$4\sqrt{f_c'}\, b_w\, d = 0.219(12)(26.4) = 69.4 \text{ kips}$$

$$69.4 > 55.62$$

Therefore, $s_{max} = d/2$ or 24 in., where $d/2 = 13.2$ in. Also, check

$$s_{max} = \frac{A_v f_y}{50 b_w} = \frac{0.22\,(40,000)}{50\,(12)} = 14.7 \text{ in.}$$

Therefore, $s_{max} = 13.2$ in.

6. Spacing requirements based on shear strength:

$$V_s = V_{s\,(max)} - mx$$

$$= 55.62 - 6.94\,(x)$$

$$\text{Required } s = \frac{A_v f_y d}{V_s}$$

Therefore,

$$\text{Required } s = \frac{0.22\,(40)(26.4)}{55.62 - 6.94\,(x)} = \frac{232.3}{55.62 - 6.94\,(x)}$$

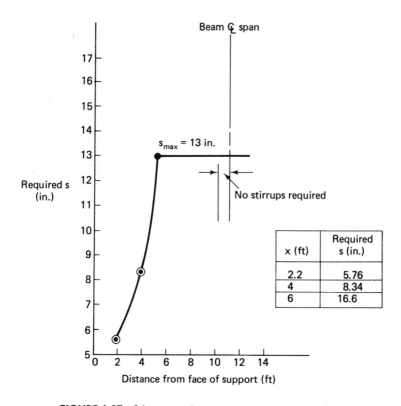

FIGURE 6-17 Stirrup spacing requirements—Example 6–1

x (ft)	Required s (in.)
2.2	5.76
4	8.34
6	16.6

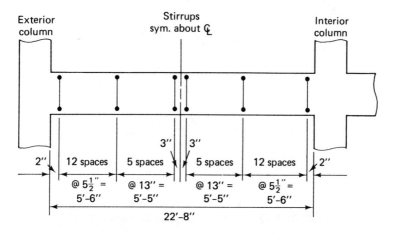

FIGURE 6-18 Stirrup spacing for Example 6–1, end span (interior spans similar)

181

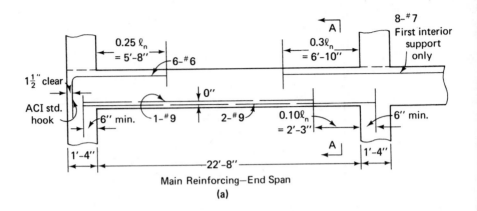

Main Reinforcing—End Span

(a)

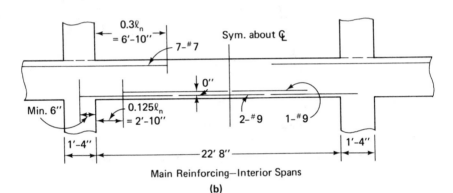

Main Reinforcing—Interior Spans

(b)

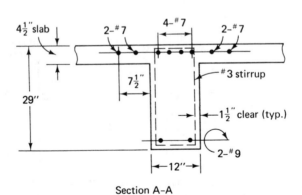

Section A-A

(c)

FIGURE 6-19 Design sketches for Example 6–1

182

The results for several arbitrary values of x are shown tabulated and plotted in Fig. 6–17.

7. Utilizing Fig. 6–17, the stirrup pattern shown in Fig. 6–18 is developed. Despite the lack of symmetry in the shear diagram, the stirrup pattern is symmetrical with respect to the centerline of the span. This is conservative and will be used for all spans.

The design sketches are shown in Fig. 6–19. As with the slab design, the typical bar cutoff points of Fig. 6–3 are used for this beam. Therefore, all bars terminate in compression zones, and the requirements of ACI Code Section 12.11.5 need not be checked.

PROBLEMS

6–1. For the one-way slab shown, determine all moments and shears for service loads of 100 psf dead load (includes the slab weight) and 300 psf live load.

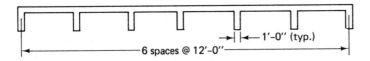

FIGURE P6-1

6–2. For the continuous beam shown, determine all moments and shears for service loads of 2.00 kips/ft dead load (includes weight of beam and slab) and 3.00 kips/ ft live load.

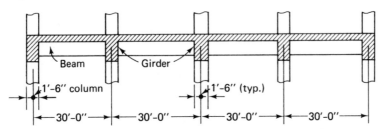

FIGURE P6-2

6–3. A floor system is to consist of beams, girders, and a slab. Service loads are to be 30 psf dead load (*does not* include the weight of the floor system) and 150 psf live load. $f'_c = 4,000$ psi and $f_y = 60,000$ psi.
   a. Design the continuous one-way slab.
   b. Design the continuous beam along column line 2.
Be sure to include complete design sketches.

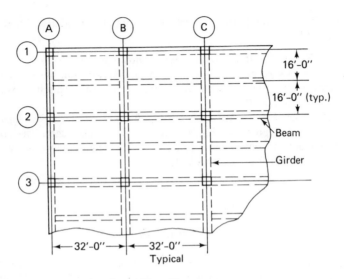

Partial Floor Plan

FIGURE P6-3

6–4. Same as Problem 6–3, except that the beam spacing is 15 ft, 0 in., the beam span is 28 ft, 0 in. center-to-center of columns, the service loads are 30 psf dead load and 200 psf live load, $f'_c = 3,000$ psi, and $f_y = 40,000$ psi.

7

Serviceability

7-1 INTRODUCTION

The ACI Code, Section 9.1, requires that bending members have structural strength adequate to support the anticipated factored design loads and that they have adequate performance at service load levels. Adequate performance, or *serviceability*, relates to deflections and cracking in reinforced concrete beams and slabs. It is important to realize that serviceability is to be assured at *service load levels*, not at ultimate strength. At service loads, deflections should be held to specified limits because of many considerations, among which are esthetics, effects on nonstructural elements such as windows and partitions, undesirable vibrations, and proper functioning of roof drainage systems. Any cracking should be limited to small numbers of hairline cracks for reasons of both appearance and prevention of corrosion of the reinforcing steel.

7-2 DEFLECTIONS

Guidelines for the control of deflections are found in the ACI Code, Section 9.5. In addition, Table 9.5(b) indicates the maximum permissible deflections. For the purpose of following the Code guidelines, either of two methods may be used: (1) utilization of the minimum thickness (or depth of member) criteria as established in Table 9.5(a) of the Code, which will result in sections which are sufficiently deep and stiff so that deflections will not be excessive; and (2) calculation of expected deflections using standard deflection formulas in combination with the Code provisions for moment of inertia and the effects of the load/time history of the member.

Minimum thickness (depth) guidelines are simple and direct and should be used whenever possible. Note that the tabulated minimum thicknesses apply to nonprestressed, one-way members that do not support, and are not attached to, partitions or other construction likely to be damaged by large deflections. For the latter members, deflections *must* be calculated.

For the second method, in which deflections are calculated, the Code stipulates that the members should have their deflections checked at *service load levels*. Therefore, the *properties* at service load levels must be used. Under service loads, concrete flexural members still exhibit generally elastic-type behavior (see Fig. 1–1) but will have been subjected to cracking in tension zones at any point where the applied moment is large enough to produce tensile stress in excess of concrete tensile strength. The cross section for moment-of-inertia determination then has a shape as shown in Fig. 7–1.

The moment of inertia of the cracked section of Fig. 7–1 is designated I_{cr} and is determined based on the assumption that the concrete is cracked to the neutral axis. In other words, the concrete is assumed to have no tensile strength, and the small tension zone below the neutral axis and above the upper limit of cracking is neglected.

The moment of inertia of the cracked section described represents one end of a range of values that may be used for deflection calculations. At the opposite end, as a result of a small bending moment and a maximum flexural stress less than the modulus of rupture, the full uncracked section may be considered in determining the moment of inertia to resist deflection. This is termed the *moment of inertia of the gross cross section* and is designated I_g.

In reality, both I_{cr} and I_g occur in a bending member where the maximum moment is in excess of the cracking moment. The I_{cr} occurs at or near the cracks, whereas the I_g occurs between the cracks. Research has indicated that due to the presence of the two extreme conditions, a more realistic value of a moment of inertia lies somewhere between these values. The ACI Code recommends that deflections be calculated using an *effective moment of inertia I_e*, where

$$I_g > I_e > I_{cr}$$

The 1971 ACI Code introduced the effective moment of inertia in an attempt to provide a basis for more realistic deflection calculations. This approach has also been included in the 1977 Code. Prior to 1971 and in accordance with the 1963 ACI Code, the moment of inertia value used was either I_{cr} or I_g.

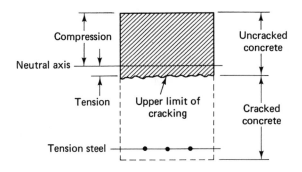

FIGURE 7-1 Typical cracked cross section

Once the effective moment of inertia is determined, the member deflection may be calculated utilizing standard deflection expressions. The effective moment of inertia, of course, will depend on the values of the moment of inertia of the gross section and the moment of inertia of the cracked section. The gross (or uncracked) moment of inertia may be easily calculated by

$$I_g = \frac{bh^3}{12}$$

This expression neglects the presence of any reinforcing steel.

7-3 CALCULATION OF I_{cr}

The moment of inertia of the cracked cross section can also be calculated in the normal way once the problem of the differing materials (steel and concrete) is overcome. To accomplish this, the steel area will be replaced by an equivalent area of concrete A_{eq}. This is a fictitious concrete that *can* resist tension. The determination of the magnitude of A_{eq} is based on the theory (from strength of materials) that when two differing elastic materials are subjected to equal strains, the stresses in the materials will be in proportion to their moduli of elasticity.

In Fig. 7–2, ϵ_1 is the compressive strain in the concrete at the top of the beam and ϵ_2 is a tensile strain at the level of the steel.

Using the notation

f_s = tensile steel stress

$f_{c\,(tens)}$ = theoretical tensile concrete stress at the level of the steel

E_s = steel modulus of elasticity

E_c = concrete modulus of elasticity

187

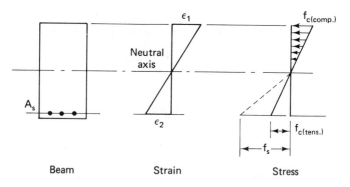

Beam Strain Stress

FIGURE 7-2 Beam bending—elastic theory

the following relationships can be established:

$$\epsilon_2 = \frac{f_s}{E_s} \quad \text{and} \quad \epsilon_2 = \frac{f_{c\,(\text{tens})}}{E_c}$$

Equating these two expressions and solving for f_s yields

$$\frac{f_s}{E_s} = \frac{f_{c\,(\text{tens})}}{E_c}$$

$$f_s = \frac{E_s}{E_c}[f_{c\,(\text{tens})}]$$

The ratio $\dfrac{E_s}{E_c}$ is normally called the *modular ratio* and is denoted n.
Therefore,

$$f_s = nf_{c\,(\text{tens})}$$

The ACI Code, Section B.5.4, indicates that values of n may be taken as the nearest whole number. Values of n are tabulated in Table A–5 in Appendix A.

Since we are replacing the steel (theoretically) with an equivalent concrete area, the equivalent concrete area A_{eq} must provide the same tensile resistance as that provided by the steel. Therefore,

$$A_{\text{eq}}\, f_{c\,(\text{tens})} = f_s A_s$$

Substituting, we obtain

$$A_{\text{eq}}\, f_{c\,(\text{tens})} = nf_{c\,(\text{tens})}\, A_s$$

from which

$$A_{\text{eq}} = nA_s$$

188

This defines the equivalent area of concrete with which we are replacing the steel. Another way of visualizing this is to consider the steel to be transformed into an equivalent concrete area of nA_s. The resulting *transformed concrete cross section* is composed of a single (though hypothetical) material and may be dealt with in the normal fashion for neutral-axis and moment-of-inertia determinations. Figure 7–3 depicts the transformed section. Since the steel is normally assumed to be concentrated at its centroid which is a distance d from the compression face, the replacing equivalent concrete area must also be assumed to act at the same location. Therefore, the representation of this area is shown as a thin rectangle extending out past the beam sides. Recalling that the cross section is assumed cracked up to the neutral axis, the resulting effective area is as shown in Fig. 7–4, and the neutral axis will be located a distance $\overline{y}$ down from a reference axis at the top of the section.

The neutral axis may be determined by taking a summation of moments of the effective areas about the reference axis:

$$\overline{y} = \frac{\sum(Ay)}{\sum A} = \frac{(b\overline{y})\dfrac{\overline{y}}{2} + nA_s d}{b\overline{y} + nA_s}$$

$$b(\overline{y}^2) + nA_s\overline{y} = \frac{b(\overline{y}^2)}{2} + nA_s d$$

$$\frac{b(\overline{y}^2)}{2} + nA_s\overline{y} - nA_s d = 0$$

This is a quadratic equation of the form $ax^2 + bx + c = 0$ and may be solved either by completion of the square as was done in Example 3–7

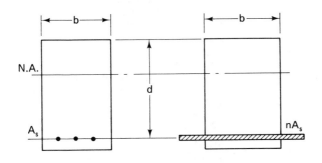

Beam Cross Section   Transformed Cross Section

FIGURE 7-3 Method of transformed section

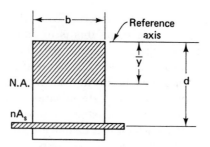

FIGURE 7-4   Effective beam

or by using the formula for roots of a quadratic equation which will result in the following useful expression:

$$\bar{y} = \frac{nA_s\left[\sqrt{1 + 2\dfrac{bd}{nA_s}} - 1\right]}{b}$$

Once the neutral axis is located, the moment of inertia (I_{cr}) may be found using the familiar transfer formula from engineering mechanics.

EXAMPLE 7-1

Find the cracked moment of inertia for the cross section shown in Fig. 7-5(a). $A_s = 2.00$ in.^{2}, $n = 9$ (Table A–5), and $f'_c = 3,000$ psi. With reference to the transformed section, shown in Fig. 7-5(b), the neutral axis is located as follows:

$$\bar{y} = \frac{nA_s\left[\sqrt{1 + 2\dfrac{bd}{nA_s}} - 1\right]}{b}$$

$$= \frac{9(2.0)\left[\sqrt{1 + 2\dfrac{(8)(17)}{9(2.0)}} - 1\right]}{8}$$

$$= 6.78 \text{ in.}$$

The moment of inertia of the cracked section may now be found (note that all units are inches):

$$I_{cr} = \frac{b\bar{y}^3}{3} + nA_s(d - \bar{y})^2$$

$$= \frac{8(6.78)^3}{3} + 18(17 - 6.78)^2$$

$$= 831.1 + 1880$$

$$= 2,711 \text{ in.}^4$$

190

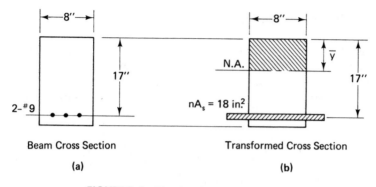

Beam Cross Section

(a)

Transformed Cross Section

(b)

FIGURE 7-5 Sketches for Example 7-1

If the beam cross section contains compression steel, this steel may also be transformed and the neutral-axis location and cracked moment-of-inertia calculations carried out as before. Since the compression steel actually displaces concrete which is in compression, it should theoretically be transformed using $(n - 1)A'_s$ rather than nA'_s. However, for deflection calculations, which are only approximate, the use of nA'_s will not detract from the accuracy expected. The resulting transformed section will appear as shown in Fig. 7–6.

The neutral axis location may be determined from the following expression:

$$\frac{b}{2}\,\overline{y}^2 + nA'_s\,\overline{y} - nA'_s\,d' - nA_s\,d + nA_s\overline{y} = 0$$

and the moment of inertia with respect to the neutral axis from the following expression:

$$I_{cr} = \frac{b\,\overline{y}^3}{3} + nA_s(d - \overline{y})^2 + nA'_s(\overline{y} - d')^2$$

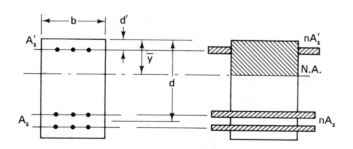

Beam Cross Section

Transformed Cross Section

FIGURE 7-6 Doubly reinforced beam

191

7–4 IMMEDIATE DEFLECTION

This is the deflection that occurs as soon as load is applied on the member. For all practical purposes, the member is elastic. The ACI Code, Section 9.5.2.3, states that this deflection may be calculated using a concrete modulus of elasticity E_c as specified in Section 8.5.1 and an effective moment of inertia I_e computed as follows:

$$I_e = \left\{ \left(\frac{M_{cr}}{M_a} \right)^3 I_g + \left[1 - \left(\frac{M_{cr}}{M_a} \right)^3 \right] I_{cr} \right\} \le I_g$$

where I_e = effective moment of inertia

I_{cr} = moment of inertia of the cracked section transformed to concrete

I_g = moment of inertia of the gross (uncracked) concrete cross section about the centroidal axis, neglecting all steel reinforcement

M_a = maximum moment in the member at the stage for which the deflection is being computed

M_{cr} = that moment which would initially crack the cross section computed from

$$M_{cr} = \frac{f_r I_g}{y_t}$$

where f_r = modulus of rupture for the concrete.

This is $7.5\sqrt{f_c'}$ for normal-weight concrete.
Values are tabulated in Table A–5; modifications for f_r for lightweight aggregate concretes are indicated in the ACI Code, Sections 9.5.2.3a and b

y_t = distance from the neutral axis of the uncracked cross section (neglecting steel) to the extreme tension fiber

Inspection of the formula for the effective moment of inertia will show that if the maximum moment is low with respect to the cracking moment M_{cr}, the moment of inertia of the gross section I_g will be the dominant factor. However, if the maximum moment is large with respect to the cracking moment, the moment of inertia of the cracked section I_{cr} will be dominant. In any case, I_e will lie somewhere between I_{cr} and I_g. For continuous beams, the use of the average value of the effective moments of inertia existing at sections of critical positive and negative moments is recommended.

The actual calculation of deflections will be made using the standard deflection methods for elastic members. Deflection formulas of the type found in standard handbooks may be suitable, or one may utilize more rigorous techniques when necessary.

7-5 LONG-TIME DEFLECTION

In addition to deflections that occur immediately, reinforced concrete members are subject to added deflections that occur gradually over long periods of time. These additional deflections are due mainly to creep and shrinkage and may eventually become excessive. The additional (or long-time) deflections are computed based on two items: (1) amount of sustained dead and live load, and (2) the ratio of compressive to tensile reinforcement in the beam, and are stated in terms of a factor times the immediate deflection due to the sustained loads.

$$\Delta_{LT} = \Delta_i \left(2 - 1.2 \frac{A'_s}{A_s} \right)$$

where Δ_{LT} = long-time deflection
Δ_i = immediate deflection due to *sustained* loads
A'_s = compression steel area
A_s = tensile steel area

The long-time deflection factor $(2 - 1.2 A'_s/A_s)$ shall not be less than 0.6. Some judgment will be required in determining just what portion of the live loads should be considered *sustained*. In a residential application, 20% sustained live load might be a logical estimate, while in storage facilities 100% sustained live load would be reasonable.

The calculated deflections must not exceed the maximum permissible deflections which are found in the ACI Code, Table 9.5(b). This table sets permissible deflections in terms of fractions of span length. Basically, these limitations guard against damage to the various parts of the system (both structural and nonstructural parts) as a result of excessive deflection. In the case of attached nonstructural elements, only the deflection that takes place after such attachment need be considered.

EXAMPLE 7-2

Determine which of the ACI Code deflection criteria will be satisfied by the nonprestressed reinforced concrete beam, having a cross section shown in Fig. 7-7, subject to the stated conditions. Assume a 50% sustained live load. Assume normal-weight concrete. $f'_c = 3,000$ psi and $f_y = 40,000$ psi. The beam is on a simple span of 30 ft.

1. The maximum *service* moments at midspan are

$$M_{DL} = 20 \text{ ft-kips} \quad \text{and} \quad M_{LL} = 15 \text{ ft-kips}$$

2. Check the beam depth based on ACI Code criteria (Table 9.5a).

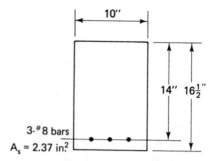

FIGURE 7-7 Sketch for Example 7–2

$$\text{Minimum } h = \frac{\ell}{16}\left(0.4 + \frac{f_y}{100,000}\right)$$

$$= \frac{30(12)}{16}(0.8) = 18 \text{ in.}$$

Since $18 > 16.5$, deflections must be calculated.

3. Effective moment of inertia will be calculated using

$$I_e = \left\{\left(\frac{M_{cr}}{M_a}\right)^3 I_g + \left[1 - \left(\frac{M_{cr}}{M_a}\right)^3\right] I_{cr}\right\}$$

Therefore, we first must compute the various terms within the expression. The moment of inertia of the cracked transformed section will be determined with reference to Fig. 7–8. The steel area of 2.37 in.² and a modular ratio n of 9 (see Table A–5) will result in a transformed area of

$$nA_s = 9(2.37) = 21.3 \text{ in.}^2$$

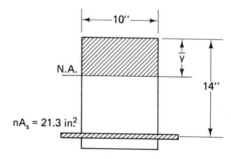

FIGURE 7-8 Transformed section for Example 7–2

The neutral-axis location is determined as follows:

$$\bar{y} = \frac{nA_s\left[\sqrt{1 + 2\dfrac{bd}{nA_s}} - 1\right]}{b}$$

$$= \frac{21.3\left[\sqrt{1 + 2\dfrac{(10)(14)}{21.3}} - 1\right]}{10}$$

$$= 5.88 \text{ in.}$$

The moment of inertia of the cracked transformed section is then determined:

$$I_{cr} = \frac{10(5.88)^3}{3} + 21.3(14 - 5.88)^2$$

$$= 2,082 \text{ in.}^4$$

Determination of the moment of inertia of the gross section results in

$$I_g = \frac{bh^3}{12} = \frac{1}{12}(10)(16.5)^3 = 3,743 \text{ in.}^4$$

The moment that would initially crack the cross section may be determined next:

$$M_{cr} = \frac{f_r I_g}{y_t}$$

where

$$y_t = \frac{16.5}{2} = 8.25$$

$$f_r = 7.5\sqrt{f_c'} = 0.411 \text{ ksi} \qquad \text{(from Table A–5)}$$

Therefore,

$$M_{cr} = \frac{0.411(3,743)}{8.25(12)} = 15.5 \text{ ft–kips}$$

We will assume that $M_a = 35$ ft–kips on the basis that it is the maximum moment which the beam must carry that will establish the crack pattern. This is not in strict accordance with the ACI Code, which indicates that M_a should be the maximum moment occurring at the stage the deflection is computed. It is logical that since the cracking pattern is irreversible, the use of the effective moment of inertia based on the full maximum moment is more realistic. This approach is conservative and will furnish a lower I_e, which will subsequently result in a larger computed deflection.

4. Determine the effective moment of inertia.

$$I_e = \left\{ \left(\frac{M_{cr}}{M_a} \right)^3 I_g + \left[1 - \left(\frac{M_{cr}}{M_a} \right)^3 \right] I_{cr} \right\}$$

$$= \left\{ \left(\frac{15.5}{35} \right)^3 3{,}743 + \left[1 - \left(\frac{15.5}{35} \right)^3 \right] (2{,}082) \right\} = 2{,}226 \text{ in.}^4$$

5. Compute the immediate dead load deflection ($M_{DL} = 20$ ft–kips).

$$\Delta = \frac{5w\ell^4}{384 E_c I_e} = \frac{5M\ell^2}{48 E_c I_e}$$

where M is the moment due to a uniform load and E_c may be found in Table A–5.

$$\Delta = \frac{5(20)(30)^2(1{,}728)}{48(3{,}120)(2{,}226)} = 0.466 \text{ in.}$$

6. Compute the immediate live load deflection ($M_{LL} = 15$ ft–kips). By proportion:

$$\Delta = \frac{15}{20}(0.466) = 0.35 \text{ in.}$$

7. The total immediate DL + LL deflection is

$$0.466 + 0.35 = 0.816 \text{ in.}$$

8. The long-time deflection (DL + sustained LL) multiplier is

$$2 - 1.2 \frac{A_s'}{A_s} = 2 - (1.2)\left(\frac{0}{2.37} \right) = 2.0$$

$$2.0 > 0.6 \qquad\qquad\qquad\text{(OK)}$$

Since $M_{DL} = 20$ ft–kips and 50% $M_{LL} = 7.5$ ft–kips, the sustained moment for long-time deflection = 27.5 ft–kips.

$$\Delta_{LT} = \frac{27.5}{20}(0.466)(2.0) = 1.28 \text{ in.}$$

9. A comparison of actual deflections to maximum permissible deflections may now be made. In the comparison, made in Table 7–1, the maximum permissible deflections are from the ACI Code, Table 9.5(b). Since the allowable deflection is not exceeded in cases *a* or *b*, this beam is limited in usage to floors and flat roofs which do not support and are not attached to nonstructural elements likely to be damaged by large deflections.

7–6 DEFLECTIONS FOR CONTINUOUS SPANS

To compute deflections for continuous spans subject to uniformly distributed loads such as the beam in Fig. 7–9, the following approximate approach may be used:

TABLE 7–1
Permissible vs. Actual Deflections (Example 7–2)

Maximum Permissible Deflection	Actual Computed Deflection
$\dfrac{\ell}{180} = \dfrac{30\,(12)}{180} = 2$ in.	$\Delta_{LL} = 0.35$ in. (immediate LL)
$\dfrac{\ell}{360} = \dfrac{30\,(12)}{360} = 1$ in.	$\Delta_{LL} = 0.35$ in. (immediate LL)
$\dfrac{\ell}{480} = \dfrac{30\,(12)}{480} = 0.75$ in.	$\Delta_{LL} + \Delta_{LT} = 0.35 + 1.28 = 1.63$ in.
$\dfrac{\ell}{240} = \dfrac{30\,(12)}{240} = 1.5$ in.	$\Delta_{LL} + \Delta_{LT} = 0.35 + 1.28 = 1.63$ in.

$$\Delta = \frac{5w\ell_n^4}{384E_cI_e} - \frac{M\ell_n^2}{8E_cI_e}$$

where M = negative moment at supports (based on service loads) for span being investigated; if values are different, use average moment

ℓ_n = clear span

E_c = modulus of elasticity for concrete (see Table A–5)

w = uniformly distributed service load

I_e = effective moment of inertia—use average value of I_e at positive moment area and I_e at negative moment area

In a similar manner, for long-time deflections the long-time deflection multipler should be averaged for the different locations.

7-7 CRACK CONTROL

With the advent of the higher-strength reinforcing steels, where more strain is required to produce the higher stresses, cracking of reinforced concrete flexural members has presented problems.

Of primary interest when considering cracking is the width of the cracks that form. Rather than a small number of large cracks, it is more

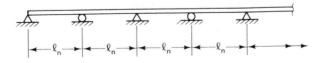

FIGURE 7-9 Continuous beam

197

desirable to have only hairline cracks and accept more numerous cracks if necessary. The present Code methods for crack control are based largely on research which has shown that crack width is generally proportional to the amount of stress in the steel and the amount of cover on the steel, and that crack width is minimized if the steel is well distributed in the tension zone of the concrete. This problem is discussed in the ACI Code, Section 10.6, and applies to beams and one-way slabs.

The ACI Code directs that only deformed reinforcement is to be used and that the tension reinforcement must be *well distributed* in zones of maximum concrete tension. The problem of flanges in tension (negative moment in T-beams) is specially noted.

Where the design yield strength f_y of the steel exceeds 40,000 psi, a special check must be performed to ensure that there is proper distribution of the tension steel. For a particular beam or slab, a quantity Z is calculated for the check on the steel distribution with upper limits of Z being set depending upon the type of exposure to which the structure will be subjected. Essentially, these limits are values of Z corresponding to crack widths of 0.016 in. for interior exposures and 0.013 in. for exterior exposures, which the ACI committee has judged as reasonable maximums.

$$Z = f_s \sqrt[3]{d_c A}$$

where f_s = calculated stress in the steel
　　　　at service loads (and may be taken as $0.6f_y$) (ksi)
　　d_c = thickness of concrete cover measured from the extreme
　　　　tension fiber to the *center of the closest bar* (in.)
　　A = effective tension area of concrete surrounding the main
　　　　tension reinforcing bars and having the same centroid as that
　　　　reinforcement, divided by the number of bars (in.^{2})
　　Z = quantity limiting distribution of flexural reinforcement with
　　　　maximum limits of 175 kips/in. for interior exposures
　　　　and 145 kips/in. for exterior exposures

EXAMPLE 7–3

Check the steel distribution for the beam shown in Fig. 7–10 to establish whether reasonable control of flexural cracking is accomplished in accordance with the ACI Code, Section 10.6. f_y = 60,000 psi.

1. The concrete cover from the extreme tension fiber to the center of the closest bar is

$$d_c = 1.5 + 0.38 + \frac{1.128}{2} = 2.44 \text{ in.}$$

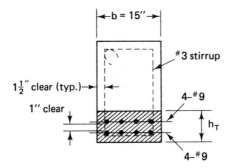

FIGURE 7-10 Sketch for Example 7–3

2. The effective tension area per bar A may be expressed as follows:

$$A = \frac{bh_T}{\text{number of bars}}$$

where h_T is a linear distance equal to twice the distance from the extreme tension fiber to the centroid of the tensile reinforcement

$$h_T = (1.5 + 0.38 + 1.128 + 0.5)2 = 7.0 \text{ in.}$$

Therefore,

$$A = \frac{15(7.0)}{8} = 13.13 \text{ in.}^2 \text{ per bar}$$

3.
$$Z = f_s \sqrt[3]{d_c A}$$

where $f_s = 0.60 f_y = 0.60(60) = 36$ ksi

$$Z = 36 \sqrt[3]{2.44(13.13)} = 114 \text{ kips/in.}$$

$$114 < 145 \text{ and } 175 \tag{OK}$$

Therefore, the steel distribution is satisfactory for either interior or exterior exposure.

PROBLEMS

7–1. Locate the neutral axis and calculate the moment of inertia for the cracked transformed cross sections shown. $f'_c = 4,000$ psi and $f_y = 60,000$ psi.

7–2. Find I_g and I_{cr} for the T-beam. The effective flange width is 60 in. and $f'_c = 4,000$ psi.

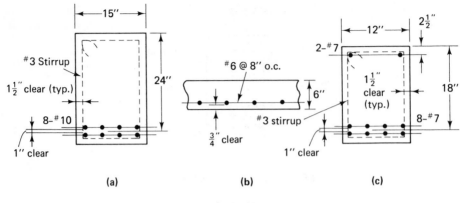

FIGURE P7-1

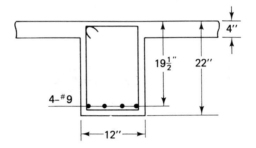

FIGURE P7-2

7–3. The beam of cross section shown is on a simple span of 20 ft and carries service loads of 1.5 kip/ft dead load (includes beam weight) and 1.0 kip/ft live load. $f'_c = 3,000$ psi and $f_y = 40,000$ psi. Compute:

   a. The immediate deflection due to dead load and live load.
   b. The long-time deflection due to the dead load.

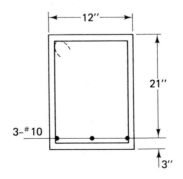

FIGURE P7-3

7–4. A floor beam that supports structural elements likely to be damaged by large deflections is on a simple span of 16 ft. The service loads are 0.6 kip/ft dead load (does not include the beam weight) and 1.40 kips/ft live load. Assume that the live load is 60% sustained.

$$f'_c = 3,000 \text{ psi and} f_y = 40,000 \text{ psi.}$$

   a. Check the beam for deflections.
   b. If the beam is unsatisfactory, redesign it so that it meets both flexural strength and deflections requirements.

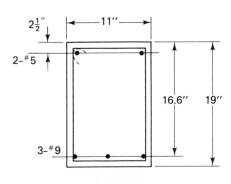

FIGURE P7-4

7–5. Calculate the expected long-time deflection due to dead load and sustained live load for (a) the slab, and (b) the beam of the floor system designed for Problem 6–3. Assume 10% sustained live load.

7–6. Check the cross sections of Fig. P7–1(a) and (b) for acceptability under the Code provisions for distribution of flexural reinforcement (crack control).

7–7. Check the distribution of flexural reinforcement for the members designed in Problems 2–12 and 2–14. If necessary, redesign the steel.

8

Walls

8-1 INTRODUCTION

Walls are generally utilized to provide lateral support for an earth fill, embankment, or some other material, and/or to support vertical loads. Some of the more common types of walls are shown in Fig. 8–1.

One of the primary purposes for these walls is to maintain a difference in the elevation of the ground surface on each side of the wall. The earth whose ground surface is at the higher elevation is commonly called the *backfill*, and the wall is said to retain this backfill.

All the walls shown in Fig. 8–1 have applications in either building or bridge projects. They do not necessarily behave in an identical manner under load, but still serve the same basic function of providing lateral support for a mass of earth or other material which is at a higher elevation behind the wall than the earth or other material in front of the wall. Hence they all may be broadly termed *retaining structures* or *retaining walls* even though they may be subject to vertical loading which in effect will change the behavior of the wall.

The gravity wall [Fig. 8–1(a)] depends mostly on its own weight for stability. It is usually made of plain concrete and is used for walls up to approximately 10 ft in height. The semigravity wall is a modification of the gravity wall in which small amounts of reinforcing steel are introduced. This, in effect, reduces the massiveness of the wall.

The cantilever wall [Fig. 8–1(b)] is the most common type of retaining structure and generally is used for walls in the range 10–25 ft in height. It derives its name from the fact that its individual parts (toe,

Retaining wall under construction—Seabrook Station, New Hampshire. (Courtesy Public Service Company of New Hampshire)

heel, and stem) behave as, and are designed as, cantilever beams. Aside from its stability, the capacity of the wall is a function of the strength of its individual parts.

The counterfort wall [Fig. 8–1(c)] may be economical when the wall height is in excess of 25 ft. The counterforts are spaced at intervals and act as tension members to support the stem. The stem is then designed as a continuous member spanning horizontally between the counterforts.

The buttress wall [Fig. 8–1(d)] is similar to the counterfort wall

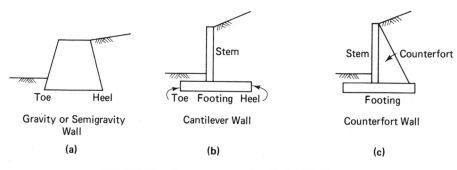

FIGURE 8-1 Common types of walls (a) (b) (c)

except that the buttresses are located on the side of the stem opposite to the retained material and act as compression members to support the stem. The counterfort wall is more commonly used because it has a clean, uncluttered exposed face and allows for more efficient use of space in front of the wall.

The basement or foundation wall [Fig. 8–1(e)] may act as a cantilever retaining wall. However, the first floor may provide an additional horizontal reaction similar to the basement floor slab, thereby making the wall act as a vertical beam. This wall would then be designed as a simply supported member spanning the first floor and the basement floor slab.

The bridge abutment [Fig. 8–1(f)] is similar, in some respects, to the basement wall. The bridge superstructure induces horizontal as well as vertical loads, thus altering the normal cantilever behavior.

The bearing wall [Fig. 8–1(g)] may exist with or without lateral loads. "The American Standard Building Code Requirements for Masonry"[1] defines a bearing wall as a wall that supports any vertical load in addition to its own weight. Depending on the magnitudes of the vertical and lateral loads, the wall may have to be designed for combined bending and axial compression. Bearing walls and basement walls will be further discussed later in this chapter.

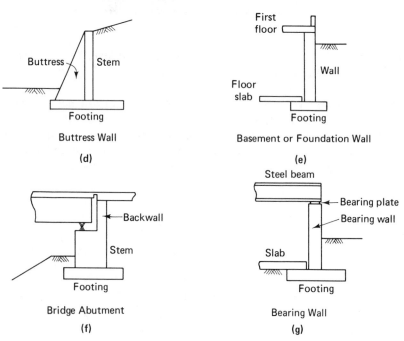

FIGURE 8-1 Common types of walls (continued)

8-2 LATERAL FORCES ON RETAINING WALLS

The design of a retaining wall must account for all of the applied loads. The load that presents the greatest problem and is of primary concern is the lateral earth pressure induced by the retained soil. The comprehensive earth pressure theories evolving from the original Coulomb and Rankine theories may be observed in almost any textbook on soil mechanics.

The magnitude and direction of the pressures as well as the pressure distribution exerted by a soil backfill upon a wall are highly indeterminate, due to many variables. These variables include, but are not limited to, type of backfill used, drainage of the backfill material, level of the water table, slope of the backfill material, added loads applied on the backfill, degree of soil compaction, and movement of the wall caused by the action of backfill.

An important consideration is that water must be prevented from accumulating in the backfill material. Walls are rarely designed to retain saturated material, which means that proper drainage must be provided. It is generally agreed that the best backfill material behind a retaining structure is a well-drained cohesionless material. Hence it is the condition that one usually specifies and for which one designs. Materials that contain combinations of types of soil will act like the predominant material.

The lateral earth pressure can exist and/or develop in three different categories: active state, at rest, and passive state. If a wall is absolutely rigid, earth pressure at rest will develop. If the wall should deflect or move a very small amount away from the backfill, active earth pressure will develop and in effect reduce the lateral earth pressure occurring in the at-rest state. Should, for some reason, the wall be forced to move toward the backfill, passive earth pressure will develop and increase the lateral earth pressure appreciably above that occurring in the at-rest state. As indicated, the magnitude of earth pressure at rest lies somewhere between active and passive earth pressures.

Under normal conditions, earth pressure at rest is of such a magnitude that the wall deflects slightly, thus relieving itself of the at-rest pressure. The active pressure results. For this reason, retaining walls are generally designed for active earth pressure due to the retained soil.

Because of the involved nature of a rigorous analysis of an earth backfill and the variability of the material and conditions, assumptions and approximations are made with respect to the nature of lateral pressures on a retaining structure. It is common practice to assume a linear active and passive earth pressure distribution. The pressure intensity is assumed to increase with depth as a function of the weight of the

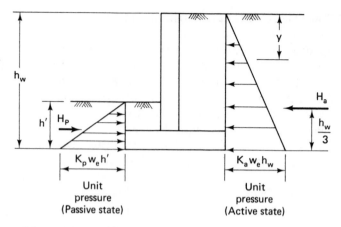

FIGURE 8-2 Analysis of forces acting on walls—level backfill

soil in a manner similar to that which would occur in a fluid. Hence this horizontal pressure of the earth against the wall is frequently called an *equivalent fluid pressure*. Experience has indicated that walls designed on the basis of these assumptions and those of the following discussion are safe and relatively economical.

Level Backfill

If we consider a level backfill (of well-drained cohesionless soil), the assumed pressure diagram is shown in Fig. 8–2. The unit pressure intensity p_y in any plane a distance y down from the top is

$$p_y = K_a w_e y$$

Therefore, the total active earth pressure acting on a 1-ft width of wall may be calculated as the product of the average pressure on the total wall height h_w and the area on which this pressure acts:

$$H_a = \tfrac{1}{2} K_a w_e h_w{}^2 \tag{8–1}$$

where K_a, the *coefficient of active earth pressure*, has been established by both Rankine and Coulomb to be

$$K_a = \frac{1 - \sin\phi}{1 + \sin\phi} = \tan^2\left(45° - \frac{\phi}{2}\right) \tag{8–2}$$

and

$$w_e = \text{unit weight of earth (lb/ft}^3)$$

$$\phi = \text{angle of internal friction (soil on soil)}$$

206

K_a usually varies from 0.27 to 0.40. The term $K_a w_e$ in Eq. (8–1) is generally called an *equivalent fluid weight*, since the resulting pressure is identical to that which would occur in a fluid of that weight (units are lb/ft³).

In a similar manner, the total passive earth pressure force may be established as

$$H_p = \tfrac{1}{2} K_p w_e (h')^2 \tag{8–3}$$

where h' is the height of earth and K_p the *coefficient of passive earth pressure*.

$$K_p = \frac{1 + \sin\phi}{1 - \sin\phi} = \tan^2\left(45° + \frac{\phi}{2}\right) = \frac{1}{K_a}$$

K_p usually varies from 2.5 to 4.0.

The total force in each case is assumed to act at one-third the height of the triangular pressure distribution as shown in Fig. 8–2.

Sloping Backfill

If we consider a sloping backfill, the assumed active earth pressure distribution is shown in Fig. 8–3, where $H_s = \tfrac{1}{2} K_a w_e h_b^2$, h_b is the height of the backfill at the back of the footing, and K_a is the coefficient of active earth pressure.

$$K_a = \cos\theta \left(\frac{\cos\theta - \sqrt{\cos^2\theta - \cos^2\phi}}{\cos\theta + \sqrt{\cos^2\theta - \cos^2\phi}}\right)$$

where θ is the slope angle of the backfill and ϕ is as previously defined. Note that H_s is shown acting parallel to the slope of the backfill.

For walls approximately 20 ft in height or less, it is recommended that the horizontal force component H_H simply be assumed equal to H_s and assumed to act at $h_b/3$ above the bottom of the footing, as shown in Fig. 8–3. The effect of the vertical force component H_V is neglected. This is a conservative approach.

Assuming a well-drained cohensionless soil backfill which has a unit weight of 110 lb/ft³ and an internal friction angle ϕ of 33°40′, values of equivalent fluid weight for sloping backfill may be determined as listed in Table 8–1.

Level Backfill with Surcharge

Loads are often imposed on the backfill surface behind a retaining wall. They may be either live loads or dead loads. These loads are generally termed a *surcharge* and theoretically may be transformed into an equivalent height of earth.

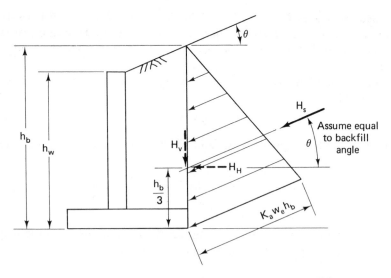

FIGURE 8-3 Analysis of forces acting on walls—sloping backfill

A uniform surcharge over the adjacent area adds the same effect as an additional (equivalent) height of earth. This equivalent height of earth h_{su} may be obtained as follows:

$$h_{su} = \frac{w_s}{w_e}$$

where w_s = surcharge load (psf)

w_e = unit weight of earth (lb/ft³)

In effect, this adds a rectangle of pressure behind the wall with a total lateral surcharge force assumed acting at its midheight, as shown in Fig. 8–4. Surcharge loads far-enough removed from the wall cause no additional pressure acting on the wall.

TABLE 8–1
$K_a w_e$ **Values for Sloping Backfill**

θ (deg)	$K_a w_e$ (lb/ft³)
0	32
10	33
20	38
30	54

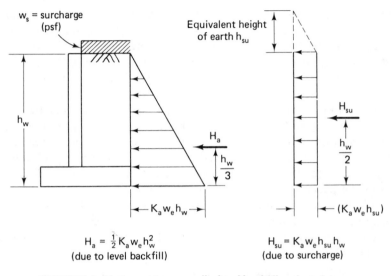

$$H_a = \tfrac{1}{2} K_a w_e h_w^2$$
(due to level backfill)

$$H_{su} = K_a w_e h_{su} h_w$$
(due to surcharge)

FIGURE 8-4 Forces acting on wall—level backfill and surcharge

8-3 DESIGN OF REINFORCED CONCRETE CANTILEVER RETAINING WALLS

A retaining wall must be stable as a whole, and it must have sufficient strength to resist the forces acting on it. Four possible *modes of failure* will be considered. *Overturning about the toe*, point *O*, as shown in Fig. 8–5, could occur due to lateral loads. The stabilizing moment must be

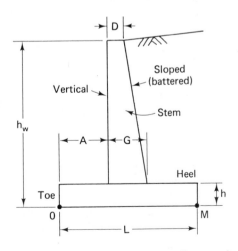

FIGURE 8-5 Cantilever retaining-wall proportions

sufficiently in excess of the overturning moment so that an adequate factor of safety against overturning is provided. The factor of safety should never be less than 1.5 and should preferably be 2.0 or more. *Sliding on the base of the footing*, surface *OM* in Fig. 8–5, could also occur due to lateral loads. The resisting force is based on an assumed coefficient of friction of concrete on earth. The factor of safety against sliding should never be less than 1.5 and should preferably be 2.0 or more. *Excessive soil pressure* under the footing will lead to undesirable settlements and possible rotation of the wall. Actual soil pressures should not be allowed to exceed specified maximum pressures, which depend on the characteristics of the underlying soil. The *structural failure of component parts of the wall* such as stem, toe, and heel, each acting as a cantilever beam, could occur. These must be designed to have sufficient strength to resist all anticipated loads.

A *general design procedure* for specific, known conditions may be summarized as follows:

1. Establish the general shape of wall based on the desired height and function.

2. Establish the site soil conditions, loads and other design parameters. This includes the determination of allowable soil pressure, earth-fill properties for active and passive pressure calculations, amount of surcharge, and the desired factors of safety.

3. Establish the tentative proportions of the wall.

4. Analyze the stability of the wall. Check factors of safety against overturning and sliding and compare actual soil pressure with allowable soil pressure.

5. Assuming that all previous steps are satisfactory, design the component parts of the cantilever retaining wall, stem, toe, and heel, as cantilever beams.

In a manner similar to one-way slabs, the analysis and design of cantilever retaining walls is based on a twelve inch (one foot) wide strip measured along the length of wall.

The tentative proportions of a cantilever retaining wall may be obtained from the following rules of thumb (see Fig. 8–5):

1. Footing width L: use ½ to ⅔h_w.

2. Footing thickness h, and bottom-of-stem thickness G: use $\frac{1}{10}h_w$.

3. Toe width A: use ¼ to ⅓L.

4. Use a minimum wall batter of ¼ in./ft to improve the efficiency of the stem as a bending member and decrease the quantity of concrete required.

5. Top of stem thickness D: use a minimum of 10 in.

EXAMPLE 8–1

Design a retaining wall for the conditions shown in Fig. 8–6. Use f'_c = 3,000 psi and f_y = 40,000 psi.

1. The general shape of the wall, as shown, is that of a cantilever wall, since the overall height of 18 ft is within the range in which this type of wall is normally economical.

2. Design data: unit weight of earth w_e = 100 lb/ft³, allowable soil pressure = 4,000 psf, equivalent fluid weight $K_a w_e$ = 30 lb/ft³, and surcharge load w_s = 400 psf. The desired factor of safety against overturning is 2.0 and against sliding is 1.75.

3. Establish tentative proportions for the wall:

   a. Footing width:

   ½ to ⅔ of wall height

   ½(18) to ⅔(18) = 9 to 12 ft

   Use 12 ft. *[handwritten: ft.]*

   b. Footing thickness and bottom of stem thickness:

   $$\frac{1}{10}(18) = 1.8 \text{ ft}$$ *[handwritten: ✷ this may be low]*

   Use 1 ft 9 in. *[handwritten: 1.75']* *[handwritten: ft.]*

   c. Toe width:

   ¼ to ⅓ footing width

   ¼(12) to ⅓(12) = 3 to 4 ft

   Use 3 ft 6 in. *[handwritten: ft.]* *[handwritten: ¼ to]*

   d. Use a batter for the rear face of wall approximately ½ in./ft.

[handwritten: not less than 10"]

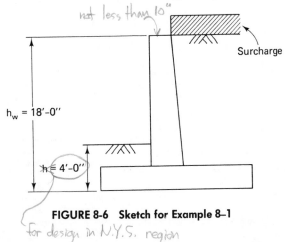

Surcharge

h_w = 18'–0"

h = 4'–0" *[handwritten, circled]*

FIGURE 8-6 Sketch for Example 8–1

[handwritten: for design in N.Y.S. region]

e. Top of stem thickness, based on G and a batter of ½ in./ft:

$$D = G - (h_w - h)½$$

$$= 21 - (18 - 1.75)½ = 12.88 \text{ in.}$$

Use 1 ft 0 in. Therefore, the actual batter is

$$\frac{21 - 12}{18 - 1.75} = 0.55 \text{ in./ft}$$

The preliminary wall proportions are shown in Fig. 8–7.

4. For the stability analysis, use actual unfactored weights and loads in accordance with the ACI Code, Section 15.2.4.

a. Factor of safety against overturning

The tendency of the wall to overturn is a result of the horizontal loads acting on the wall. An assumption is made that in overturning, the wall will rotate about the toe and the horizontal loads are said to create an *overturning moment* about the toe. Any vertical loads will tend to create rotation about the toe in the opposite direction and are therefore said to provide a *stabilizing moment*.

The factor of safety against overturning is then expressed as

$$F.S. = \frac{\text{stabilizing moment}}{\text{overturning moment}}$$

The minimum required factor of safety against overturning is normally governed by the applicable building code. A minimum of 1.5 is generally

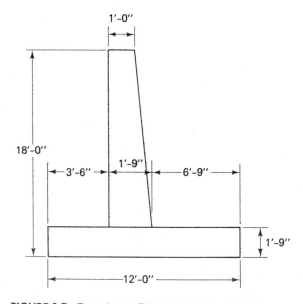

FIGURE 8-7 Tentative wall proportions—Example 8–1

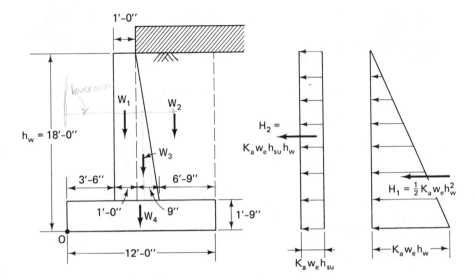

FIGURE 8-8 Stability analysis diagram

considered to be good practice. (The New York State Building Code requires a minimum of 1.5.) The passive earth resistance of the soil in front of the wall is generally neglected in stability computations because of the possibility of its removal by erosion or excavation.

As discussed previously, the surcharge may be converted into an equivalent height of earth,

$$h_{su} = \frac{w_s}{w_e} = \frac{400}{100} = 4 \text{ ft}$$

thus adding a rectangle of earth pressure behind the wall.

The various vertical and horizontal forces and their associated moments are now calculated (see Fig. 8–8). Note that soil on the toe is neglected. Both stabilizing and overturning moments are calculated with respect to point O in Fig. 8–8.

Stabilizing Moments (Vertical Forces)

Force	Magnitude (lb)	Lever Arm (ft)	Moment (ft-lb)
W_1	1.0 (16.25) (150) = 2,437.5	4.0	9,750
W_2	7.5 (16.25 + 4.0) (100) = 15,187.5	8.25	125,297
W_3	½ (16.25) (0.75) (50) = 304.7	4.75	1,447
W_4	12.0 (1.75) (150) = 3,150.0	6.0	18,900
	$\Sigma W = 21,080$ lb		$\Sigma M = 155,394$ ft-lb

213

Note that W_2 is the weight of soil and surcharge on the heel from the back edge of the heel to the vertical dashed line in the stem. W_3 represents weight difference between reinforced concrete and soil (50 lb/ft³) in the triangular portion which is the back of the stem.

Overturning Moments (Horizontal Forces)

Force	Magnitude (lb)	Lever Arm (ft)	Moment (ft-lb)
H_1	½ (30) (18)² = 4,860	6.0	29,160
H_2	4 (30) (18) = 2,160	9.0	19,440
	ΣH = 7,020 lb		ΣM = 48,600 ft-lb

The factor of safety against overturning is

$$\text{F.S.} = \frac{155,394}{48,600} = 3.20 > 2.0 \qquad \text{(OK)}$$

b. Factor of safety against sliding

The tendency of the wall to slide is primarily a result of the horizontal forces, while the vertical forces cause the frictional resistance against sliding.

The total frictional force available (or resisting force) may be expressed as

$$F = f(\Sigma W)$$

where f = coefficient of friction between the concrete and soil and ΣW = summation of vertical forces (see Fig. 8–8). A typical value for the coefficient of friction is f = 0.50. Then the resisting force F can be calculated:

$$F = f(\Sigma W)$$

$$0.50(21,080) = 10,540 \text{ lb}$$

The factor of safety against sliding may be expressed as

$$\text{F.S.} = \frac{\text{resisting force } F}{\text{actual horizontal force } \Sigma H}$$

$$= \frac{10,540}{7,020} = 1.5$$

The required factor of safety for this problem is 1.75; hence the resistance against sliding is *inadequate*.

One solution to this problem is to reproportion the wall until the requirements are met. Rather than change the wall, we will utilize a base

shear key

shear key to mobilize the passive resistance of the soil and in effect increase the resisting force F and subsequently increase the factor of safety. The design of the base shear key will be one of the last steps in the design of the retaining wall.

c. Soil pressures and location of resultant force

The soil pressure under the footing of the wall is a function of the location of the resultant force, which in turn is a function of the vertical forces and horizontal forces. It is generally desirable and usually required that for walls on soil, the resultant of all the forces acting on the wall must lie within the middle third of the base. When this occurs, the resulting pressure distribution could be either triangular, rectangular, or trapezoidal in shape, with the soil in compression under the entire width of the footing. The resulting maximum foundation pressure must not exceed the safe bearing capacity of the soil.

The first step is to locate the point at which the resultant of the vertical forces ΣW and the horizontal forces ΣH intersects the bottom of the footing. This may be visualized with reference to Fig. 8–9. ΣW and ΣH are first combined to form the resultant force R. Since R may be moved anywhere along its line of action, it is moved to the bottom of the footing, where it is then resolved back into its components ΣW and ΣH. The moment about point O due to the resultant force acting at the bottom of the footing must be the same as the moment effect about point O of the components ΣW and ΣH. Therefore, using the components in the plane of the bottom of footing, the location of the resultant which is a distance x from point O may be

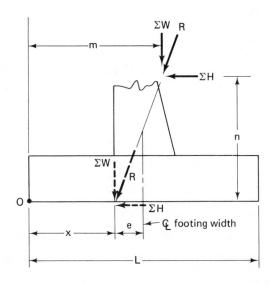

FIGURE 8-9 Resultant of forces acting on wall

determined. Note that the moment due to ΣH at the bottom of the footing is zero.

$$\Sigma W(x) = \Sigma Wm - \Sigma Hn$$

As may be observed, ΣWm is the stabilizing moment of the vertical forces with respect to point O, and ΣHn is the overturning moment of the horizontal forces with respect to the same point. Hence this expression may be rewritten as

$$\sum W(x) = \text{stabilizing } M - \text{overturning } M$$

$$x = \frac{\text{stabilizing } M - \text{overturning } M}{\sum W}$$

With reference to Fig. 8–8,

$$x = \frac{155,394 - 48,600}{21,080} = 5.07 \text{ ft from toe}$$

Therefore, the eccentricity e with respect to the centerline of footing is

$$e = 6.0 - 5.07 = 0.93 \text{ ft}$$

With the footing length of 12 ft, the middle third lies 4 ft in from each end of footing. Therefore, the resultant does lie within the middle third of the footing.

With an eccentricity of 0.93 ft, the resulting pressure distribution is trapezoidal. If the resultant intersected the base at the edge of the middle third, the pressure distribution would have been triangular, and if $e = 0$, the pressure distribution would have been rectangular, indicating a uniform soil pressure distribution.

The pressures may now be calculated considering ΣW applied as an eccentric load on a rectangular section 12 ft long by 1 ft wide (this is a *typical* 1-ft-wide strip of the wall footing). The pressures are obtained by utilizing the basic equations for bending and axial compression

$$p = \frac{P}{A} \pm \frac{Mc}{I}$$

where p = unit soil pressure intensity under the footing

P = total vertical load $(\sum W)$

A = footing cross-sectional area $[(L))(1.0)]$

M = moment due to eccentric load $[(\sum W))(e)]$

c = distance from centerline of footing to outside edge (L/2)

I = moment of inertia of footing with respect to its centerline

$$\left[\frac{(1.0)(L)^3}{12} \right]$$

The expression given previously may be rewritten as

$$p = \frac{\sum W}{(L)(1.0)} \pm \frac{(\sum W))(e))(L/2)}{1.0(L)^3/12}$$

Simplifying and rearranging yields

$$p = \frac{\sum W}{L}\left(1 \pm \frac{6e}{L}\right)$$

Substituting, we obtain

$$p = \frac{21,080}{12}\left[1 \pm \frac{6(0.93)}{12}\right]$$

$$= 1,757(1 \pm 0.465)$$

Maximum $p = 1,757(1.465) = 2,574$ psf

Minimum $p = 1,757(0.535) = 940$ psf

allowable soil pressure

The resulting pressures are less than 4,000 psf and are therefore satisfactory. The resulting pressure distribution beneath the footing may be observed in Fig. 8–10.

5. Design of component parts:

a. Design of heel

The load on the heel is primarily earth dead load and surcharge, if any, acting vertically downward. With the assumed straight-line pressure distribution under the footing, the downward load is reduced somewhat by the upward-acting pressure. For design purposes the heel is assumed to be a cantilever beam 1 ft in width and with a span length equal to 6 ft 9 in. fixed at the rear face of wall (point *A* in Fig. 8–10).

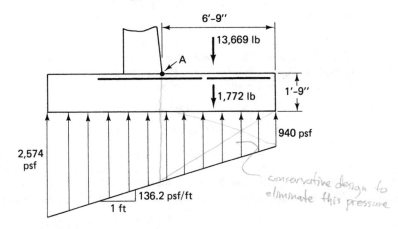

conservative design to eliminate this pressure

FIGURE 8-10   Presure distribution under footing

Weight of footing:

$$1.75(150)(6.75) = 1,772 \text{ lb}$$

Earth and surcharge weight:

$$6.75(20.25)(100) = 13,669 \text{ lb}$$

Slope of pressure distribution under footing:

$$\frac{2,574 - 940}{12} = 136.2 \text{ psf/linear foot}$$

Thus far, all analysis has been on the basis of service (actual) load because allowable soil pressures are determined using a certain factor of safety against reaching pressures that will cause unacceptable settlements. Reinforced concrete, however, is designed on the basis of factored loads. The ACI Code, Section 9.2.4, specifies that where lateral earth pressure H must be included in design, the strength U shall be at least equal to $1.4D + 1.7L + 1.7H$.

A conservative approach to the design of the heel is to use factored loads, as required, and to ignore the relieving effect of the upward pressure under the heel. This is the unlikely condition that would exist if there occurred a lateral force overload (and no associated increased vertical load) causing uplift of the heel.

The *maximum moment* may be obtained by taking a summation of moments about point A (utilizing service loads).

$$M = (13,669 + 1,772)\frac{6.75}{2} = 52,113 \text{ ft-lb}$$

The bending of the heel is such that tension occurs in the top of the footing.

The *maximum shear* may be obtained by taking a summation of vertical forces on the heel side of point A (utilizing service loads).

$$V = 13,669 + 1,772 = 15,441 \text{ lb}$$

The maximum moment and shear must be modified for strength design. Since the loads on the heel are predominantly dead load, a load factor of 1.4 is used. This, in effect, considers the surcharge to be a dead load. However, this is acceptable because of the conservative nature of the design.

$$M_u = 52,113(1.4) = 72,958 \text{ ft-lb}$$

$$V_u = 15,441(1.4) = 21,617 \text{ lb}$$

The *footing size and reinforcement* (for heel) will next be determined. Since it commonly occurs that shear strength will be the controlling factor with respect to footing thickness, the shear will be checked first. The heel effective depth available, assuming 2 in. cover (ACI 7.7.1) and No. 8 bars, is

$$d = 21 - 2 - 0.5 = 18.5 \text{ in.}$$

The shear strength ϕV_n of the heel, if no shear reinforcing is provided, is the shear strength of the concrete alone:

$$\phi V_n = \phi V_c = \phi \left(2\sqrt{f_c'} \right) bd$$
$$= 0.85(2)\sqrt{3{,}000}\,(12)(18.5)$$
$$\phi V_n = 20{,}757 \text{ lb}$$

Therefore,

$$V_u > \phi V_n$$

This is unsatisfactory. But since it is not desirable or economical to reinforce a footing for shear, one solution is to increase the footing thickness. Since, as a limit, V_u cannot exceed ϕV_n, the preceding expression can be rewritten and rearranged as follows:

$$\text{Required } d = \frac{V_u}{(\phi)(2\sqrt{f_c'})\,b} = \frac{21{,}617}{0.85(110)(12)}$$
$$= 19.27 \text{ in.}$$

Therefore, increase the footing thickness from 1 ft 9 in. to 2 ft 0 in. This increases the effective depth d as follows:

$$d = 24 - 2 - 0.5 = 21.5 \text{ in.} > 19.27 \text{ in.} \qquad \text{(OK)}$$

The footing thickness could be reduced to 22 in.; however, since it is the opinion of many designers that wall heights in excess of 16 ft should have a footing thickness of not less than 2 ft 0 in., the thickness of 24 in. will be used.

The increase of the footing thickness will affect the stability and soil pressures previously calculated very little, since there will be an associated decrease of equal amount in the height of the earth fill. Previous soil pressures will be used for the rest of the design, and the stem thickness of 1 ft 9 in. at top of footing will be maintained.

The *tensile reinforcement* requirement may now be determined in the normal way:

$$\text{Required } \bar{k} = \frac{M_u}{\phi bd^2} = \frac{72.958(12)}{0.9(12)(21.5)^2}$$
$$= 0.1754 \text{ ksi}$$

Therefore, from Table A–6, the required $\rho = 0.0046$.

Utilizing the ACI Code minimum ρ for beams as applicable for the component parts of the wall,

$$\rho_{min} = \frac{200}{f_y} = \frac{200}{40{,}000} = 0.0050$$

Since required $\rho < \rho_{min}$, USE ρ_{min}.

$$\text{Required } A_s = \rho bd$$
$$= 0.0050(12)(21.5)$$
$$= 1.29 \text{ in.}^2$$

Use No. 9 bars at 9 in. o.c. ($A_s = 1.33$ in.^{2}).

b. Design of toe

The load on the toe is primarily a result of the soil pressure distribution on the bottom of the footing acting in an upward direction. For design purposes the toe is assumed to be a cantilever beam 1 ft in width and with a span length equal to 3 ft 6 in. fixed at the front face of the wall (point B in Fig. 8–11). The soil on the top of the toe is conservatively neglected. Forces and pressures are calculated with reference to Fig. 8–11.

Since the reinforcing steel will be placed in the bottom of the footing, the effective depth available, assuming a 3-in. cover and No. 8 bars, is

$$d = 24 - 3 - 0.5 = 20.5 \text{ in.}$$

The weight of footing for the toe design is

$$2.0(3.5)(150) = 1,050 \text{ lb}$$

The soil pressure in the plane of point B, recalling that the slope of the pressure diagram is 136.2 psf/ft, is

$$2,574 - 3.5(136.2) = 2,097 \text{ psf}$$

The *design moment* M_u may be obtained by a summation of moments about point B in Fig. 8–11. Both the moment and the shear must be modified

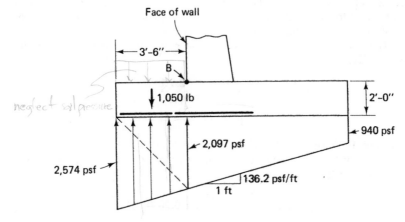

FIGURE 8-11   Pressure distribution for toe design

for strength design. According to the ACI Code, Section 9.2.4, a load factor of 1.7 should be used for horizontal earth pressure and for live load, while 1.4 should be used for dead load. However, since the soil pressure under the footing is primarily the result of horizontal earth pressure, a conservative procedure of using 1.7 is recommended. In addition, since the weight of the footing reduces the effect of the horizontal earth pressure, a factor of 0.9 is used for the footing dead load, as recommended by the ACI Code.

The design moment M_u is then calculated as

$$M_u = 1.7(\tfrac{1}{2})(2{,}574)(3.5)^2(\tfrac{2}{3}) + 1.7(\tfrac{1}{2})(2{,}097)(3.5)^2(\tfrac{1}{3}) - 0.9(1{,}050)\left(\frac{3.5}{2}\right)$$

$$= 23{,}492 \text{ ft-lb}$$

The bending of the toe is such that tension occurs in the bottom of the footing.

The *design shear* V_u is obtained by a summation of vertical forces on the toe side of point B and applying load factors as previously discussed.

$$V_u = 1.7(\tfrac{1}{2})(2{,}574)(3.5) + 1.7(\tfrac{1}{2})(2{,}097)(3.5) - 0.9(1{,}050)$$

$$= 12{,}951 \text{ lb}$$

Footing size and reinforcement, based on the requirements of the toe, are treated as they were for the heel. The shear strength ϕV_n of the toe, if no shear reinforcing is provided, is the shear strength of the concrete alone.

$$\phi V_n = \phi V_c = \phi\left(2\sqrt{f'_c}\right)bd$$

$$= 0.85(2)\sqrt{3{,}000}(12)(20.5)$$

$$\phi V_n = 23{,}001 \text{ lb}$$

$$V_u < \phi V_n$$

Therefore, no shear reinforcement is required and the thickness of the footing is satisfactory.

Based on the determined M_u,

$$\text{Required } \bar{k} = \frac{M_u}{\phi bd^2} = \frac{23.492(12)}{0.9(12)(20.5)^2}$$

$$= 0.0621 \text{ ksi}$$

Therefore, from Table A–6,

$$\text{Required } \rho = 0.0016$$

Since this is less than ρ_{min} of 0.0050, use ρ_{min}. Then

$$\text{Required } A_s = \rho bd$$

$$= 0.0050(12)(20.5)$$

$$= 1.23 \text{ in.}^2$$

Use No. 9 bars at 9 in. o.c.($A_s = 1.33$ in.^{2}).

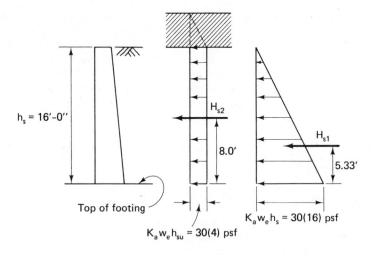

FIGURE 8-12 Forces acting on stem

c. Design of stem

The load on the stem is primarily lateral earth pressure acting (in this problem) horizontally. For design purposes the stem is assumed to be a vertical cantilever beam 1 ft in width and with a span length equal to 16 ft 0 in. fixed at the top of footing.

The loads acting on the stem from the top of wall to the top of footing are depicted in Fig. 8–12. The magnitudes of the horizontal forces are follows:

$$H_{s1} = \frac{1}{2}(30)(16.0)^2 = 3{,}840 \text{ lb}$$

$$H_{s2} = 4(30)(16.0) = 1{,}920 \text{ lb}$$

The *design moment* M_u in the stem may be obtained by a summation of moments about the top of the footing. Both shears and moments must be modified for strength design. Since the forces are due to lateral earth pressure, a factor of 1.7 should be used. M_u may then be calculated as

$$M_u = 1.7(3{,}840)(5.33) + 1.7(1{,}920)(8.0)$$

$$= 60{,}906 \text{ ft-lb}$$

The *design shear* V_u is obtained by a summation of the horizontal forces acting on the stem above the top of the footing.

$$V_u = 1.7(3{,}840) + 1.7(1{,}920)$$

$$= 9{,}792 \text{ lb}$$

The *stem size and reinforcement* requirements will next be determined. Because the stem is assumed to be a cantilever slab, V_u as a limit may equal ϕV_n where $V_n = V_c$. It is not practical to reinforce the stem for shear:

222

therefore, if shear strength is inadequate, stem thickness must be increased. The effective depth available at the top of the footing, assuming 2 in. cover and No. 8 bars, is

$$d = 21 - 2 - 0.5 = 18.5 \text{ in.}$$

The shear strength ϕV_n of the stem (at top of footing) if no shear reinforcement is provided is the shear strength of the concrete alone.

$$\phi V_n = \phi V_c = \phi 2\sqrt{f'_c}bd$$

$$= 0.85(2)\sqrt{3,000}(12)(18.5)$$

$$\phi V_n = 20,757 \text{ lb}$$

$$V_u < \phi V_n$$

Therefore, no shear reinforcement is required. Based on M_u at the bottom of the stem,

$$\text{Required } \bar{k} = \frac{M_u}{\phi bd^2} = \frac{60.906(12)}{0.9(12)(18.5)^2}$$

$$= 0.1977 \text{ ksi}$$

Therefore, from Table A–6, the required $\rho = 0.0052$. This is greater than $\rho_{min} = 0.0050$. Therefore,

$$\text{Required } A_s = \rho bd$$

$$= 0.0052(12)(18.5)$$

$$= 1.15 \text{ in.}^2$$

Use No. 8 bars at 8 in. o.c.($A_s = 1.18$ in.²).

The *stem reinforcement pattern* will be investigated more closely. Since the moment and shear vary along the height of the wall, it is only logical that the steel requirements also vary. A pattern is usually created whereby some of the stem reinforcement is cut off where it is no longer required. The actual pattern used is the choice of the designer, based on various practical constraints.

A typical approach to the creation of a stem reinforcement pattern is first to draw the complete M_u diagram for the stem and then to determine bar cutoff points in the same manner as was discussed in Chapter 5. Stem moments due to the external loads will be calculated at 5 ft from the top of the wall and at 10 ft from the top of the wall. These distances are arbitrary and should be chosen so that a sufficient number of points will be available to draw the M_u diagram.

With reference to Fig. 8–13, the moment at 5 ft from the top of the wall is calculated by first determining the horizontal forces.

$$H_{s1} = \frac{1}{2}(30)(5.0)^2 = 375 \text{ lb}$$

$$H_{s2} = 4(30)(5.0) = 600 \text{ lb}$$

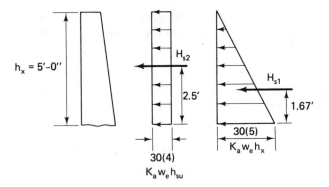

FIGURE 8-13 Stem analysis

Summing the moments about a plane 5 ft below the top of the wall and introducing the appropriate load factor,

$$M_u = 1.7(375)\left(\frac{5.0}{3}\right) + 1.7(600)\left(\frac{5.0}{2}\right)$$

$$= 3{,}613 \text{ ft-lb}$$

With reference to Fig. 8–14, the moment at 10 ft below the top of the wall may be calculated.

$$H_{s1} = \tfrac{1}{2}(30)(10.0)^2 = 1{,}500 \text{ lb}$$

$$H_{s2} = 4(30)(10.0) = 1{,}200 \text{ lb}$$

and

$$M_u = 1.7(1{,}500)\left(\frac{10.0}{3}\right) + 1.7(1{,}200)\left(\frac{10.0}{2}\right)$$
$$= 18{,}700 \text{ ft-lb}$$

The three M_u values may now be plotted as the dark solid curve in Fig. 8–15. This diagram will reflect both the design moment and moment strength with respect to the distance from the top of the wall and must be plotted to scale.

The moment strength M_R of the stem at various locations may be computed from the expression

$$M_R = \phi b d^2 \overline{k}$$

Since we are primarily attempting to establish a cutoff location for some vertical stem reinforcement, the moment strength will be based on a pattern of alternate vertical bars being cut off. With alternate bars cut off, the remaining reinforcement pattern is No. 8 bars at 16 in. o.c., which furnishes an $A_s = 0.59$ in.² Caution must be exercised that the spacing of the remaining bars does not exceed three times the wall thickness or 18 inches, whichever is less.

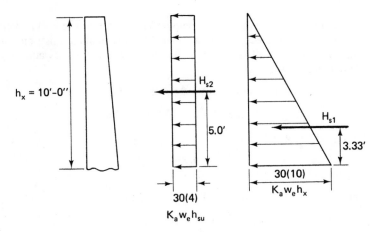

FIGURE 8-14   Stem analysis

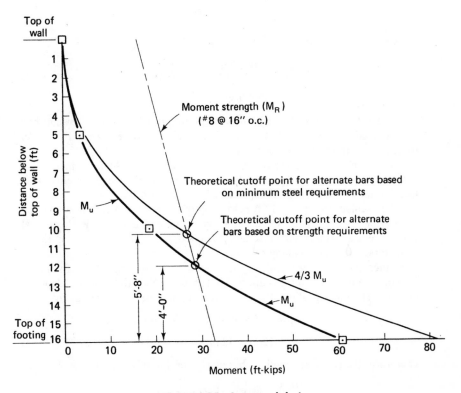

FIGURE 8-15   Stem steel design

The moment strength will be computed at the top of the wall and at the top of the footing, neglecting minimum steel requirements at this time. Since the rear face of the wall is battered, the effective depth varies and may be computed as follows:

$$\text{Top of wall: } d = 12 - 2 - 0.5 = 9.5 \text{ in.}$$

$$\text{Top of footing: } d = 21 - 2 - 0.5 = 18.5 \text{ in.}$$

At the top of the wall:

$$\rho = \frac{A_s}{bd} = \frac{0.59}{12(9.5)} = 0.0052$$

Therefore, from Table A–6, $\bar{k} = 0.1995$ ksi. Then

$$M_R = \frac{0.9(12)(9.5)^2}{12}(0.1995)$$

$$= 16.2 \text{ ft-kips}$$

At the top of the footing,

$$\rho = \frac{A_s}{bd} = \frac{0.59}{12(18.5)} = 0.0027$$

Therefore, from Table A–6, $\bar{k} = 0.1057$ ksi. Then

$$M_R = \frac{0.9(12)(18.5)^2}{12}(0.1057)$$

$$= 32.6 \text{ ft-kips}$$

These two values (M_R) may now be plotted in Fig. 8–15. A straight line will be used to approximate the variation of M_R for No. 8 bars at 16 in. o.c. from the top to the bottom of the wall.

The intersection of the M_R and the M_u curves establishes a theoretical cutoff point for alternate vertical bars based on strength requirements. This location may be determined graphically by scaling on Fig. 8–15.

The scaled distance is 12 ft below top of the wall or 4 ft 0 in. above the top of the footing. The No. 8 bar must be extended past this theoretical cutoff point a distance equal to the effective depth of the member, or 12 bar diameters, whichever is larger.

Since the effective depth of the stem steel at the top of the footing is 18.5 in. and the wall batter is approximately 0.55 in./ft, the effective depth of the stem steel at 4 ft above the top of footing may be calculated as follows:

$$d = 18.5 - 0.55(4) = 16.3 \text{ in.}$$

$$12d_b = 12(1) = 12 \text{ in.}$$

Therefore, the required length of bar above the top of the footing would be

$$4 + \frac{16.3}{12} = 5.36 \text{ ft} \qquad \text{Use } 5 \text{ ft } 5 \text{ in.}$$

However, the ACI Code, Section 10.5.2, stipulates that not less than minimum reinforcement shall be used at any section unless the area of steel furnished is one-third greater than that required. Recall that the minimum steel ratio $\rho = 200/f_y = 0.0050$. Checking the actual steel ratio at the theoretical cutoff point 4 ft above the top of footing:

$$\rho = \frac{A_s}{bd} = \frac{0.59}{12(16.3)} = 0.0030$$

Since $0.0030 < 0.0050$, alternate bars may *not* be stopped at 5 ft 5 in. above the top of the footing.

A simple approach to establish the theoretical cutoff point for alternate bars and conform to the ACI Code is to plot another moment curve using values equal to $1\frac{1}{3}$ times the design moment M_u. The intersection of this curve and the moment strength curve for No. 8 bars at 16 in. will furnish the theoretical cutoff point for alternate bars which will then conform to the Code requirement of an additional one-third of the required steel if ρ_{min} is not satisfied.

Multiplying the design moment M_u by 1.333, the following moments are obtained:

At 5 ft below the top of the wall:

$$M_u = 3,613(1.333) = 4,816 \text{ ft-lb}$$

At 10 ft below the top of the wall:

$$M_u = 18,700(1.333) = 24,927 \text{ ft-lb}$$

At the top of the footing:

$$M_u = 60,906(1.333) = 81,188 \text{ ft-lb}$$

Plotting this moment curve and scaling the distance above the top of the footing, we arrive at a value of 5 ft 8 in., or 5.67 ft. The bars must extend past this theoretical cutoff point a distance equal to the effective depth of the member or 12 bar diameters, whichever is larger.

$$d = 18.5 - 0.55(5.67) \approx 15.4 \text{ in.}$$

$$12d_b = 12(1) = 12 \text{ in.}$$

Therefore, the length of bar required above the top of footing is

$$5.67 + \frac{15.4}{12} = 6.95 \text{ ft} \qquad \text{say, 7 ft 0 in.}$$

The development length of the stem reinforcing steel into the stem above the top of the footing and also above the theoretical cutoff point of the alternate bars must be a minimum of ℓ_d where $\ell_d = (23)(0.8) = 18.4$ in. The actual development length furnished is far in excess of that required.

The ACI Code, Section 12.11.5, stipulates that flexural reinforcement must not be terminated in a tension zone unless one of several conditions is

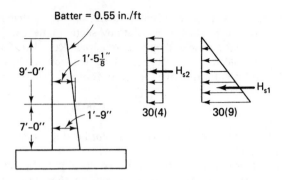

FIGURE 8-16 Shear at bar cutoff point

satisfied. One of these conditions is that the shear at the cutoff point does not exceed two-thirds of the shear permitted. Therefore, check the shear at the actual cutoff point, 7.0 ft above the top of the footing (see Fig. 8–16).

The effective depth of the wall in a plane 7.0 ft above the top of the footing is: $d = 17.125 - 2 - 0.5 = 14.62$ in.

$$H_{s1} = \frac{1}{2}(30)(9.0)^2 = 1{,}215 \text{ lb}$$

$$H_{s2} = 4(30)(9.0) = \underline{1{,}080 \text{ lb}}$$

$$\text{Total} = 2{,}295 \text{ lb}$$

$$\text{Design } V_u = 2{,}295(1.7) = 3{,}902 \text{ lb}$$

$$\tfrac{2}{3}\phi V_n = \tfrac{2}{3}\phi V_c = \tfrac{2}{3}\phi(2\sqrt{f'_c})\,bd$$

$$= \tfrac{2}{3}(0.85)(2)\sqrt{3{,}000}(12)(14.62)$$

$$\tfrac{2}{3}\phi V_n = 10940 \text{ lb}$$

$$V_u < \tfrac{2}{3}\phi V_n \qquad\qquad\qquad\qquad (\text{OK})$$

In summary, alternate vertical stem reinforcing bars may be stopped at 7 ft 0 in. above the top of the footing (see the typical wall section, Fig. 8–19).

d. Additional design details

The *stem steel anchorage* into the footing may be computed using Table A–12 and applicable modification factors. The basic development length for the No. 8 bar is 23 in. Modification factors are as follows:

$$\text{Spacing in excess of 6 in.} = 0.8 \quad \text{sparse bar mod. factor}$$

$$\frac{\text{required } A_s}{\text{provided } A_s} = \frac{1.15}{1.18} = 0.97$$

therefore the development length required is:

$$\ell_d = (23)(0.8)(0.97) = 17.8 \text{ in.}$$

228

With the footing thickness equal to 24 in. and a minimum of 3 in. of cover required for the steel at the bottom of the footing, the anchorage length available is $24 - 3 = 21$ in. This is adequate. Therefore, a straight embedment length is satisfactory and no hook is required.

The *length of splice* required for main stem reinforcing steel (see the typical wall section, Fig. 8–19) may be calculated recognizing that the class B splice is applicable for this condition (see Chapter 5). Therefore, the length of the splice required equals

$$1.3(23.0)(0.8)(0.97) = 23.2 \text{ in.} \quad \text{Use 24 in.}$$

Stem face steel in the form of horizontal and vertical reinforcement will be provided as per the ACI Code, Sections 14.2.10 and 14.2.11. The minimum recommended steel is as follows:

Horizontal bars (per foot of height of wall):

$$A_s = 0.0025b \text{ (thickness)}$$

Vertical bars (per foot of wall horizontally):

$$A_s = 0.0015b \text{ (thickness)}$$

The Code also stipulates that walls more than 10 in. thick, except basement walls, must have reinforcement for each direction in each face of the wall. The exposed face must have a minimum of one-half and a maximum of two-thirds the total steel required for each direction. The maximum spacing for the steel must not exceed three times the wall thickness nor 18 in. (ACI Code, Section 10.15) for both vertical bars and horizontal bars. For the front face of the wall (exposed face), use two-thirds of the total steel required and an average stem thickness = 16.5 in.

For horizontal steel:

$$0.0025(12)(16.5) = 0.50 \text{ in.}^2/\text{ft of height}$$

$$= 0.50(0.67) = 0.34 \text{ in.}^2$$

Use No. 5 bars at 11 in. ($A_s = 0.34 \text{ in.}^2$).

For vertical steel:

$$0.0015(12)(16.5) = 0.30 \text{ in.}^2/\text{ft of horizontal length}$$

$$= (0.30)(0.67) = 0.20 \text{ in.}^2$$

Use No. 4 bars at 12 in. ($A_s = 0.20 \text{ in.}^2$).

Rear face of wall, not exposed, for horizontal steel:

$$0.50(0.33) = 0.17 \text{ in.}^2$$

Use No. 4 bars at 11 in. ($A_s = 0.22 \text{ in.}^2$).

Longitudinal reinforcemment in the footing should provide a *minimum* steel area of 0.1% of footing cross-sectional area. A common practice is to use an area somewhere between 0.1 and 0.2% of the footing area.

$$0.001(12)(2.0)(144) = 3.46 \text{ in.}^2$$

Use 15 No. 5 bars ($A_s = 4.65$ in.^{2}) as shown in Fig. 8–19. This provides a steel percentage between 0.1% and 0.2%.

Transverse reinforcement in the footing need not run the full width of footing. However proper anchorage length must be provided from the point of maximum tensile stress.

The basic development length for the No. 9 bars is 29 in. For the top bars in the heel, the applicable modification factors are as follows:

$$\text{Top bars} = 1.4$$

$$\text{Spacing in excess of 6 in.} = 0.8$$

$$\frac{\text{Required } A_s}{\text{Provided } A_s} = \frac{1.29}{1.33} = 0.97$$

The development length required for the top bars, therefore, is

$$\ell_d = 29(1.4)(0.8)(0.97) = 31.5 \text{ in.} \qquad \text{Use 32 in.}$$

For the bottom bars in the toe, the applicable modification factors are as follows:

$$\text{Spacing in excess of 6 in.} = 0.8$$

$$\frac{\text{Required } A_s}{\text{Provided } A_s} = \frac{1.23}{1.33} = 0.93$$

The development length required for the bottom bars, therefore, is

$$\ell_d = 29(0.8)(0.93) = 21.6 \text{ in.} \qquad \text{Use 22 in.}$$

These details are included in the typical wall section, Fig. 8–19.

A *shear key* will be provided between the stem and the footing. We will use a depressed key formed by a 2 × 6 plank (dressed dimensions 1½ × 5½), as shown in Fig. 8–19.

The need for a shear key is questionable, since considerable slip is required to develop the key for purposes of lateral force transfer. However, it may be considered as an added mechanical factor of safety.

The shear-friction method of the ACI Code Section 11.7 should be used to design for horizontal force transfer between the stem and footing. This approach eliminates the need for the shear key, provided that the contact surface is both adequately roughened and crossed by properly anchored reinforcement. According to the Code, the surface must be intentionally roughened to an amplitude of ¼ in. or more by raking or some other means.

The shear-friction approach assumes a shear plane already cracked. Since the stem is poured against hardened concrete, the plane between the two may be considered analogous to a cracked plane. Sliding along this plane is resisted by the two contact surfaces. As slip begins to occur, the roughness of the contact surfaces forces the two surfaces to separate. This separation is resisted by the reinforcement A_{vf} which is placed approximately perpendicular to the surfaces. Thus, tension is induced in the reinforcement and contact pressure develops between the stem and the footing.

The ideal shear strength (or friction force which resists sliding) may be computed from

$$V_n = A_{vf} f_y \mu$$

where A_{vf} = Area of shear-friction reinforcement
μ = Coefficient of friction in accordance with ACI 11.7.5. It may be taken as 1.0 for concrete placed against hardened concrete roughened as described previously.

Note that A_{vf} is steel that is provided for the shear-friction development and that it is in addition to any other steel already provided.

As a limit

may as well be =

$$V_u \gtrless \phi V_n$$

where V_u is shear force applied at the cracked plane. Substituting for V_n

$$V_u \leq \phi A_{vf} f_y \mu$$

Solving for A_{vf}

$$\text{Required } A_{vf} = \frac{V_u}{\phi f_y \mu}$$

$$= \frac{9792}{(0.85)(40,000)(1.0)}$$

$$= 0.29 \text{ in.}^2 \text{ per ft length of wall}$$

Since it is desirable to distribute the shear-friction reinforcement across the width of the contact surface, use 2 No. 4 dowels per foot, one in each face of stem.

$$\text{Provided } A_{vf} = 0.20(2) = 0.40 \text{ in.}^2$$

actual — $\text{Shear strength } V_n = A_{vf} f_y \mu$

$$= (0.40)(40,000)(1.0)$$

$$= 16,000 \text{ lb}$$

The ACI Code Section 11.7.4 stipulates that the maximum V_n shall not exceed

upper limit

$$0.2 f'_c A_c \quad \text{or} \quad 800 A_c$$

where A_c is the area of the contact surface resisting shear transfer.

$$\text{Max } V_n = (0.2)(3000)(12)(21) = 151,200 \text{ lb}$$

$$\text{Max } V_n = (800)(12)(21) = 201,600 \text{ lb}$$

Therefore max V_n = 16,000 lb and since V_u = 9,792 lb

lowest of Max V_n controls

$$9,792 < (0.85)(16,000)$$

$$9,792 < 13,600 \tag{OK}$$

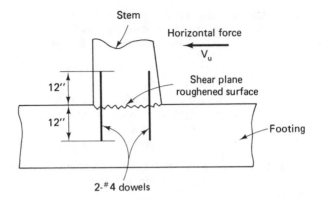

FIGURE 8-17   Shear-friction reinforcement

The development length of the No. 4 dowels must be furnished into both the stem and the footing. These are considered to be tension bars. The dowels are 12 in. on centers.

Basic development length for the No. 4 bars = 8 in.

The modification factors are as follows:

Spacing in excess of 6 in. = 0.8

$$\frac{\text{Required } A_s}{\text{Provided } A_s} = \frac{0.29}{0.40} = 0.73$$

Required development length $\ell_d = (8)(0.8)(0.73) = 4.7$ in. Use the required minimum $\ell_d = 12$ in.

These bars are depicted in Fig. 8–17.

6. Footing base shear key

The *footing base shear key* (sometimes called a *bearing lug*) is primarily used to prevent a sliding failure. The magnitude of the additional resistance to sliding offered by the key is questionable and is a function of the subsoil material. In addition, the excavation for a key will generally disturb the subsoil during construction and conceivably will do more harm than good. Hence the use of the key for the purpose intended is controversial.

An acceptable design approach is to utilize the passive earth resistance H_p in front of the wall (from the top of the footing to the bottom of the key) as an additional resistance to sliding (see Fig. 8–18).

The passive earth resistance may be expressed in terms of an equivalent fluid weight. Since $K_a w_e = 30$ lb/ft³ and $w_e = 100$ lb/ft³, then

$$K_a = \frac{30}{100} = 0.3$$

The coefficient of passive earth resistance is

$$K_p = \frac{1 + \sin \phi}{1 - \sin \phi} = \frac{1}{K_a} = \frac{1}{0.3} = 3.33$$

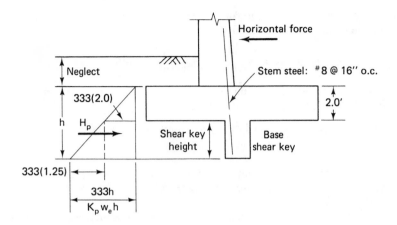

FIGURE 8-18 Base shear key force analysis

Therefore,

$$K_p w_e = 3.33(100) = 333 \text{ lb/ft}^3$$

The unfactored actual horizontal force $\Sigma H = 7{,}020$ lb. Therefore, the required resistance to sliding that will furnish a factor of safety of 1.75 is

$$1.75(7{,}020) = 12{,}285 \text{ lb}$$

The actual frictional resistance furnished is 10,540 lb (see the discussion of stability analysis). Therefore, the resistance that must be furnished by the base shear key and the passive earth resistance is

$$12{,}285 - 10{,}540 = 1{,}745 \text{ lb}$$

With reference to Fig. 8–18, the height h required to furnish this resistance may be obtained by establishing horizontal equilibrium ($\Sigma H = 0$).

$$\tfrac{1}{2}(333)(h)^2 = 1{,}745$$

$$h^2 = 10.48$$

Required $h = 3.24$ ft $= 3$ ft $- 3$ in.

Since the footing thickness $= 2$ ft 0 in., the base shear key height may be taken as 1 ft 3 in., thereby furnishing a required h. The key must be designed for moment and shear. Assuming the key as a vertical cantilever beam, taking a summation of moment about the plane of the bottom of footing, and applying a load factor of 1.7:

$$M_u = 1.7(333)(2.0)(1.25)\left(\frac{1.25}{2}\right) + 1.7(\tfrac{1}{2})(333)(1.25)^2(\tfrac{2}{3})(1.25)$$

$$= 1{,}253 \text{ ft-lb}$$

Assume a 10-in. width of key and a No. 8 bar, which provides an effective depth $d = 10 - 3 - 0.5 = 6.5$ in.

$$\text{Required } \bar{k} = \frac{M_u}{\phi bd^2} = \frac{1.253(12)}{0.9(12)(6.5)^2} = 0.0330 \text{ ksi}$$

From Table A–6, the required $\rho = 0.0010$ Since this is less than ρ_{min} of 0.0050, use ρ_{min}.

$$\text{Required } A_s = \rho bd = 0.0050(12)(6.5)$$

$$= 0.39 \text{ in.}^2$$

This is a very small amount of steel and it will be more practical to extend some existing stem bars into the key, therefore: use No. 8 bars at 16 in. ($A_s = 0.59$ in.²).

A convenient pattern would consist of extending alternate stem bars into the key (see the typical wall section, Fig. 8–19). The shear strength of the key is

$$\phi V_n = \phi V_c$$

$$= \phi(2\sqrt{f_c'})bd$$

$$= 0.85(2)(\sqrt{3,000})(12)(6.5)$$

$$\phi V_n = 7,293 \text{ lb}$$

The design shear is

$$V_u = 1.7(1,745) = 2,967 \text{ lb}$$

$$V_u < \phi V_n \qquad\qquad\qquad\qquad \text{(OK)}$$

The anchorage of the stem steel into the base shear key should also be checked. The basic development length for the No. 8 bar $= 23$ in., and the applicable modification factors are as follows:

$$\text{Spacing in excess of 6 in.} = 0.8$$

$$\frac{\text{Required } A_s}{\text{Provided } A_s} = \frac{0.39}{0.59} = 0.66$$

The development length required is

$$\ell_d = 23(0.8)(0.66) = 12.1 \text{ in.}$$

Available embedment $= 12$ in. assuming a 3 in.-cover on the end of the bar. Since 12.1 in. ≈ 12 in. no hook is required. The entire design is shown in Fig. 8–19.

8–4 DESIGN CONSIDERATIONS FOR BEARING WALLS

Bearing walls [Fig. 8–1(g)] were briefly described at the beginning of this chapter as those walls that carry vertical load in addition to their own weight. Recommendations for the empirical design of such walls are

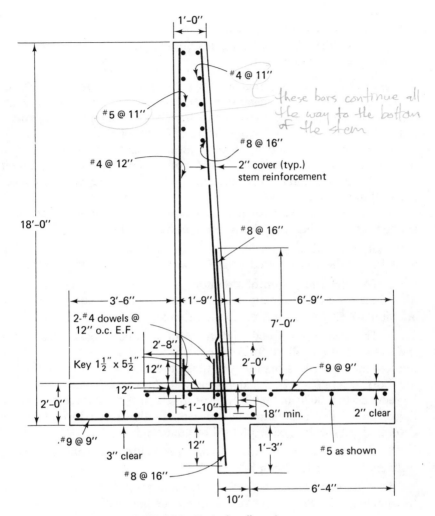

FIGURE 8-19 Typical wall section

presented in Chapter 14 of the ACI Code (318–77) and apply primarily to walls spanning vertically, that is, supported at top and bottom. (Also see the ACI Code, Section 10.15.)

The ACI Code stipulates that reinforced concrete walls carrying reasonably concentric loads may be designed utilizing an empirical method. "Reasonably concentric" implies that the resultant factored load falls within the middle third of the cross section.

The design axial load strength or capacity of such a wall will be

$$\phi P_{nw} = 0.55\phi f_c' A_g \left[1 - \left(\frac{\ell_c}{40h} \right)^2 \right] \quad \text{[ACI Eq. (14-1)]}$$

235

where $\phi = 0.70$

  h = thickness of wall (in.)

  ℓ_c = vertical distance between supports (in.)

  A_g = gross area of section (in.^{2})

Research studies have indicated that this expression is applicable for walls that are restrained against rotation at one or both ends and are braced top and bottom against lateral translation. This in effect overestimates the wall capacity for the theoretical case of pin-ended walls.

Where the wall is subject to concentrated loads, the effective length of wall for each concentration must not exceed the center-to-center distance between loads, nor exceed the width of bearing plus four times the wall thickness.

The following requirements applicable to bearing walls are prescribed, among others, by the ACI Code, Section 14.2.

1. Reinforced concrete bearing walls must have a thickness of at least 1/25 of the unsupported height or width, whichever is shorter.

2. The uppermost 15 ft of a reinforced concrete wall must be at least 6 in. thick, and the minimum thickness must be increased 1 in. for each additional 25 ft downward, or fraction thereof.

3. The area of horizontal reinforcement must be at least 0.0025 times the area of the wall ($0.0025bh$ per foot) and the area of vertical reinforcement not less than 0.0015 times the area of the wall ($0.0015bh$ per foot), where b = 12 in. and h is the wall thickness. These values may be reduced to 0.0020 and 0.0012, respectively, if the reinforcement is not larger than ⅝ in. in diameter and consists of either welded wire fabric or deformed bars with f_y of 60,000 psi or greater.

4. Exterior basement walls, foundation walls, fire walls, and party walls must not be less than 8 in. thick.

5. Reinforced concrete walls must be anchored to floors, roofs, or to columns, pilasters, buttresses, and intersecting walls.

6. Walls more than 10 in. thick, except for basement walls, must have reinforcement in each direction for each face. The exterior surface shall have a minimum of one-half and a maximum of two-thirds of the total steel required, with the interior surface having the balance of the reinforcement. Bars shall not be less than No. 3 placed 18 in. on centers.

EXAMPLE 8–2

Design a reinforced concrete bearing wall to support a series of steel wide-flange beams at 8 ft 0 in. o.c. Each beam rests on a bearing plate 6 in. × 12 in. Wall is braced top and bottom against lateral translation. Assume bottom end

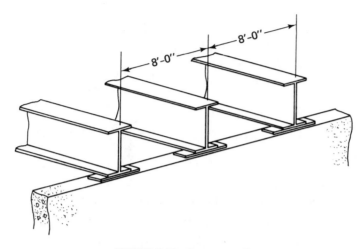

FIGURE 8-20 Bearing wall

fixed against rotation. The wall height is 15 ft and the design (factored) load P_u from each beam is 125 kips. $f'_c = 3{,}000$ psi and $f_y = 40{,}000$ psi. See Fig. 8–20.

1. Assume an 8-in.-thick wall with full concentric bearing.

2. The bearing strength of the concrete under the bearing plate, as governed by the ACI Code, Section 10.16.1, is

$$\phi(0.85)f'_c A_1 = 0.70(0.85)(3{,}000)(6)(12)$$

$$= 128{,}520 \text{ lb}$$

Factored bearing load $P_u = 125{,}000$ lb

$$128{,}520 > 125{,}000 \qquad \text{(OK)}$$

3. The effective length of the wall (ACI Code, Section 14.2.4) must not exceed the center-to-center distance between loads nor the width of bearing plus four times the wall thickness. Beam spacing = 96 in.

$$12 + 4(8) = 44 \text{ in.}$$

Therefore, use 44 in.

4. The minimum thickness required is 1/25 times the shorter of the unsupported height or width. Assume, for our case, that width does not control.

$$h_{\min} = \frac{\ell_c}{25} = \frac{15(12)}{25} = 7.2 \text{ in.}$$

Also, $h_{\min}$ in any case is 6 in. Therefore, the 8-in. wall is satisfactory.

237

5. Capacity of wall:

$$\phi P_{nw} = 0.55\phi f'_c A_g \left[1 - \left(\frac{\ell_c}{40h} \right)^2 \right]$$

$$= 0.55(0.70)(3)(44)(8)\left\{ 1 - \left[\frac{15(12)}{40(8)} \right]^2 \right\}$$

$$= 406.6(1 - 0.316)$$

$$= 278 \text{ kips}$$

$$\phi P_{nw} > P_u \qquad\qquad\qquad\qquad \text{(OK)}$$

6. Reinforcing steel (ACI Code, Section 14.2):
   Vertical reinforcement per foot of wall length:

$$\text{Required } A_s = 0.0015bh$$

$$A_s = 0.0015(12)(8) = 0.144 \text{ in.}^2$$

Horizontal reinforcement per foot of wall height:

$$\text{Required } A_s = 0.0025bh$$

$$A_s = 0.0025(12)(8) = 0.24 \text{ in.}^2$$

The maximum spacing of reinforcement must not exceed three times the wall thickness nor 18 in. (ACI Code, Section 7.6.5).

$$3(8) = 24 \text{ in.} \quad \text{Use 18 in.}$$

Minimum size reinforcing bar: No. 3 (ACI 14.2.12)

$$\text{Vertical steel: use No. 4 bars at 16 in.}(A_s = 0.15 \text{ in.}^2)$$

$$\text{Horizontal steel: use No. 5 bars at 15 in. } (A_s = 0.25 \text{ in.}^2)$$

The steel may be placed in one layer, since the wall is less than 10 in. thick. (see Fig. 8–21).

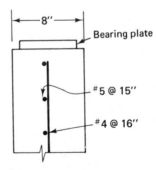

FIGURE 8-21   Section of wall

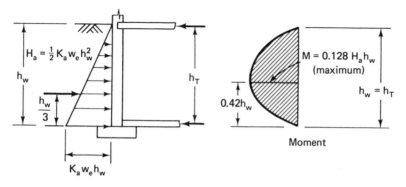

FIGURE 8-22 Basement wall with full height backfill—forces and moment diagram

8-5 DESIGN CONSIDERATIONS FOR BASEMENT WALLS

A basement wall is a type of retaining wall in which there is lateral support assumed to be provided at bottom and top by the basement floor slab and first-floor construction, respectively. As previously mentioned, the wall would be designed as a simply supported member with a loading diagram and moment diagram as shown in Fig. 8–22.

If the wall is part of a bearing wall, the vertical load will relieve some of the tension in the vertical reinforcement. This may be neglected since its effect may be small compared to the uncertainties in the assumption of loads. If the vertical load is of a permanent nature and of significant magnitude, its effect should be considered in the design.

When a part of the basement wall is above ground, the lateral bending moment may be small and may be computed as shown in Fig.

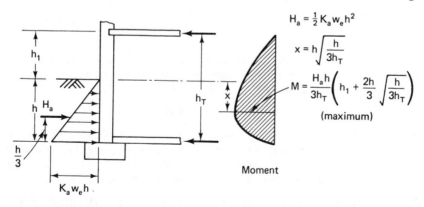

FIGURE 8-23 Basement wall with partial backfill—forces and moment diagram

8–23. This assumes that the wall is spanning in a vertical direction. Depending upon the type of construction, the basement wall may also span in a horizontal direction and may behave as a slab reinforced in either one or two directions. If the wall design assumes two horizontal reactions as shown, caution must be exercised that the two supports are in place prior to backfilling behind the wall.

PROBLEMS

8–1 Compute the active earth pressure horizontal force on the wall shown for the following conditions $w_e = 100$ lb/ft³.

Case	ϕ	θ	w_s (psf)	h_w (ft)
a	25	0	400	15
b	28	10	0	18
c	30	0	200	20
d	33	20	0	25

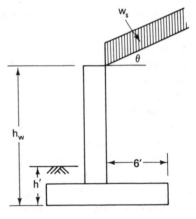

FIGURE P8-1

8–2. Find the passive earth pressure force in front of the wall for Problem 8–1(a) if $h' = 4$ ft.

8–3 For the wall shown, determine the factors of safety against overturning and sliding and determine the soil pressures under the footing. $K_a = 0.3$ and $w_e = 100$ lb/ft³. Coefficient of friction $f = 0.50$

8–4 Same as Problem 8–3 but the toe is 2 ft 0 in.

8–5 In Problem 8–1(c), assume a footing depth of 2 ft. Design the stem steel for M_u at the top of the footing. Check the anchorage into the footing. Disregard other stem steel details. $f'_c = 4,000$ psi and $f_y = 60,000$ psi. Use thickness at top of stem = 12 in. and thickness at bottom of stem = 1 ft 9 in.

8–6 Completely design the cantilever retaining wall shown. The height of wall h_w is 22 ft, the backfill is level, the surcharge is 600 psf, $w_e = 100$ lb/ft³, $h' = 4$ ft, $K_a = 0.30$, and the allowable soil pressure is 4 ksf. Use a coefficient of friction

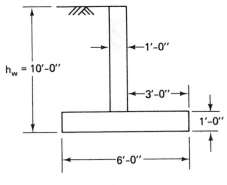

FIGURE P8-3

of concrete on soil of 0.5, f'_c = 3,000 psi, and f_y = 40,000 psi. The required factor of safety against overturning is 2.0 and against sliding is 1.5. The wall batter should be between ¼ and ½ in./ft. The design is to be in accordance with the ACI Code (318–77).

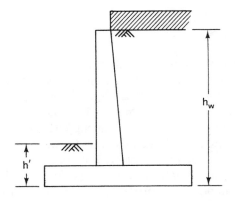

FIGURE P8-6

REFERENCE

1. "The American Standard Building Code Requirements for Masonry," U.S. Department of Commerce, National Bureau of Standards, American National Standards Institute, A41.1–1953. Available from American National Standards Institute, 1430 Broadway, New York, N.Y. 10018.

9

Columns

9-1 INTRODUCTION

The main vertical load-carrying members in buildings are called columns. The ACI Code defines a column as a member used primarily to support axial compressive loads and with a height at least three times its least lateral dimension. The Code further defines as a *pedestal* an upright compression member having a ratio of unsupported height to least lateral dimension of 3 or less. The Code definition for columns will be extended to include members which are subjected to combined axial compression and bending moment (in other words, eccentrically applied compressive loads), since, for all practical purposes, no column is truly axially loaded.

There are three basic types of reinforced concrete columns. These are shown in Fig. 9-1. *Tied columns* [Fig. 9-1(a)] are reinforced with longitudinal bars which are enclosed by horizontal, or lateral, ties placed at specified spacings. *Spiral columns* [Fig. 9-1(b)] are reinforced with longitudinal bars enclosed by a continuous, rather closely spaced, steel spiral. The spiral is made up of either wire or bar and is formed in the shape of a helix. A third type of reinforced concrete column is a *composite column* and is shown in Fig. 9-1(c). This type encompasses compression members reinforced longitudinally with structural steel shapes, pipes or tubes with or without longitudinal bars. Code requirements for composite compression members are found in Section 10.14. Our discussion will be limited to the first two types—tied and spiral columns. Tied columns are generally square or rectangular while spiral columns are normally circu-

Foundation Columns, Turbine Generator Building—Seabrook
Station, New Hampshire.

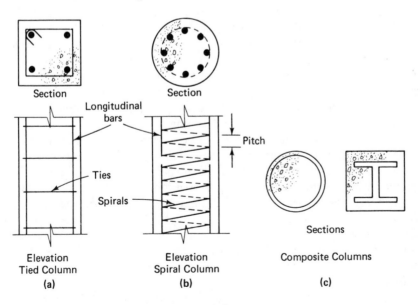

Section   Section

Longitudinal
bars

Pitch

Ties

Spirals

Elevation   Elevation   Sections
Tied Column   Spiral Column   Composite Columns
(a)   (b)   (c)

FIGURE 9-1   Column types

243

lar. However, this is not a hard-and-fast rule, since square spirally re-inforced columns and circular tied columns do exist as well as do other shapes, such as octagonal and L-shaped columns.

We will initially discuss the analysis and design of columns which are *short*. A column is said to be short when its length is such that lateral buckling need not be considered. The ACI Code does, however, require that the length of columns be a design consideration. It is recognized that as length increases, the usable strength of a given cross section is decreased because of the buckling problem. Concrete columns, by their very nature, are more massive and therefore stiffer than their structural steel counterparts. For this reason, slenderness is less of a problem in reinforced concrete columns. It has been estimated that more than 90% of typical reinforced concrete columns existing in braced frame buildings may be classified as short columns and slenderness effects may be neglected.

9-2 STRENGTH OF REINFORCED CONCRETE COLUMNS—SMALL ECCENTRICITY

If a compressive load P is applied coincident with the longitudinal axis of a symmetrical column, it theoretically induces a uniform compressive stress over the cross-sectional area. If, however, the compressive load is applied a small distance e away from the longitudinal axis, there is a tendency for the column to bend due to the moment $M = Pe$. This distance e is called the *eccentricity*. Unlike the zero eccentricity condition, the compressive stress is not uniformly distributed over the cross section but is greater on one side than the other. This is analogous to our previously discussed eccentricity of applied loads with respect to retaining wall footings, where the eccentrically applied load resulted in a nonuniform soil pressure under the footing.

We will consider an *axial load* to be a load that acts parallel to the longitudinal axis of a member but *need not* be applied at any particular point on the cross section, such as a centroid or a geometric center. The column that is loaded with a compressive axial load at zero eccentricity is probably nonexistent and even the axial load/small eccentricity (axial load/small moment) combination is relatively rare. Nevertheless, we will first consider the case of columns that are loaded with compressive axial loads at *small eccentricities*, further defining this situation as that in which the induced moments, although they are present, are so small that they are of little significance.

The fundamental assumptions for the calculation of column axial load strength (small eccentricities) are that at ultimate load the concrete is stressed to $0.85f'_c$ and the steel is stressed to f_y. For the cross sections shown in Fig. 9–1(a) and (b), the nominal axial load strength at small eccentricity will be a straightforward sum of the forces existing in the concrete and longitudinal steel when each of the materials is stressed to its maximum. The following ACI notation will be used:

A_g = gross area of the column section (in.^{2})

A_{st} = total area of *longitudinal* reinforcement (in.^{2})

P_0 = nominal, or theoretical, axial load strength at zero eccentricity

P_n = nominal, or theoretical, axial load strength at given eccentricity

P_u = factored applied axial load at given eccentricity

For convenience we will define a steel ratio ρ_g for columns:

$$\rho_g = \frac{A_{st}}{A_g}$$

The nominal, or theoretical, axial load strength for the special case of zero eccentricity may be written

$$P_0 = 0.85f_c'(A_g - A_{st}) + f_y A_{st}$$

This theoretical strength must be further reduced to a maximum usuable axial load strength utilizing two different strength reduction factors.

Extensive testing has shown that spiral columns are tougher than tied columns, as depicted in Fig. 9–2. Both types behave similarly up to the column yield point, at which time the outer shell spalls off. At this point the tied column fails through crushing and shearing of the concrete and through outward buckling of the bars between the ties. The spiral column, however, has a core area within the spiral which is effectively laterally supported and continues to withstand load. Failure occurs only when the spiral steel yields following large deformation of the column.

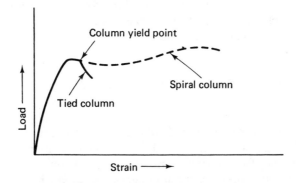

FIGURE 9-2 Load–strain relationship for columns

Naturally, the size and spacing of the spiral steel will affect the final load at failure. Although both columns have exceeded their usable strengths once the outer shell spalls off, the code recognizes the greater tenacity of the spiral column in Section 9.3.2, where it directs that a strength reduction factor of 0.75 be used for a spiral column while the strength reduction factor for the tied column must be 0.70. In justifying the use of a spirally reinforced column with respect to strength and economics (code requirements for ductility notwithstanding), one should consider that the use of the spiral will increase the usable strength by a factor of 0.75/0.70 = 1.07, or only 7%.

The Code directs that the basic load–strength relationship be

$$P_u \leq \phi P_n$$

where P_n is the nominal axial load strength at a given eccentricity. Logically, for the case of zero eccentricity, if it could exist, P_n would equal P_0. However, the ACI Code recognizes that no practical column can be loaded with zero eccentricity. Therefore, in addition to imposing the strength reduction factor ϕ, the Code directs that the nominal strengths be further reduced by factors of 0.80 and 0.85 for tied and spiral columns, respectively. This results in the following expressions for usable axial load strengths:

$$\phi P_{n\,(\text{max})} = 0.85\phi[0.85f'_c(A_g - A_{st}) + f_y A_{st}] \quad \text{ACI Eq. (10–1)}$$

for spiral columns and

$$\phi P_{n(\text{max})} = 0.80\phi[0.85f'_c(A_g - A_{st}) + f_y A_{st}] \quad \text{ACI Eq. (10–2)}$$

for tied columns. These expressions provide the magnitude of the maximum usable axial load strength that may be realized from any column cross section. This will be the axial load strength at small eccentricity. Should the eccentricity (and the associated moment) become larger, ϕP_n will have to be reduced, as will be shown in Section 9–9. It may be recognized that the Code equations for $\phi P_{n\,(\text{max})}$ provide for an extra margin of axial load strength. This will, in effect, provide some reserve strength to carry small moments.

9–3 CODE REQUIREMENTS CONCERNING COLUMN DETAILS

Main (longitudinal) reinforcing should have a cross-sectional area so that ρ_g will be between 0.01 and 0.08. The minimum number of longitudinal bars must be six for a circular arrangement and four for a rectangular arrangement. The foregoing requirements are stated in the ACI Code,

Section 10.9. Although not mentioned in the present Code, the 1963 code recommended a minimum bar size of No. 5.

The clear distance between longitudinal bars must not be less than 1.5 times the nominal bar diameter nor 1½ in. (ACI Code, Section 7.6). This requirement also holds true where bars are spliced. Table A–14 may be used to determine the maximum number of bars allowed in one row around the periphery of circular or square columns.

Cover shall be 1½ in. minimum over primary reinforcement, ties or spirals (ACI Code, Section 7.7.1).

Tie requirements are discussed in detail in the ACI Code, Section 7.10.5. The minimum size is No. 3 for longitudinal bars No. 10 and smaller; otherwise, minimum tie size is No. 4. (See Table A–14 for a suggested tie size.) Usually, No. 5 is a maximum. The center-to-center spacing of ties should not exceed the smaller of 16 longitudinal bar diameters, 48 tie-bar diameters, or the least column dimension. Furthermore, the ties shall be so arranged that every corner and alternate longitudinal bar will have lateral support provided by the corner of a tie having an included angle of not more than 135° and no bar shall be farther than 6 in. clear on each side from such a laterally supported bar. See Fig. 9–3 for typical tie arrangements.

Spiral requirements are discussed in the ACI Code, Section 7.10.4. The minimum spiral size is ⅜ in. in diameter (⅝ in. is usually maximum). Clear space between spirals must not exceed 3 in. or be less than 1 in. The *spiral steel ratio* ρ_s must not be less than the value given by

$$\rho_{s(\min)} = 0.45 \left(\frac{A_g}{A_c} - 1 \right) \frac{f'_c}{f_y} \qquad \text{(see ACI Code, Section 10.9.3.)}$$

where $\rho_s = \dfrac{\text{volume of spiral steel in one turn}}{\text{volume of column core in height }(s)}$

s = center-to-center spacing of spiral (sometimes called the *pitch*)

A_g = gross cross-sectional area of the column

A_c = cross-sectional area of the core (out-to-out of spiral)

f_y = *spiral* steel yield point

This particular spiral steel ratio will result in a spiral that will make up the strength lost due to the spalling of the outer shell (see Fig. 9–2).

An approximate formula for the actual spiral steel ratio in terms of physical properties of the column cross section may be derived from the preceding definition of ρ_s. In Fig. 9–4, we will denote the overall core diameter (out-to-out of spiral) as D_c and the spiral diameter (center to center) as D_s. Further, the cross-sectional area of the spiral bar will be denoted A_{sp}. From the definition of ρ_s, an expression may be written:

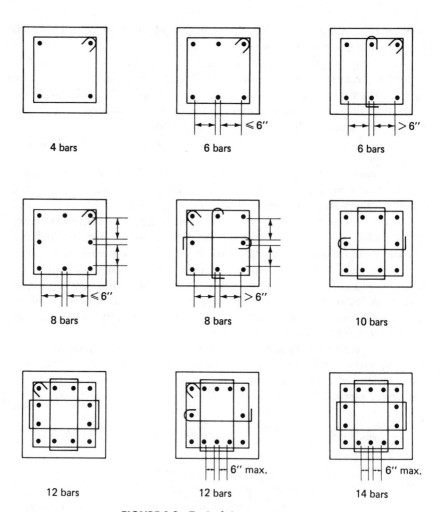

FIGURE 9-3 Typical tie arrangements

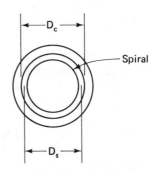

FIGURE 9-4 Definition of D_c and D_s

$$\text{Actual } \rho_s = \frac{A_{sp} \pi D_s}{(\pi D_c^2/4)(s)}$$

If the small difference between D_c and D_s is neglected, then, in terms of D_c:

$$\text{Actual } \rho_s = \frac{4 A_{sp}}{D_c s}$$

9–4 ANALYSIS OF SHORT COLUMNS—SMALL ECCENTRICITY

The analysis of short columns carrying axial loads that have small eccentricites involves checking the maximum usable strength and the various details of the reinforcing. The procedure is summarized in Section 9–6.

EXAMPLE 9–1

Find the maximum usable axial load strength for the tied column of cross section shown in Fig. 9–5. Check the ties. Assume a short column. $f'_c = 4,000$ psi and $f_y = 60,000$ psi for both longitudinal steel and ties.

1. Check the steel ratio for the longitudinal steel.

$$\rho_g = \frac{A_{st}}{A_g} = \frac{8.00}{(16)^2} = 0.0313$$

$$0.01 < 0.0313 < 0.08 \qquad \text{(OK)}$$

2. From Table A–14, using a 13-in. core (column size less cover on each side), the maximum number of No. 9 bars is eight. Therefore, the number of longitudinal bars is satisfactory.

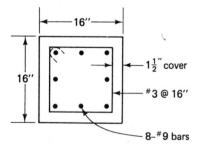

FIGURE 9-5 Sketch for Example 9–1

3. The maximum usable strength may now be calculated.

$$\phi P_{n(max)} = 0.80\phi[0.85f'_c(A_g - A_{st}) + f_y A_{st}]$$

$$= 0.80(0.70)[0.85(4)(256 - 8) + 60(8)]$$

$$= 741 \text{ kips}$$

4. Check the ties. Tie size of No. 3 is acceptable for longitudinal bar size up to No. 10. The spacing of the ties must not exceed the smaller of

48 tie bar diameters = 48($\frac{3}{8}$) = 18 in.

16 longitudinal bar diameters = 16(1.128) = 18 in.

Least column dimension = 16 in.

Therefore, the tie spacing is OK. The tie arrangement for this column may be checked by ensuring that the clear distance between longitudinal bars does not exceed 6 in. Clear space in excess of 6 in. would require additional ties in accordance with the ACI Code, Section 7.10.5.3.

$$\text{Clear distance} = \frac{16 - 2(1\frac{1}{2}) - 2(\frac{3}{8}) - 3(1.128)}{2}$$

$$= 4.4 \text{ in.} < 6 \text{ in.}$$

Therefore, no extra ties are needed.

EXAMPLE 9–2

Determine whether the column of cross section shown in Fig. 9–6 is adequate to carry a design axial load (P_u) of 550 kips. Assume small eccentricity. Check the spiral. f'_c = 4,000 psi and f_y = 60,000 psi.

1. From Table A–2, A_{st} = 5.53 in.², and from Table A–14 a diameter of 15 in. results in a circular area A_g = 176.7 in.² Therefore,

$$\rho_g = \frac{5.53}{176.7} = 0.0313$$

$$0.01 < 0.0313 < 0.08 \qquad \qquad \text{(OK)}$$

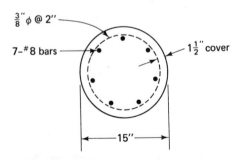

$\frac{3}{8}" \phi$ @ 2"

7-#8 bars

$1\frac{1}{2}"$ cover

15"

FIGURE 9-6   Sketch for Example 9–2

2. From Table A–14, using a 12-in. core, 7 No. 8 bars are satisfactory.

3. Find the maximum usable strength.

$$\phi P_{n\,(\text{max})} = 0.85\phi[0.85f_c'\,(A_g - A_{st}) + f_yA_{st}]$$

$$= 0.85(0.75)[0.85(4)(176.7 - 5.53) + 60(5.53)]$$

$$= 583 \text{ kips}$$

The strength is OK since 583 kips > 550 kips.

4. Check the spiral steel: the ⅜-in. diameter is OK (ACI Code, Section 7.10.4.2; and Table A–14). The minimum ρ_s is calculated using Table A–14 for the value of A_c:

$$\rho_{s\,(\text{min})} = 0.45\left(\frac{A_g}{A_c} - 1\right)\frac{f_c'}{f_y}$$

$$= 0.45\left(\frac{176.7}{113.1} - 1\right)\frac{4}{60} = 0.0169$$

$$\text{Actual } \rho_s = \frac{4A_{sp}}{D_cS} = \frac{4(0.11)}{12(2)} = 0.0183$$

$$0.0183 > 0.0169 \tag{OK}$$

The clear distance between spirals must not be in excess of 3 in. nor less than 1 in.:

$$\text{Clear distance} = 2.0 - \tfrac{3}{8} = 1\tfrac{5}{8} \text{ in.} \tag{OK}$$

The column is satisfactory for the specified conditions.

9–5 DESIGN OF SHORT COLUMNS—SMALL ECCENTRICITY

The design of reinforced concrete columns involves the proportioning of the steel and concrete areas and the selection of properly sized and spaced ties or spirals. Since the ratio of steel to concrete area must fall within a given range ($0.01 \leq \rho_g \leq 0.08$), the strength equation given in Section 9–2 is modified to include this term. For a tied column,

$$\phi P_{n\,(\text{max})} = 0.80\phi[0.85f_c'\,(A_g - A_{st}) + f_y(A_{st})]$$

$$\rho_g = \frac{A_{st}}{A_g}$$

from which

$$A_{st} \doteq \rho_gA_g$$

Therefore,

$$\phi P_{n\,(\text{max})} = 0.80\phi[0.85f_c'(A_g - \rho_gA_g) + f_y\,\rho_gA_g]$$

$$= 0.80\phi A_g[0.85f_c'(1 - \rho_g) + f_y\,\rho_g]$$

Since

$$P_u \leqslant \phi P_{n\,(max)}$$

an expression can be written for required A_g in terms of the material strengths, P_u and ρ_g. For tied columns,

$$\text{Required } A_g = \frac{P_u}{0.80\phi[0.85f'_c(1 - \rho_g) + f_y\rho_g]}$$

Similarly for spiral columns:

$$\text{Required } A_g = \frac{P_u}{0.85\phi[0.85f'_c(1 - \rho_g) + f_y\rho_g]}$$

It should be recognized that there can be many valid choices as to size of column that will provide the necessary strength to carry any load P_u. A low ρ_g will result in a larger required A_g, and vice versa. Other considerations will normally affect the practical choice of column size. Among them are architectural requirements and the desirability of maintaining column size from floor to floor so that forms may be reused.

The procedure for design of short columns for loads at small eccentricities is summarized in Section 9–6.

EXAMPLE 9–3

Design a square tied column to carry axial service loads of 320 kips dead load and 190 kips live load. There is no identified applied moment. Assume that the column is short. Use ρ_g about 0.03. $f'_c = 4,000$ psi and $f_y = 60,000$ psi.

1. Material strengths and approximate ρ_g are given.

2. The factored design load is

$$P_u = 1.7(190) + 1.4(320) = 771 \text{ kips}$$

3. The required gross column area is

$$\text{Required } A_g = \frac{P_u}{0.80\phi[0.85f'_c(1 - \rho_g) + f_y\,\rho_g]}$$

$$= \frac{771}{0.80(0.70)[0.85(4)(1 - 0.03) + 60(0.03)]}$$

$$\text{Required } A_g = 270 \text{ in.}^2$$

4. The required size of square column would be

$$\sqrt{270} = 16.4 \text{ in.}$$

Use a 16-in. square column. This choice will require that the actual ρ_g be slightly in excess of 0.03.

$$\text{Actual } A_g = (16 \text{ in.})^2 = 256 \text{ in.}^2$$

5. Load on the concrete area (this is approximate since ρ_g will increase slightly):

$$\text{Load on concrete} = 0.80\phi(0.85f_c')A_g(1 - \rho_g)$$

$$= 0.80(0.70)(0.85)(4)(256)(1 - 0.03)$$

$$= 473 \text{ kips}$$

Therefore, the load to be carried by the steel is

$$771 - 473 = 298 \text{ kips}$$

Since the maximum usable strength of the steel is $(0.80\phi A_{st}f_y)$, the required steel area may be calculated as

$$\text{Required } A_{st} = \frac{298}{0.80(0.70)(60)} = 8.87 \text{ in.}^2$$

We will distribute bars of the same size evenly around the perimeter of the column and must therefore select bars in multiples of four. Use 8 No. 10 bars ($A_{st} = 10.2$ in.²). Table A–14 indicates a maximum of 8 No. 10 bars for a 13-in. core (OK).

6. Design the ties. From Table A–14, select a No. 3 tie. The spacing must not be greater than

$$48 \text{ tie diameters} = 48(\tfrac{3}{8}) = 18 \text{ in.}$$

$$16 \text{ longitudinal bar diameters} = 16(1.27) = 20.3 \text{ in.}$$

$$\text{Least column dimension} = 16 \text{ in.}$$

Use No. 3 ties spaced 16 in. o.c. Check the arrangement with reference to Fig. 9–7. The clear space between adjacent bars in the same face is

$$\frac{16 - 3 - 0.75 - 3(1.27)}{2} = 4.22 \text{ in.} < 6.0 \text{ in.}$$

Therefore, no additional ties are required by the ACI Code, Section 7.10.5.3.

7. The design sketch is shown in Fig. 9–7.

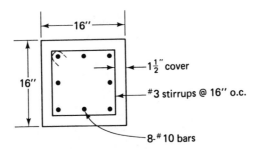

FIGURE 9-7 Design sketch for Example 9–3

EXAMPLE 9–4

Redesign the column of Example 9–3 as a circular spirally reinforced column.

1. Use $f'_c = 4{,}000$ psi, $f_y = 60{,}000$ psi, and ρ_g approximately 0.03.

2. $P_u = 771$ kips, as in Example 9–3.

3.
$$\text{Required } A_g = \frac{P_u}{0.85\,\phi\,[0.85f'_c(1 - \rho_g) + f_y\rho_g]}$$

$$= \frac{771}{0.85\,(0.75)\,[0.85\,(4)\,(1 - 0.03) + 60\,(0.03)]}$$

$$= 237 \text{ in.}^2$$

4. From Table A–14, use an 18 in.-diameter column.

$$\text{Actual } A_g = 254.5 \text{ in.}^2$$

5.
$$\text{Load on concrete} = 0.85\phi(0.85)f'_c A_g(1 - \rho_g)$$

$$= 0.85(0.75)(0.85)(4)(254.5)(1 - 0.03)$$

$$= 535 \text{ kips}$$

$$\text{Load on steel} = 771 - 535 = 236 \text{ kips}$$

$$\text{Required } A_{st} = \frac{236}{0.85\phi f_y}$$

$$= \frac{236}{0.85(0.75)60} = 6.17 \text{ in.}^2$$

Use 8 No. 8 bars ($A_{st} = 6.32$ in.²). Table A–14 indicates a maximum of twelve No. 8 bars for a circular core 15 in. in diameter (OK).

6. Design the spiral. From Table A–14, select a ⅜" diameter spiral. The spacing will be based on the required spiral steel ratio. A_c is from Table A–14.

$$\rho_{s(min)} = 0.45\left(\frac{A_g}{A_c} - 1\right)\frac{f'_c}{f_y}$$

$$= 0.45\left(\frac{254.5}{176.7} - 1\right)\frac{4}{60} = 0.0132$$

The maximum spiral spacing may be found by setting the actual ρ_s equal to $\rho_{s(min)}$.

$$\text{Actual } \rho_s = \frac{4A_{sp}}{D_c s}$$

from which

$$s_{max} = \frac{4A_{sp}}{D_c\,\rho_{s(min)}}$$

$$= \frac{4(0.11)}{15(0.0132)} = 2.22 \text{ in.}$$

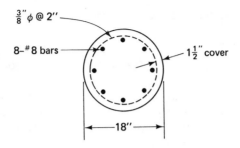

 is positioned here with labels:

$\frac{3}{8}"\phi$ @ 2"

8-#8 bars

$1\frac{1}{2}"$ cover

18"

FIGURE 9-8 Design sketch for Example 9-4

Use a spiral spacing of 2 in. The clear space between spirals must not be less than 1 in. nor more than 3 in.

$$\text{Clear space} = 2 - \frac{3}{8} = 1\frac{5}{8} \text{ in.} \qquad \text{(OK)}$$

7. The design sketch is shown in Fig. 9–8.

9-6 SUMMARY OF PROCEDURE FOR ANALYSIS AND DESIGN OF SHORT COLUMNS WITH SMALL ECCENTRICITIES

Analysis

1. Check ρ_g within acceptable limits.

$$0.01 \leqslant \rho_g \leqslant 0.08$$

2. Check the number of bars within acceptable limits for the clear space (see Table A–14). The minimum number is four for tied columns and six for spiral columns.

3. Calculate the maximum usable axial load strength $\phi P_{n\,(max)}$. See Section 9–2.

4. Check the lateral reinforcing. For ties, check size, spacing, and arrangement. For spirals, check size, ρ_s, and clear distance.

Design

1. Establish the material strengths. Establish the desired ρ_g (if any).

2. Establish the factored design load P_u.

3. Determine the required gross column area A_g.

4. Select the column dimensions. Use full-inch increments. ~or two inch

5. Find the load carried by the concrete and the load carried by the longitudinal steel. Determine the required longitudinal steel area. Select the longitudinal steel.

6. Design the lateral reinforcing—ties or spiral.

7. Sketch the design.

255

9-7 THE LOAD-MOMENT RELATIONSHIP

The equivalency between an eccentrically applied load and an axial load–moment combination may be shown with reference to Fig. 9-9. Assume that a force P_u is applied to a cross section at a distance e (eccentricity) from the centroid as shown in Fig. 9-9(a) and (b). Add equal and opposite forces P_u at the centroid of the cross section [Fig. 9-9(c)]. The original eccentric force P_u may now be combined with the upward force P_u to form a couple $P_u e$ which is a pure moment. This will leave remaining one force P_u acting downward at the centroid of the cross section. It can therefore be seen that if a force P_u is applied with an eccentricity e, the situation that results is identical to the case where an axial load of P_u and a moment of $P_u e$ are simultaneously applied [Fig. 9-9(d)]. If we define M_u as the factored design moment to be applied on a *compression* member along with a factored design load of P_u, the relationship between the two is

$$e = \frac{M_u}{P_u}$$

We have previously spoken of a column's nominal axial load strength at zero eccentricity P_0. Temporarily neglecting the strength reduction factors that must be applied to P_0, let us assume that we may apply a load P_u with zero eccentricity on a column of axial load strength P_0 (where $P_u = P_0$). If we now move the load P_u away from the zero eccentricity position a distance of e, the column must resist the load P_u and, in addition, a moment $P_u e$. Since $P_u = P_0$ (in our hypothetical case), it is clear that there is no additional strength to carry the moment and that the column is overloaded. In order that this particular column not be overloaded when subjected to load at eccentricity e, we must reduce P_u to the point where the column can carry both P_u and $P_u e$. The amount of the required decrease in P_u will depend on the magnitude of the eccentricity.

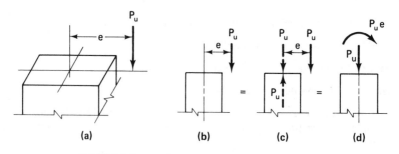

FIGURE 9-9 Load–moment–eccentricity relationship

256

The preceding discussion, of course, must be modified due to the fact that the ACI Code imposes $\phi P_{n(max)}$ as the upper limit of axial load strength for any column. Nevertheless, the strength of any column cross section is such that it will support a broad spectrum of load and moment (or load and eccentricity) combinations. In other words, we may think of a column cross section as having many different axial load strengths, each with its own related moment strength.

9-8 COLUMNS SUBJECTED TO AXIAL LOAD AT LARGE ECCENTRICITY

The 1971 ACI Code stipulated that compression members must be designed for an eccentricity e of not less than $0.05h$ for spirally reinforced columns or $0.10h$ for tied columns, but at least 1 in. in any case. h is defined as the overall dimension of the column. These specified minimum eccentricities were originally intended to serve as a means of reducing the axial load design strength of a section in pure compression. The effect of the minimum eccentricity requirement was to limit the *maximum* axial load strength of a compression member.

Under the 1977 ACI Code, as we have discussed, the maximum axial load strength $\phi P_{n(max)}$ is given by ACI Eqs. (10–1) and (10–2). These two equations apply when eccentricities are not in excess of *approximately* the $0.10h$ and $0.05h$ minimum eccentricity limits of the 1971 Code. Therefore, *small eccentricities* may be considered as those eccentricities up to about $0.10h$ and $0.05h$ for tied and spiral columns, respectively. We will consider cases of *large eccentricities* as those in which ACI Eqs. (10–1) or (10–2) no longer apply and where ϕP_n must be reduced below $\phi P_{n(max)}$.

The occurrence of columns subjected to eccentricities sufficiently large so that moment must be a design consideration is common. Even interior columns supporting beams of equal spans will receive unequal loads from the beams due to applied live load patterns. These unequal loads could mean that the column must carry both load and moment as shown in Fig. 9–10(a), and the resulting eccentricity of the loads could be appreciably in excess of our definition of small eccentricity. Another example of a column carrying both load and moment is shown in Fig. 9–10(b). In both cases, the rigidity of the joint will require the column to rotate along with the end of the beam that it is supporting. The rotation will induce moment in the column. A third and very practical example can be found in precast work [Fig. 9–10(c)], where the beam reaction can clearly be seen to be eccentrically applied on the column through the column bracket.

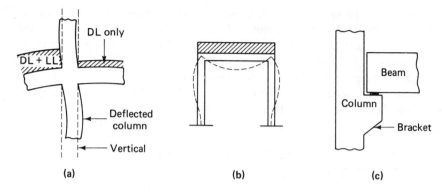

FIGURE 9-10 Eccentrically loaded columns

9-9 ANALYSIS OF SHORT COLUMNS—LARGE ECCENTRICITY

The first step in our investigation of short columns carrying loads at large eccentricity will be to determine the strength of a given column cross section which carries loads at various eccentricities. This may be thought of as an analysis process. For this development we will find the usable axial load strength ϕP_n, where P_n is defined as the nominal, or theoretical, axial load strength at a given eccentricity.

EXAMPLE 9-5

Find the usable strength ϕP_n for the tied column for the following conditions: (a) small eccentricity, (b) pure moment, (c) $e = 5$ in., and (d) the balanced condition. The column cross section is shown in Fig. 9-11. Assume a short column. Bending is about the Y-Y axis. $f'_c = 4,000$ psi and $f_y = 60,000$ psi.

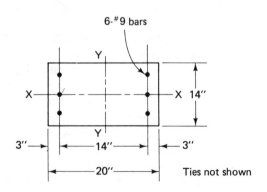

FIGURE 9-11 Column cross section for Example 9-5

258

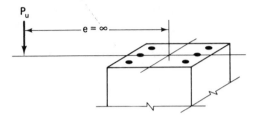

FIGURE 9-12 Columns of Example 9–5 loaded with pure moment

a. Small eccentricity

This is the case of analysis similar to Examples 9–1 and 9–2.

$$\phi P_n = \phi P_{n(max)}$$

$$= 0.80\phi \, [0.85f'_c(A_g - A_{st}) + f_y A_{st}]$$

$$= 0.80(0.70) \, [0.85(4)(280 - 6) + 60(6)]$$

$$= 723 \text{ kips}$$

b. Pure moment

This is the case where eccentricity e is infinite. The situation is shown in Fig. 9–12. We will find the usable moment strength M_R, since P_u and ϕP_n will both be zero.

With reference to Fig. 9–13(d), notice that for pure moment, the bars on the load side of the column are in compression while the bars on the side away from the load are in tension. The total tensile and compressive forces must be equal to each other. Since $A_s = A'_s$, A'_s *must* be at a stress less than yield. Assume A_s is at yield stress.

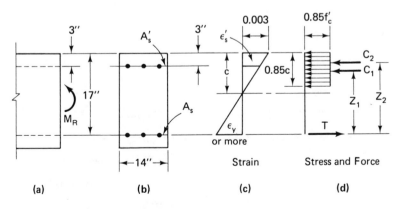

FIGURE 9-13 Example 9–5b, $e = \infty$

259

$$C_1 = \text{concrete compressive force}$$

$$C_2 = \text{steel compressive force}$$

$$T = \text{steel tensile force}$$

Referring to the compressive-strain-diagram portion of Fig. 9–13(c) and noting that the basic units are kips and inches:

$$\epsilon'_s = 0.003 \, \frac{c - 3}{c}$$

Since

$$f'_s = E_s \epsilon'_s$$

Substituting yields

$$f'_s = 29{,}000 \, (0.003) \left(\frac{c - 3}{c} \right)$$

$$= 87 \left(\frac{c - 3}{c} \right)$$

for equilibrium in Fig. 9–13(d),

$$C_1 + C_2 = T$$

Substituting into the foregoing and accounting for the concrete displaced by the compression steel, we obtain

$$(0.85f'_c)(0.85c)(b) + f'_s A'_s - 0.85f'_c(A'_s) = f_y A_s$$

$$(0.85)(4)(0.85c)(14) + 87 \left(\frac{c - 3}{c} \right)(3) - 0.85(4)(3) = 3(60)$$

Solving the preceding equation for the one unknown quantity c yields

$$c = 3.62 \text{ in.}$$

Therefore,

$$f'_s = 87 \left(\frac{3.62 - 3}{3.62} \right) = 14.90 \text{ ksi (compression)}$$

Summarizing the forces:

$$C_1 = 0.85f'_c \,(0.85)\,cb = 0.85(4)(0.85)(3.62)(14) = \quad 146.5 \text{ kips}$$

$$\text{Displaced concrete} = 0.85(f'_c)A'_s = 0.85(4)(3) = \quad -10.2 \text{ kips}$$

$$C_2 = f'_s A'_s = 14.90(3) = \quad \underline{\quad 44.7 \text{ kips}}$$

$$181.0 \text{ kips}$$

$$T = f_y A_s = 60(3.0) = 180 \text{ kips}$$

The slight error between T and $(C_1 + C_2)$ will be neglected.

Summarizing the internal couples:

$$M_{n(1)} = C_1 Z_1$$

$$= \frac{146.5}{12}\left[17 - \frac{0.85(3.62)}{2}\right]$$

$$= 188.8 \text{ ft-kips}$$

$$M_{n(2)} = C_2 Z_2$$

$$= (44.7 - 10.2)\left(\frac{14}{12}\right)$$

$$= 40.3 \text{ ft-kips}$$

$$M_n = M_{n(1)} + M_{n(2)}$$

$$= 188.8 + 40.3$$

$$= 229 \text{ ft-kips}$$

Applying the tied column strength reduction factor of 0.70 and neglecting, for now, ACI Code, Section 9.3.2(c), the usable strength becomes

$$M_R = \phi M_n = 0.70(229)$$

$$= 160 \text{ ft-kips}$$

c. e = 5 in.

This situation is shown in Fig. 9–14. In part **a**, all steel was in compression. In part **b**, the steel on the side of column away from the load was in tension. Therefore, there is some value of eccentricity at which this steel will change from tension to compression. Since this value of eccentricity is not known, the strain situation shown in Fig. 9–15 is *assumed* and will be verified (or disproved) by calculation.

The assumptions at ultimate load are:

1. Maximum concrete strain = 0.003.

2. $\epsilon'_s > \epsilon_y$. Therefore, $f'_s = f_y$.

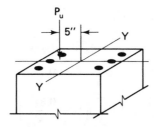

FIGURE 9-14   Example 9–5c, e = 5 in.

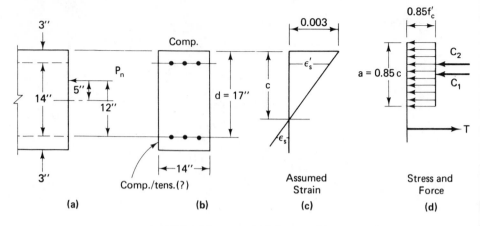

FIGURE 9-15   Example 9–5c, e = 5 in.

3. ϵ_s is tensile.

4. $\epsilon_s < \epsilon_y$. Therefore, $f_s < f_y$.

The unknown quantities are P_n and c.

Using basic units of kips and inches, the tensile and compressive forces are evaluated. Force C_2 is the force in the compressive steel accounting for the concrete displaced by the steel. T is with reference to Fig. 9–15(c):

$$C_1 = 0.85f_c'ab = 0.85(4)(0.85c)(14) = 40.46c$$

$$C_2 = f_yA_s' - 0.85f_c'A_s' = A_s'(f_y - 0.85f_c')$$

$$= 3[60 - 0.85(4)] = 169.8$$

$$T = f_sA_s = \epsilon_sE_sA_s = 87\left(\frac{d-c}{c}\right)A_s$$

$$= 87\left(\frac{17-c}{c}\right)3 = 261\left(\frac{17-c}{c}\right)$$

From Σ forces $= 0$ in Fig. 9–15:

$$P_n = C_1 + C_2 - T$$

$$= 40.46c + 169.8 - 261\left(\frac{17-c}{c}\right)$$

From Σ moments $= 0$, taking moments about T in Fig. 9–15:

$$P_n(12) = C_1\left(d - \frac{a}{2}\right) + C_2(14)$$

$$P_n = \frac{1}{12}\left[40.46(c)\left(17 - \frac{0.85c}{2}\right) + 169.8(14)\right]$$

The preceding two equations for P_n may be equated and the resulting cubic equation solved for c by trial or by some other iterative method. The solution will yield $c = 14.86$ in., which will result in a value for P_n (from either equation) of 733 kips. Therefore,

$$\phi P_n = 0.70(733)$$

$$= 513 \text{ kips}$$

Check the assumptions that were made.

$$\epsilon_s' = \left(\frac{14.86 - 3}{14.86}\right)(0.003) = 0.0024$$

$$\epsilon_y = 0.00207$$

Since $\epsilon_s' > \epsilon_y$,

$$f_s' = f_y \tag{OK}$$

Based on the location of the neutral axis, the steel away from the load *is* in tension and

$$f_s = 87\left(\frac{17 - 14.86}{14.86}\right) = 12.53 \text{ ksi} < 60 \text{ ksi} \tag{OK}$$

All assumptions are verified.

We may also determine the moment strength for an eccentricity of 5 in. We will term this M_R.

$$M_R = \phi P_n e = \frac{513(5)}{12}$$

$$= 214 \text{ ft-kips}$$

Therefore, the given column has a usable load–moment combination strength of 513 kips axial load and 214 ft-kips moment. This assumes that the moment is applied about the $Y-Y$ axis.

d. The balanced condition

Recall that the balanced condition exists when the concrete reaches a strain of 0.003 at the same time that the tension steel reaches its yield strain, as shown in Fig. 9–16(c). P_b is defined as nominal, or theoretical, axial load strength at the balanced condition, e_b is the associated eccentricity, and c_b is the distance from the compression face to the balanced neutral axis.

From Eq. (2–1) in Chapter 2, we may calculate the value of c_b:

$$c_b = \frac{87,000}{87,000 + f_y}(d) = \frac{87,000}{147,000}(17) = 10.06 \text{ in.}$$

We then may determine ϵ_s':

$$\epsilon_s' = \frac{7.06}{10.06}(0.003) = 0.0021$$

Since $0.0021 > 0.00207$, the compression steel has yielded and $f_s' = f_y = 60$ ksi.

Summarize the forces in Fig. 9–16(d) and let C_2 account for the force in the displaced concrete.

$$C_1 = 0.85(4)(0.85)(10.06)(14) = 407 \text{ kips}$$

$$C_2 = 60(3) - 0.85(4)(3) = 170 \text{ kips}$$

$$T = 60(3) = 180 \text{ kips}$$

$$P_b = C_1 + C_2 - T = 407 + 170 - 180$$

$$= 397 \text{ kips}$$

The value of e_b may be established by summing moments about T:

$$P_b(e_b + 7) = C_1\left(d - \frac{0.85c_b}{2}\right) + C_2(14)$$

$$397(e_b + 7) = 407\left[17 - \frac{0.85(10.06)}{2}\right] + 170(14)$$

from which $e_b = 12.0$ in. Therefore, at the balanced condition,

$$\phi P_b = 0.70(397) = 278 \text{ kips}$$

$$\phi P_b e_b = M_R = \frac{278}{12}(12) = 278 \text{ ft-kips}$$

The results of the four parts of Example 9–5 are tabulated and plotted in Fig. 9–17. All usable axial load strengths are denoted ϕP_n and all usable moment strengths are denoted $\phi P_n e$. This plot is commonly called an *interaction diagram*. It applies *only* to the column analyzed. However, it is a representation of *all* combinations of axial load and moment strengths for that column cross section.

In Fig. 9–17, any point *on* the solid line represents an allowable combination of load and moment. Any point *within* the solid line represents a load–moment combination which is also allowable, but for

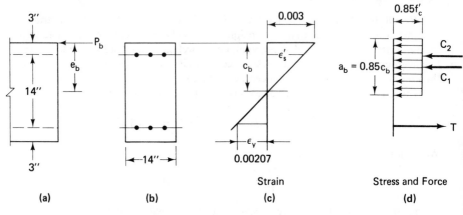

FIGURE 9-16   Example 9–5d, balanced condition

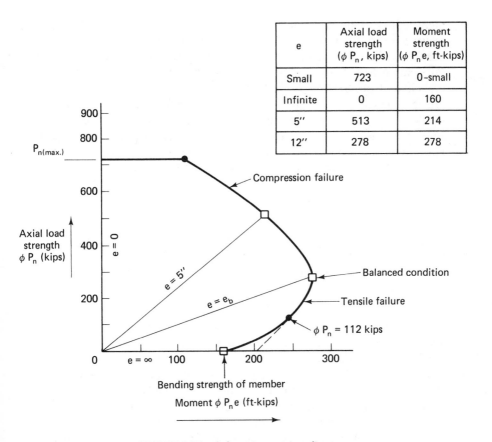

e	Axial load strength $(\phi P_n$, kips)	Moment strength $(\phi P_n e$, ft-kips)
Small	723	0-small
Infinite	0	160
5"	513	214
12"	278	278

FIGURE 9-17 Column interaction diagram

which this column is *overdesigned*. Any point *outside* the solid line represents an unacceptable load–moment combination or a load–moment combination for which this column is *underdesigned*. The value of $\phi P_{n(\text{max})}$ (which we calculated in part **a**) is superimposed on the plot as a horizontal line.

Radial lines from the origin represent various eccentricities. (Actually, the slopes of the radial lines are equal to $\phi P_n/\phi P_n e$, or $1/e$.) The intersection of the $e = e_b$ line with the solid line represents the balanced condition. Any eccentricity *less* than e_b will result in compression controlling the column, and any eccentricity *greater* than e_b will result in tension controlling the column.

It is apparent that the calculations involved with column loads at large eccentricities are involved and tedious. The previous examples were analysis examples. Design of a cross section using the calculation approach would be a trial-and-error method and would become exceedingly tedious. Therefore, design and analysis aids have been developed which shorten the process to a great extent. These aids may be found in the form of tables and/or charts. A chart approach is developed in ACI Publication SP-17A(78), *Design Handbook, Volume 2—Columns*. Several charts from SP-17A(78) have been reproduced in Appendix A. These are in accordance with provisions of the 1977 ACI Code concerning $\phi P_{n(\text{max})}$. The charts take on the basic form of the interaction diagram in Fig. 9–17. But, in order to generalize the charts so that they are applicable to more situations, some new terms are introduced. Referring to Chart A-25, the following definitions will be useful:

$$\rho_g = \frac{A_{st}}{A_g}$$

h = column dimension perpendicular to the bending axis (see sketch included on the chart)

γ = ratio of distance between centroid of outer rows of bars and thickness of the cross section, in the direction of bending.

Note that the load (vertical) axis and the moment (horizontal) axis scales are in terms of ϕP_n. The charts may be used to find the usable axial load strength of a column along with its associated moment strength. It will be recalled that the primary load–strength relationship for columns is that $P_u \leq \phi P_n$. The charts, then, are basically analysis aids. For design, they may be used for a trial-and-error procedure.

Charts for the three types of cross sections shown in Fig. 9–18 are included in Appendix A. These are for material strengths of $f_c' = 4{,}000$ psi and $f_y = 60{,}000$ psi only. For other combinations, see SP-17A(78). For those columns of the type shown in Fig. 9–18(a), the steel is assumed to be distributed equally in each face, that is, one-fourth of the total steel per face. ϕ factors of 0.70 or 0.75 (as applicable) are reflected in the charts and need not be included in any calculations.

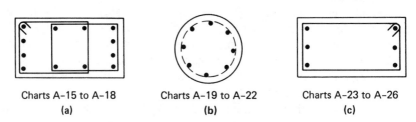

Charts A–15 to A–18   Charts A–19 to A–22   Charts A–23 to A–26
(a)   (b)   (c)

FIGURE 9-18   Design aid column types

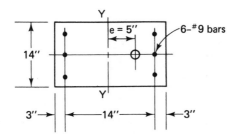

FIGURE 9-19 Sketch for Example 9-6

EXAMPLE 9-6

Using the charts of Appendix A, find the usable axial load strength ϕP_n for the column cross section shown in Fig. 9-19 at an eccentricity of 5 in. $f'_c = 4{,}000$ psi and $f_y = 60{,}000$ psi. Compare the results with Example 9-5(c).

First determine which chart to use based on the type of cross section, the material strengths and the factor γ.

$$\gamma h = 14 \text{ in.}$$

$$\gamma = \frac{14}{20} = 0.70$$

Chart A-24 is for $\gamma = 0.60$ and Chart A-25 is for $\gamma = 0.75$. For purposes of this example, use Chart A-25 rather than an interpolation between the two charts.

$$\rho_g = \frac{A_{st}}{A_g} = \frac{6.00}{20(14)} = 0.0214$$

$$0.01 < 0.0214 < 0.08$$

$$\frac{e}{h} = \frac{5.0}{20} = 0.25 \qquad\qquad \text{(OK)}$$

The quantities ρ_g and e/h are intersected on Chart A-25, as shown in Fig. 9-20. From the intersection, reading horizontally to the vertical axis on the left, one may read a value of

$$\frac{\phi P_n}{A_g} = 1.85$$

from which

$$\phi P_n = 1.85 A_g = 1.85(14)(20) = 518 \text{ kips}$$

The associated moment strength may be determined:

$$\phi P_n e = (518)\left(\frac{5}{12}\right) = 216 \text{ ft-kips}$$

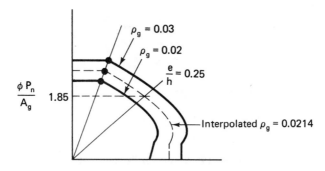

FIGURE 9-20 Use of Chart A-25

This compares favorably with the results of Example 9–5c. The use of interpolation between Charts A-24 and A-25 would produce an even closer correlation with the results of Example 9–5(c). The usable load–moment strengths for other eccentricities for this column may be easily found, since the ρ_g term remains constant.

EXAMPLE 9–7

Using the column charts, find the value of $\phi P_{n(\max)}$ for the column of Example 9–6.
From Example 9–6, $\rho_g = 0.0214$. Interpolating at the left vertical axis ($e/h = 0$), between lines representing $\rho_g = 0.02$ and 0.03 on either Chart A-24 or A-25 results in:

$$\frac{\phi P_n}{A_g} = 2.6$$

Therefore,

$$\phi P_{n(\max)} = 2.6 A_g = 2.6(14)(20) = 728 \text{ kips}$$

This compares favorably with 723 kips calculated in Example 9–5(a).
Note in Charts A-24 and A-25 that the horizontal lines which represent $\phi P_{n(\max)}$ intersect their respective strength curves at about e/h of 0.10. This confirms the fact that ACI Eqs. (10–1) and 10–2) give approximately the same results for maximum usable strength as did the minimum eccentricity requirements of the 1971 Code.

9–10 ϕ FACTOR CONSIDERATIONS

Columns discussed so far have had strength reduction factors applied in a straightforward manner. That is, $\phi = 0.75$ for spiral columns and $\phi = 0.70$ for tied columns. Eccentrically loaded columns, however, carry both axial load and moment. It will be recalled that beams (flexural members) were subject to a strength reduction factor of $\phi = 0.90$ applied to the flexural (moment) strength.

The Code provides for cases where columns are subjected to very small axial loads and large moments (thereby becoming predominantly bending members, or beams, where $\phi = 0.90$ would be expected). In Section 9.3.2(c) of the Code it is stipulated that for small values of axial load, ϕ may be increased linearly to 0.90 as ϕP_n decreases from $0.10 f'_c A_g$ to zero. Additional restrictions are that f_y does not exceed 60,000 psi, the reinforcement be symmetric, and that γ be not less than 0.70. It should be noted that this is a liberalizing stipulation of the Code (use of the original ϕ would result in a conservative design) and that it applies only to cases of small axial load. In Example 9–5,

$$0.10 f'_c A_g = 0.10(4)(14)(20) = 112 \text{ kips}$$

so that if ϕP_n were 112 kips, then $\phi = 0.70$; while if ϕP_n were zero, then $\phi = 0.90$. For pure moment, as in Example 9–5(b), $\phi P_n e$ (or M_R, the moment strength of the member) would become

$$\phi P_n e = 160\left(\frac{0.9}{0.7}\right) = 206 \text{ ft-kips}$$

The strength of the member, as governed by the Code, Section 9.3.2(c) is shown by the dashed line on Fig. 9–17, for values of ϕP_n between 112 kips and zero. The allowed increase in ϕ for small axial loads is reflected in Charts A-15 through A-26.

9-11 DESIGN OF ECCENTRICALLY LOADED COLUMNS (ACI CHART APPROACH)

The ACI Charts may be used for design of an eccentrically loaded column using a trial and error approach. Recall that the charts are primarily analysis aids.

EXAMPLE 9–8

Design a short square tied column to carry a design axial load P_u of 640 kips and a design moment M_u of 330 ft-kips. Place the reinforcing uniformly in the four faces. Design the ties. Use a 1½-in. cover. $f'_c = 4,000$ psi, and $f_y = 60,000$ psi.

1. Determine the eccentricity.

$$e = \frac{M_u}{P_u} = \frac{330(12)}{640} = 6.19 \text{ in.}$$

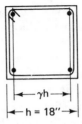

$\leftarrow \gamma h \rightarrow$

$\leftarrow h = 18'' \rightarrow$

FIGURE 9-21   Sketch for Example 9–8

2. Estimate a column size required based on the axial load P_u and $\rho_g = 0$. This will be satisfactory for an estimate.

$$\text{Required } A_g = \frac{P_u}{0.80\phi(0.85)f_c'} = \frac{640}{0.80(0.70)(0.85)(4)} = 336 \text{ in.}^2$$

Try an 18 in. $\times$ 18 in. column ($A_g = 324$ in.^{2}).

3. Determine e/h and the required $\phi P_n / A_g$ for chart use.

$$\frac{e}{h} = \frac{6.19}{18} = 0.34$$

Since, as a limit, P_u can equal ϕP_n:

$$\text{Required } \frac{\phi P_n}{A_g} = \frac{640}{18(18)} = 1.98$$

4. Determine which chart to use. In Fig. 9–21, assuming No. 3 ties and No. 8 main bars.

$$\gamma h = 18 - 2(1.5) - 2(0.38) - 2(0.5) = 13.25 \text{ in.}$$

$$\gamma = \frac{13.25}{18} = 0.736$$

Therefore, by inspection, use Chart A-17 ($\gamma = 0.75$).

5. From the chart, find the required ρ_g that is represented by the curve which passes through the intersection of $\phi P_n / A_g = 1.98$ and $e/h = 0.34$.

$$\text{Required } \rho_g = 0.05 \text{ (approximately)}$$

This is a high steel ratio and it will be difficult to fit the bars in the column. Therefore, try a 20 in. $\times$ 20 in. column.

$$\frac{e}{h} = \frac{6.19}{20} = 0.31$$

$$\text{Required } \frac{\phi P_n}{A_s} = \frac{640}{(20)(20)} = 1.60$$

$$\gamma h = 20 - 2(1.5) - 2(0.38) - 2(0.5) = 15.25$$

$$\gamma = \frac{15.25}{20} = 0.763$$

Use Chart A-17 ($\gamma = 0.75$):

Required $\rho_g = 0.027$

Required $A_{st} = \rho_g A_g = 0.027(20)(20) = 10.8$ in.^{2}

6. Select 12 No. 9 bars with $A_{st} = 12.0$ in.^{2}. Using Table A–14 check to see that 12 No. 9 bars may be used in one row in a 20-in. column. Maximum number of bars is 12 (OK).

7. Design the ties by selecting a No. 4 bar from Table A–14 and determining the spacing as follows:

$$48 \text{ tie-bar diameters} = 48(0.5) = 24 \text{ in.}$$

$$16 \text{ longitudinal bar diameters} = 16(1.128) = 18 \text{ in.}$$

$$\text{Least column dimension} = 20 \text{ in.}$$

Use No. 4 ties at 18 in. o.c.

8. The design sketch is shown in Fig. 9–22.

9–12 THE SLENDER COLUMN

Thus far, our design and analysis has been limited to short columns which require no consideration of necessary strength reduction due to the possibility of buckling. All compression members will experience the buckling phenomenon as they become longer and more flexible. These

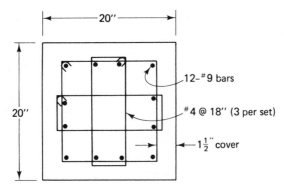

FIGURE 9-22   Design sketch for Example 9–8

are sometimes termed *slender columns*. A column may be categorized as *slender* if its cross-sectional dimensions are small in comparison to its unsupported length. The degree of slenderness may be expressed in terms of the *slenderness ratio*.

$$\frac{k\ell_u}{r}$$

where k = effective length factor for compression members (ACI Code, Section 10.11.2)
ℓ_u = unsupported length of compression member (ACI Code, Section 10.11.1)
r = radius of gyration of the cross section of the compression members; this may be taken as $0.30h$, where h is the overall dimension of a rectangular column in the direction of the moment; or $0.25D$, where D is the diameter of a circular column (ACI Code, Section 10.11.3)

The ACI Code prescribes no upper limit for the slenderness ratio of compression members as is found in other building codes. Nor does the Code define, per se, what constitutes a short column. However, it does offer an approximate approach whereby one can evaluate the effects of a high slenderness ratio. In effect, the Code recognizes that there are two categories of compression members: short members and slender members. The determination of the category to which a member belongs is not significant, but analysis to determine any strength reduction due to its slenderness is of significance and must be considered. The approximate analysis approach offered by the Code is based on a moment magnifier. This moment magnifier δ is similar to the one used for structural steel design.

The ACI Code, Section 10.11.5, stipulates that compression members shall be designed using the design axial load P_u and a magnified moment M_c:

$$M_c = \delta M_2 \qquad \text{ACI Eq. (10–6)}$$

where M_c = magnified factored design moment to be used for the design of a compression member
δ = moment magnification factor
M_2 = [for use in ACI Eq. (10–6)] the largest factored moment occurring anywhere along the member

Prior to determining the magnified design moment that results from the slenderness of a member, a relatively quick check should be made to determine whether the effects of slenderness may or may not be neglected. The Code, Section 10.11.4, stipulates that for compression members braced against sidesway, the effects of slenderness may be neglected when

$$\frac{k\ell_u}{r} < 34 - 12 \left(\frac{M_1}{M_2}\right)$$

For compression members not braced against sidesway, the effects of slenderness may be neglected when

$$\frac{k\ell_u}{r} < 22$$

M_2 is defined as the larger end moment of a column, while M_1 is the smaller end moment. M_2 is always positive, but M_1 is negative if the member is bent in double curvature and positive if it is bent in single curvature.

Compression members may be designed utilizing the approximate method of moment magnification for values of $k\ell_u/r$ up to 100. When $k\ell_u/r$ exceeds 100, a rigorous theoretical analysis must be performed and from a realistic approach is only practical with the aid of an electronic computer. This seldom occurs according to ACI-ASCE Committee 441, which surveyed typical reinforced concrete buildings to determine the normal range of variables found in the columns of such buildings. It was found that the practical upper limit on the slenderness ratio is approximately 70 in building columns.

Assuming that slenderness effects must be considered, the ACI Code, Section 10.11, provides slender column approximate design equations for the determination of the magnified moment M_c. Once M_c is established, the design (or analysis) proceeds exactly as previously outlined.

In the expression $M_c = \delta M_2$, δ is the moment magnification factor, which is empirically defined as

$$\delta = \frac{C_m}{1 - P_u/\phi P_c} \geq 1.0$$

where

$$P_c = \frac{\pi^2 EI}{(k\ell_u)^2}$$

and P_u is the factored axial design load as previously defined. C_m is a correction factor determined as follows. For members braced against sidesway and without transverse loads between supports,

$$C_m = 0.6 + 0.4 \left(\frac{M_1}{M_2}\right)$$

but never less than 0.4. For all other cases, C_m shall be taken as 1.0. For conservatism, C_m may always be taken as 1.0.

In the expression for P_c, the ACI Code, Section 10.11.5.2, allows EI to be conservatively calculated as

$$EI = \frac{E_c I_g}{2.5(1 + \beta_d)}$$

where E_c = modulus of elasticity of concrete
  I_g = gross concrete moment of inertia about centroidal axis neglecting reinforcement
  β_d = ratio of maximum factored dead load moment to maximum factored total load moment (always positive)

The question as to whether adequate bracing exists to prevent sidesway is difficult to answer and is usually a judgment factor. Sidesway itself may be described as a kind of deformation whereby one end of a member moves laterally with respect to the other. A simple example is a column fixed at one end and entirely free at the other (cantilever column or flagpole). Such a column will buckle as shown in Fig. 9–23. The upper end would move laterally with respect to the lower end. This lateral movement is termed sidesway. Since in reinforced concrete structures one is rarely concerned with single members but rather with rigid frames of various types, another basic example would be a simple portal frame as shown in Fig. 9–24. The upper end of the frame can move sideways since it is unbraced. The lower end may be theoretically

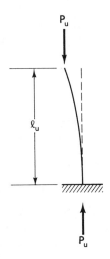

FIGURE 9-23   Fixed-free column

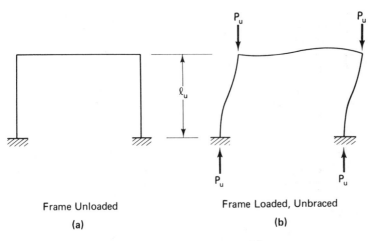

Frame Unloaded	Frame Loaded, Unbraced
(a)	(b)

FIGURE 9-24 Sidesway in portal frame

pin-connected or fully restrained or somewhere in between. Again the lateral movement is termed sidesway.

Sidesway has an effect on the loaded configuration of the compression members; hence the effective length is not only a function of the end condition and the actual length, but is also a function of whether or not sidesway exists. The effective length, by definition, is

$$k\ell_u$$

As an example, consider the simple case of a single member as shown in Fig. 9–25. The member braced against sidesway [Fig. 9–25(a)] has an effective length half that of the member without sidesway bracing [Fig. 9–25(b)] and has four times the axial load-carrying capacity.

As a rule, compression members free to buckle in a sidesway mode are appreciably weaker than when braced against sidesway. In fact, the ACI Code, Section 10.11.2.1, states that for compression members braced against sidesway, k shall be taken as 1.0 unless an analysis shows that a lower value may be used. For compression members not braced against sidesway, the effective length factor k must be greater than 1.0, depending on several variables.

Sidesway may be prevented in various ways in reinforced concrete structures. The common approach is to utilize walls or partitions sufficiently strong and rigid in their own planes to prevent the horizontal displacement. Another method is to use a rigid central core which is capable of resisting lateral loads as well as lateral displacements due to unsymmetrical loading conditions.

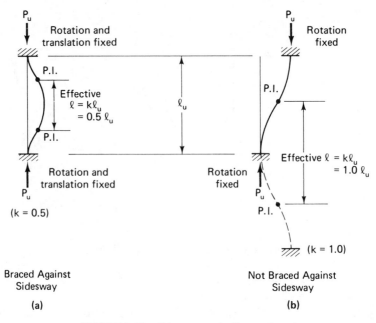

FIGURE 9-25   Sidesway and effective length

EXAMPLE 9–9

Assume that the short column designed in Example 9–8 has the following parameters:

a. Actual unbraced length $\ell_u = 15$ ft.

b. No bracing to resist sidesway.

c. The rotational restraint at the ends of the column (in combination with sidesway) is such that the effective length factor $k = 1.5$.

d. $\beta_d = 0.25$.

e. $C_m = 1.0$ (conservative).

Compute the magnified design moment M_c resulting from the slenderness of the member. $f_c' = 4,000$ psi, $f_y = 60,000$ psi, $P_u = 640$ kips, and $M_u = 330$ ft-kips.

1. Determine if slenderness must be considered.

$$r = 0.30h = 0.30(20) = 6.0 \text{ in.}$$

$$\frac{k\ell_u}{r} = \frac{1.5(15)(12)}{6.0} = 45$$

Since $45 > 22$, slenderness effects must be considered.

2. Evaluate the various terms required for the expression for δ.

$$I_g = \frac{h^4}{12} = \frac{(20)^4}{12} = 13{,}333 \text{ in.}^4$$

Therefore,

$$EI = \frac{E_c I_g}{2.5(1 + \beta_d)} = \frac{3{,}605(13{,}333)}{2.5(1 + 0.25)} = 15{,}380{,}000 \text{ kip-in.}^2$$

from which

$$P_c = \frac{\pi^2 EI}{(k\ell_u)^2} = \frac{\pi^2(15{,}380{,}000)}{[1.5(15)(12)]^2} = 2{,}082 \text{ kips}$$

3.

$$\delta = \frac{C_m}{1 - P_u/\phi P_c} = \frac{1.0}{1-[640/0.70(2082)]} = 1.78$$

4. The magnified factored design moment may now be calculated (where M_2 is taken as M_u).

$$M_c = \delta M_2 = 1.78(330) = 588 \text{ ft-kips}$$

5. The 20 in. × 20 in. column must then be checked to see if it can support this magnified moment along with the axial load of 640 kips. If not, the column must be redesigned.

EXAMPLE 9–10

Restating Example 9–8, assume that:

a. Actual unbraced length $\ell_u = 10$ ft.

b. The column is part of a frame braced against sidesway.

c. The column is bent in single curvature.

d. The moments are equal at each end of the member.

Determine if the member is satisfactory. $f_c' = 4{,}000$ psi, $f_y = 60{,}000$ psi, $P_u = 640$ kips, and $M_u = 330$ ft-kips.

1. Determine if slenderness must be considered.

$$r = 0.30h = 0.30(20) = 6.0 \text{ in.}$$

Since frame is braced against sidesway, $k = 1.0$ and

$$\frac{k\ell_u}{r} = \frac{1.0(10)(12)}{6.0} = 20$$

Since the moments are equal at each end of the member and the member is bent in single curvature, the ratio M_1/M_2 will be a positive value and

$$34 - 12\left(\frac{M_1}{M_2}\right) = 22$$

20 < 22; therefore, slenderness effects may be neglected (see the ACI Code, Section 10.11.4.1) and the column as designed in Example 9–8 is satisfactory.

PROBLEMS

9–1. Compute the maximum usable axial load strength of the tied columns shown. Assume that the columns are short. Check the tie size and spacing. $f'_c = 3,000$ psi and $f_y = 40,000$ psi.

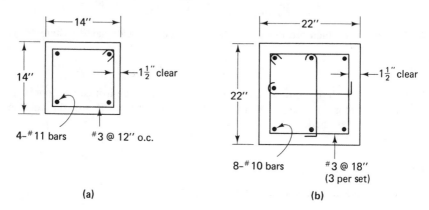

(a)   (b)

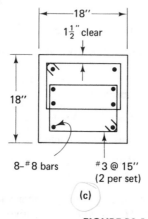

(c)

FIGURE P9-1

9–2. Compute the maximum usable axial load strength for the tied column shown. The column is short. Check the tie size and spacing. $f_c' = 4,000$ psi and $f_y = 60,000$ psi.

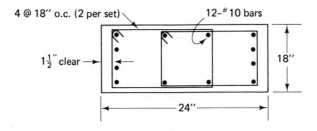

FIGURE P9-2

9–3. Find the maximum axial compressive service loads that the column of cross section shown can carry. The column is short. Assume that the service dead load and live load are equal. Check the ties. $f_c' = 4,000$ psi and $f_y = 60,000$ psi.

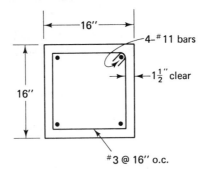

FIGURE P9-3

9–4. A short circular spiral column having a diameter of 18 in. is reinforced with 8 No. 9 bars. The cover is 1½ in. and the spiral is ⅜ in. in diameter spaced 2 in. o.c. Find the maximum usable axial load strength and check the spiral. $f_c' = 3,000$ psi and $f_y = 40,000$ psi.

9–5. Same as Problem 9–4, but $f_c' = 4,000$ psi and $f_y = 60,000$ psi.

9–6. Compute the maximum axial compressive service live load that may be placed on the column shown. The column is short and is subjected to an axial service dead load of 200 kips. Check the ties. $f_c' = 3,000$ psi and $f_y = 40,000$ psi.

9–7. Design a short, square tied column to carry a total factored design load P_u of 700 kips. Space and practical limitations require a column size of 18 in. × 18 in. Use $f_c' = 3,000$ psi and $f_y = 40,000$ psi.

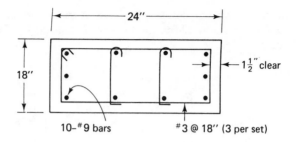

FIGURE P9-6

9–8. Design a short, square tied column for service loads of 200 kips dead load and 160 kips live load. Use ρ_g of about 0.04. $f'_c = 3,000$ psi and $f_y = 40,000$ psi. Assume that eccentricity is small.

9–9. Same as Problem 9–8, but use $f'_c = 4,000$ psi, $f_y = 60,000$ psi, and a ρ_g of about 0.03.

9–10. Design a short, circular spiral column for service loads of 175 kips dead load and 325 kips live load. Assume that the eccentricity is small. $f'_c = 4,000$ psi and $f_y = 60,000$ psi.

9–11. The short spiral column of cross section shown is to carry both axial load and moment. Find ϕP_n if $e = 6.0$ in. $f'_c = 4,000$ psi and $f_y = 60,000$ psi.

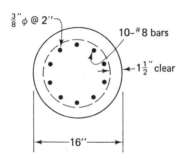

FIGURE P9-11

9–12. For the short column of cross section shown, find ϕP_n and e_b at the balanced condition using basic principles. Check by using the charts. $f'_c = 4,000$ psi and $f_y = 60,000$ psi. Bending is about the strong axis.

9–13. Design a short, square tied column to carry a factored axial design load P_u of 560 kips and a factored design moment M_u of 280 ft-kips. Place the longitudinal reinforcing uniformly in the four faces. $f'_c = 4,000$ psi and $f_y = 60,000$ psi.

9–14. Same as Problem 9–13, but design a circular spiral column.

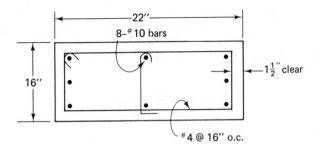

FIGURE P9-12

9–15. For the short tied column of cross section shown, find the usable axial load strength ϕP_n for an eccentricity of 14 in. Use $f'_c = 4,000$ psi and $f_y = 60,000$ psi. Assume that the ties and bracket design are adequate.

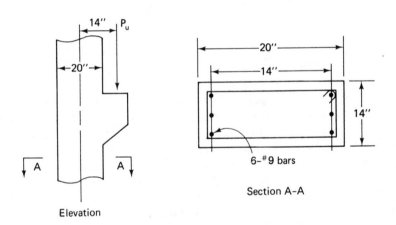

FIGURE P9-15

10

Footings

10-1 INTRODUCTION

The purpose of the structural portion of every building is to transmit
applied loads safely from one part of the structure to another. The loads
pass from their point of application into the *superstructure*, then to the
foundation, and then into the underlying supporting material. We have
discussed the superstructure and foundation walls to some extent. The
foundation is generally considered to be the entire lowermost supporting
part of the structure. Normally, a *footing* is the last, or nearly the last,
structural element of the foundation through which the loads pass. A
footing has as its function, the requirement of spreading out the superimposed load so as not to exceed the safe capacity of the underlying
material, usually soil, to which it delivers the load. Additionally, the
design of footings must take into account certain practical and, at times,
legal considerations. Our discussion will be concerned only with spread
footings depicted in Fig. 10-1. Footings supported on piles will not be
discussed.

The more common types of footings may be categorized as follows:

1. *Individual column footings* [Fig. 10-1(a)] are often termed *isolated
spread footings* and are generally square. However, if space limitations
exist, the footing may be rectangular in shape.

2. *Wall footings* support walls, which may be either bearing or
nonbearing walls [Fig. 10-1(b)].

Square Column Footing and Wall Footing.

3. *Combined footings* support two or more columns and may be either rectangular or trapezoidal in shape, [Fig. 10–1(c)(d)]. If two isolated footings are joined by a *strap beam*, the footing is sometimes called a *cantilever footing* [Fig. 10–1(e)].

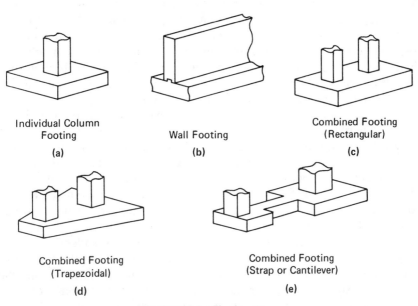

Individual Column
Footing

(a)

Wall Footing

(b)

Combined Footing
(Rectangular)

(c)

Combined Footing
(Trapezoidal)

(d)

Combined Footing
(Strap or Cantilever)

(e)

FIGURE 10-1 Footing types

4. *Mat foundations* are large continuous footings that support all columns and walls of a structure. They are commonly used where undesirable soil conditions prevail [Fig. 10–1(f)].

5. *Pile caps* or *pile footings* serve to transmit column loads to a group of piles, which will, in turn, transmit the loads to the supporting soil through friction or to underlying rock in bearing. [Fig. 10–1(g)].

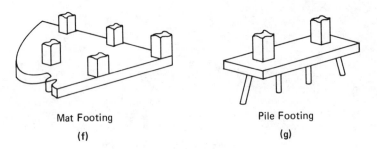

Mat Footing

(f)

Pile Footing

(g)

FIGURE 10-1   Footing types (continued)

10–2 WALL FOOTINGS

Wall footings are commonly required to support direct concentric loads. An exception to this is the footing for a retaining wall. A wall footing may be of either plain or reinforced concrete. Since it has bending in only one direction, it is generally designed in much the same manner as is a one-way slab, by considering a typical 12-in.-wide strip along the length of wall. Footings carrying relatively light loads on well-drained cohesionless soil are often made of plain concrete.

A wall footing under concentric load behaves similar to a cantilever beam, where the cantilever extends out from the wall and is loaded in an upward direction by the soil pressure. The flexural tensile stresses that are induced in the bottom of the footing are acceptable for an *unreinforced* concrete footing provided that these stresses do not exceed $5\phi\sqrt{f'_c}$ (ACI Code, Section 15.11.1). This same section stipulates that the average shear stress for one-way beam action in an unreinforced concrete footing must not exceed $2\phi\sqrt{f'_c}$.

In a *reinforced* concrete wall footing, the behavior is identical to that described above. However, reinforcing steel is placed in the bottom of the footing in a direction perpendicular to the wall, thereby resisting the induced flexural tension, similar to a reinforced concrete beam or slab.

In either case, the cantilever action is based on the maximum bending moment occurring at the face of the wall if the footing supports a concrete wall, or at a point halfway between the middle of the wall and the face of the wall if the footing supports a masonry wall. This difference stems primarily from the fact that a masonry wall is somewhat less rigid than is a concrete wall.

For each type of wall, the critical section for shear in the footing may be taken at a distance from face of wall equal to the effective depth of the footing.

EXAMPLE 10–1

Design a plain concrete wall footing to carry a 12-in. concrete block masonry wall as shown in Fig. 10–2. The service loading may be taken as 10 kips/ft dead load (which includes the weight of the wall) and 20 kips/ft live load. Use $f'_c = 3,000$ psi. The allowable soil pressure is 5,000 psf. Weight of earth $w_e = 100$ lb/ft³.

Compute the factored design load.

$$w_u = 1.4w_{DL} + 1.7w_{LL}$$

$$= 1.4(10) + 1.7(20)$$

$$= 48 \text{ kips/ft} = 48,000 \text{ lb/ft}$$

Assume a footing thickness of 3 ft 0 in. This will be checked later in the design. The footing weight is

$$150(3) = 450 \text{ psf}$$

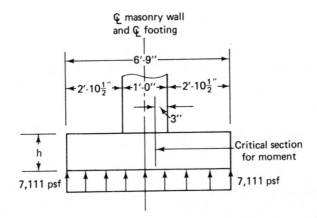

FIGURE 10-2   Plain concrete wall footing, Example 10–1

Assuming the bottom of footing to be 4 ft 0 in. below the finished ground line, the weight of the soil on top of the footing is

$$(1) (100) = 100 \text{ psf}$$

Therefore the net allowable soil pressure for superimposed service loads is

$$5,000 - 450 - 100 = 4,450 \text{ psf}$$

The maximum allowable soil pressure for strength design must now be found. It must be modified in a manner consistent with the modification of the service loads. This may be accomplished by multiplying by the ratio of the total design load (48 kips) to the total service load (30 kips). We then obtain a maximum allowable soil pressure solely for use in the strength design of the footing. This soil pressure should not be construed as an actual allowable soil pressure.

$$4,450\left(\frac{48}{30}\right) = 7,120 \text{ psf}$$

We may now determine the required footing width.

$$\frac{48,000}{7,120} = 6.74 \text{ ft} \qquad \text{Use 6 ft 9 in.}$$

Determine the design soil pressure to be used for the footing design if the footing width is 6 ft 9 in.

$$\frac{48,000}{6.75} = 7,111 \text{ psf}$$

With the design soil pressure known, the bending moment in the footing may be calculated. For concrete block masonry walls, the critical section for moment should be taken at the quarter point of wall thickness (ACI Code, Section 15.4.2b).

With reference to Fig. 10–2, the design moment is determined as follows:

$$M_u = \frac{7,111(3.125)^2}{2} = 34,722 \text{ ft-lb}$$

The unreinforced footing is a homogeneous rectangular section (12 in. in width). Therefore, the concrete tensile stress f_t at the bottom of the footing is

$$f_t = \frac{M_u}{S}$$

where S = section modulus = $\dfrac{bh^2}{6}$

$b = 12$ in.
h = total footing thickness (in.)

From the ACI Code, Section 15.11.1, the maximum tensile stress in a plain concrete footing may not exceed $5\phi\sqrt{f_c'}$, where $\phi = 0.65$ (ACI Code, Section 9.3.2f). Therefore,

$$\text{maximum } f_t = 5(0.65)\sqrt{3,000} = 178 \text{ psi}$$

The required footing thickness may now be determined.

$$\text{Required } S = \frac{M_u}{f_t}$$

Substituting for S, we obtain

$$\text{Required } \frac{bh^2}{6} = \frac{34,722(12)}{178}$$

$$\text{Required } h^2 = \frac{34,722(12)(6)}{12(178)} = 1,170 \text{ in.}^2$$

$$\text{Required } h = 34.2 \text{ in.}$$

It is common practice to assume that the bottom 1 or 2 in. of concrete placed against the ground may be of poor quality and therefore may be neglected for strength purposes. The total required footing thickness may then be determined:

$$34.2 + 2.0 = 36.2 \text{ in.} \qquad \text{say, } 36 \text{ in.}$$

This checks with our assumed footing thickness and weight. No revisions are necessary.

Shear is generally of little significance in plain concrete footings because of the large concrete footing thickness.

With reference to Fig. 10–3, if the critical section for shear is considered a distance equal to the effective depth of the member from the face of wall, and if the effective depth is taken as 34.2 in. (the required h), it may be observed that the critical section is almost at the edge of footing. Therefore, shear may be neglected, since at this section it will be very low.

From the ACI Code, Section 15.11.1, the average shear stress for unreinforced concrete footings on soil must not exceed $2\phi\sqrt{f_c'}$ for one-way beam action (the behavior exhibited by plain concrete continuous wall footings).

$$2\phi\sqrt{f_c'} = 2(0.85)\sqrt{3000} = 93.1 \text{ psi}$$

No computations for shear are necessary for this particular footing.

It is common practice to utilize some *longitudinal* steel in continuous wall footings whether or not *transverse* steel is present. This will somewhat enhance

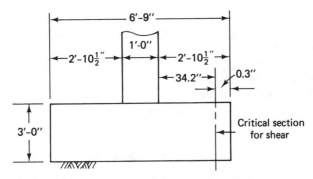

FIGURE 10-3 Plain concrete wall footing, Example 10–1

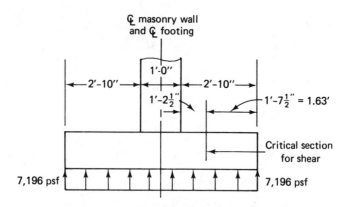

Label annotations on figure:

ℂ masonry wall and ℂ footing

2'-10'' 1'-0'' 2'-10''

1'-2½''

1'-7½'' = 1.63'

Critical section for shear

7,196 psf   7,196 psf

FIGURE 10-4 Reinforced concrete wall footing, Example 10-2

the structural integrity by limiting differential movement between parts of the footing should transverse cracking occur. It will also lend some flexural strength in the longitudinal direction. The rationale for this stems from the many uncertainties that actually exist in both the supporting soil and the applied loads.

As a guide, this longitudinal steel may be computed in a manner similar to that for temperature and shrinkage steel in a one-way slab, where

$$\text{Required } A_s = 0.0020\, bh$$

$$= 0.0020(12)(36)$$

$$= 0.86 \text{ in.}^2/\text{ft of width}$$

Number 6 bars at 6 in. o.c. would furnish 0.88 in.^{2} and would satisfy any longitudinal steel requirement. However, it is obvious that this large steel requirement does not lend itself to an economical footing design but rather leads us to conclude that one should question the use of plain concrete wall footings when superimposed loads are heavy.

A redesign utilizing a reinforced concrete wall footing is accomplished in the following problem.

EXAMPLE 10–2

Design a reinforced concrete wall footing to carry a 12-in. concrete block masonry wall as shown in Fig. 10–4. The service loading is 10 kips/ft dead load (which includes the weight of wall) and 20 kips/ft live load. $f'_c = 3,000$ psi, $f_y = 40,000$ psi, weight of earth = 100 lb/ft^3, and the allowable soil pressure = 5,000 psf. The bottom of the footing is to be 4 ft 0 in. below finished ground line.

Compute the factored design load.

$$w_u = 1.4\, w_{DL} + 1.7\, w_{LL}$$

$$= 1.4(10) + 1.7(20)$$

$$= 48 \text{ kips/ft} = 48,000 \text{ lb/ft}$$

Assume a total footing thickness = 18 in. Therefore, the footing weight is

$$150(1.5) = 225 \text{ psf}$$

Since the bottom of the footing is to be 4 ft below finished ground line, there will be 30 in. of earth on top of the footing. This depth of earth has a weight of

$$\frac{30(100)}{12} = 250 \text{ psf}$$

The net allowable soil pressure for superimposed service loads is

$$5,000 - 225 - 250 = 4,525 \text{ psf}$$

As in Example 10–1, we will use the ratio of factored load to service load to determine the soil pressure for strength design

$$\frac{48(4,525)}{30} = 7,240 \text{ psf}$$

The required footing width is

$$\frac{48,000}{7,240} = 6.63 \text{ ft} \qquad \text{Use 6 ft 8 in.}$$

The soil pressure to be used for the footing design is

$$\frac{48,000}{6.67} = 7,196 \text{ psf}$$

The actual effective depth d for the footing is determined by subtracting the concrete cover and one-half of the bar diameter (No. 8 assumed) from the total thickness:

$$d = 18 - 3 - 0.5 = 14.5 \text{ in.}$$

Since required thicknesses of reinforced concrete footings are generally controlled by shear requirements, the shear should be checked first.

With reference to Fig. 10–4, since the wall footing carries shear in a manner similar to that of a one-way slab or beam, the critical section for shear will be taken at a distance equal to the effective depth of the footing (14.5 in.) from face of wall (ACI Code, Section 11.11).

$$V_u = 1.63(1)(7,196) = 11,729 \text{ lb/ft of wall}$$

The total nominal shear strength V_n is the sum of the shear strength of the concrete V_c and the shear strength of any shear reinforcing V_s. Assuming no shear reinforcing, we obtain

$$\phi V_n = \phi V_c = \phi 2\sqrt{f_c'}bd$$

$$= 0.85(110)(12)(14.5)$$

$$= 16,269 \text{ lb}$$

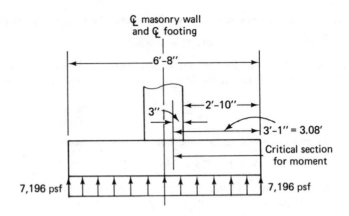

FIGURE 10-5 Reinforced concrete wall footing, Example 10–2

Therefore,

$$V_u < \phi V_n$$

Therefore, the assumed thickness of footing is satisfactory for shear, no shear reinforcing is required, and no revisions are necessary with respect to footing weight.

With reference to Fig. 10–5, the critical section for moment is taken at the quarter point of wall thickness (ACI Code, Section 15.4.2b). The maximum design moment, assuming the footing to be a cantilever beam, is

$$M_u = \frac{7,196(3.08)^2}{2} = 34,132 \text{ ft-lb}$$

The required area of tension steel is then determined in the normal way using $d = 14.5$ in. and $b = 12$ in.

$$\text{Required } \overline{k} = \frac{M_u}{\phi bd^2} = \frac{34.132(12)}{0.9(12)(14.5)^2}$$

$$= 0.1804 \text{ ksi}$$

From Table A–6, the required $\rho = 0.0047$. This is less than ρ_{min} of 0.0050, therefore, use ρ_{min}.

$$\text{Required } A_s = \rho bd = 0.0050(12)(14.5)$$

$$= 0.87 \text{ in.}^2/\text{ft of wall}$$

Use No. 7 at 8 in. o.c. ($A_s = 0.90 \text{ in.}^2$).

The development length should be checked for the bars selected. From Table A–12, the basic development length for the No. 7 bars is 18 in. Applicable modification factors are

Spacing in excess of 6 in.: 0.80

$$\frac{\text{Required } A_s}{\text{Provided } A_s} = \frac{0.87}{0.90} = 0.97$$

Development length required is

$$\ell_d = 18(0.80)(0.97) = 14 \text{ in.}$$

The development length provided, measured from the critical section and allowing for a 3-in. cover, is 34 in. Since 34 in. > 14 in., the development length is adequate.

Although not specifically required in footings by the ACI Code, longitudinal steel will be provided on the same basis as for one-way slabs (Section 7.12).

$$\text{Required } A_s = 0.0020bh = 0.0020(12)(18)$$

$$= 0.43 \text{ in.}^2/\text{ft}$$

Use No. 5 bars at 8 in. o.c. ($A_s = 0.46$ in.²). The footing design is shown in Fig. 10–6.

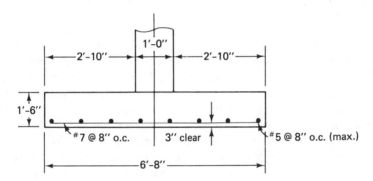

FIGURE 10-6 **Design sketch for Example 10–2**

10-3 WALL FOOTINGS UNDER LIGHT LOADS

A relatively common situation is one in which a lightly loaded wall is supported on average soil. A design as previously indicated in this chapter would result in a very small footing thickness and width.

In such a situation, experience has shown that for footings carrying plain concrete or block masonry walls, the minimum recommended dimensions shown in Fig. 10–7 should be used. The minimum depth or thickness of footing should be 8 in. but not less than the wall thickness. The minimum width of footing should equal twice the wall thickness.

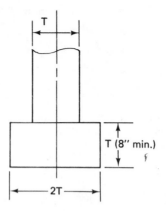

FIGURE 10-7 Recommended minimum footing dimensions for walls carrying light loads

10-4 INDIVIDUAL REINFORCED CONCRETE FOOTINGS FOR COLUMNS

This type of footing, also termed an *isolated spread footing*, is probably the most common, simplest, and most economical of the various types of footings used for structures. Individual column footings are generally square in plan. However, rectangular shapes are sometimes used where dimensional limitations exist. Basically, the footing is a slab that directly supports a column. At times a pedestal is placed between a column and a footing so that the base of the column need not be set below grade.

The footing behavior under concentric load is that of two-way cantilever action extending out from the column or pedestal. The footing is loaded in an upward direction by the soil pressure. Tensile stresses are induced in each direction in the bottom of the footing. Therefore, the footing is reinforced by two layers of steel perpendicular to each other and parallel to the edges. The required footing–soil contact area is a function of, and determined by, the allowable soil bearing pressure and the column loads being applied to the footing.

Shear

Since the footing is subject to two-way action, two different types of shear strength must be considered: two-way (punching) shear and one-way (beam) shear. The footing thickness (depth) is generally established by the shear requirements. The two-way shear is commonly termed *punching shear*, since the column or pedestal tends to punch through the footing, inducing stresses around the perimeter of the column or pedestal.

Tests have verified that if failure occurs, the fracture takes the form of a truncated pyramid with sides sloping away from the face of column or pedestal. The critical section for this two-way shear is taken perpendicular to the plane of the footing and located so that its perimeter b_0 is a minimum but does not come closer to the edge of the column or pedestal than one-half the effective depth of the footing (ACI Code, Section 11.11.1.2).

The design of the footing for two-way action is based on a shear strength V_n and is not to be taken greater than V_c unless shear reinforcement is provided. V_c may be determined from the ACI Code, Section 11.11.2:

$$V_c = \left(2 + \frac{4}{\beta_c}\right)\sqrt{f'_c}\, b_0 d \qquad \text{ACI Eq. (11–37)}$$

where β_c = ratio of long side to short side of concentrated load
or reaction area
b_0 = perimeter of critical section for two-way shear action
in footing

and V_c, f'_c, and d are as previously defined. V_c may not be in excess of $4\sqrt{f'_c}\, b_0 d$. The introduction of shear reinforcement in footings is impractical and undesirable purely on an economic basis and it is general practice to design footings based on the shear strength solely of the concrete.

The one-way (or beam) action behavior is analogous to that of a beam or one-way slab. The critical section for this one-way shear is taken on a vertical plane extending across the entire width of the footing and located at a distance equal to the effective depth of the footing from the face of concentrated load or reaction area (ACI Code, Section 11.11.1.1). As in a beam or one-way slab, the shear strength provided by the footing concrete may be taken as

$$V_c = 2\sqrt{f'_c}\, b_w d$$

For both one-way and two-way action, if we assume no shear reinforcement, the basis for the shear design will be $V_u \leq \phi V_n$, where $V_n = V_c$.

Moment and Development of Bars

The size and spacing of the footing reinforcing steel is primarily a function of the bending moment induced by the net upward soil pressure. The footing behaves as a cantilever beam in two directions. It is loaded by the soil pressure. The fixed end, or critical section for the bending moment, is located as follows (ACI Code, Section 15.4.2):

1. At the face of the column or pedestal for footings supporting a concrete column or pedestal [see Fig. 10–8(a)].

2. Halfway between the face of column and edge of steel base, for footings supporting a column with steel base plate [see Fig. 10-8(b)].

The ACI Code, Section 15.6.3, stipulates that the critical section for development length of footing reinforcement shall be assumed to be at the same location as the critical section for bending moment.

Transfer of Load from Column into Footing

All loads applied to a column must be transferred to the top of the footing (through a pedestal if there is one) by compression in the concrete and/or by reinforcement.

The bearing strength of the concrete contact area of supporting and supported member cannot exceed $\phi(0.85f'_c A_1)$ as directed by the ACI Code, Section 10.16.1. When the supporting surface is wider on all sides than is the loaded area, the design bearing strength on the loaded area may be multiplied by $\sqrt{A_2/A_1}$, where

A_2 = maximum area of the portion of the supporting
surface that is geometrically similar to
and concentric with the loaded area

A_1 = loaded area

The quantity $\sqrt{A_2/A_1}$ cannot exceed 2.0 (ACI Code, Section 10.16.1.1). Therefore, in no case can the design bearing strength for the loaded area be in excess of

$$\phi(0.85f'_c A_1)\,(2)$$

where $\phi = 0.70$ for bearing on concrete and f'_c is as previously defined.

It is quite common for the footing concrete to be of a lower strength (f'_c) than the supported column concrete. This suggests that both supporting and supported members should be considered in determining load transfer.

Where a reinforced concrete column cannot transfer the load entirely by bearing, the excess load must be transferred by reinforcement. This may be accomplished by furnishing dowels, one per column bar if necessary, or if this is still not enough, by adding extra dowels or dowels larger in diameter than the column bar size. Where dowels are used, their diameter must not exceed the diameter of the column bars by more than 0.15 in. (ACI Code, Section 15.8.4.2).

To provide a positive connection between a reinforced concrete column and footing (whether dowels are required or not), the ACI Code,

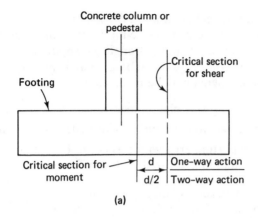

(a)

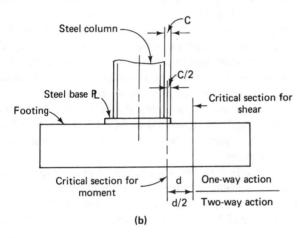

(b)

FIGURE 10-8 Critical sections for design of reinforced concrete footings supporting columns or pedestals

Section 15.8.4.1, requires a minimum area of reinforcement crossing the bearing surface of 0.005 of the column cross-sectional area, with a minimum of four bars. These four bars should preferably be dowels for the four corner bars of a square column.

The development length of the dowels must be sufficient on both sides of the bearing surface to provide the necessary development length for bars in compression (see Chapter 5).

When the dowel carries excess load into the footing, it must be spliced to the column bar using the necessary compression splice. The same procedure applies where a column rests on a pedestal and where a pedestal rests on a footing.

Where structural steel columns and column base plates are utilized, the total load is usually transferred entirely by bearing on the concrete contact area. The design bearing strength as stipulated previously also applies in this case. Where a column base detail is inadequate to transfer the total load, adjustments may be made as follows:

1. Increase column base plate dimensions.

2. Use higher-strength concrete (f_c') for pedestal or footing.

3. Increase supporting area with respect to base plate area until the ratio reaches the maximum allowed by the ACI Code.

It is common practice in building design to use a concrete pedestal between the footing and the column. The pedestal in effect distributes the column load over a larger area of the footing, thereby contributing to a more economical footing design.

Pedestals may be either plain or reinforced. If the ratio of height to least lateral dimension is in excess of three, the member is by definition a column and must be designed and reinforced as a column (see Chapter 9). If the ratio is less than three it is categorized as a pedestal and theoretically may not require any reinforcement.

The cross-sectional area of a pedestal is usually established by the concrete bearing strength as stipulated in ACI Code 10.16 or by the size of a steel column base plate or by the desire to distribute the column load over a larger footing area.

It is common practice to design a pedestal in a manner similar to a column utilizing a minimum of four corner bars (for a square or rectangular cross-section) anchored into the footing and extending up through the pedestal. Ties should be used in a manner similar to columns.

10-5 SQUARE REINFORCED CONCRETE FOOTINGS

In isolated square footings, the reinforcement should be uniformly distributed over the width of the footing in each direction. Since the bending moment is the same in each direction, the reinforcing bar size and spacing should be the same in each direction. In reality, the effective depth is not the same in both directions. However, it is common practice to use the same average effective depth for design computations for both directions. It is also common practice to assume that the minimum reinforcement ratio of $200/f_y$ is applicable to two-way footings for each of the two directions.

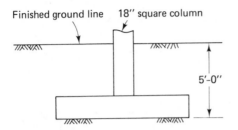

Finished ground line   18″ square column

5'-0″

FIGURE 10-9   Sketch for Example 10–3

EXAMPLE 10–3

Design a square reinforced concrete footing to support an 18-in. square tied concrete column, as shown in Fig. 10–9.

1. Design data: service dead load = 225 kips, service live load = 175 kips, allowable soil pressure = 5,000 psf, f'_c for the column = 4,000 psi and for the footing = 3,000 psi, f_y for all steel = 40,000 psi, and longitudinal column steel consists of No. 8 bars.

2. Since the footing thickness is unknown at this point, one can use an average weight for the 5 ft 0 in. from bottom of footing to finished ground line of 125 lb/ft³. The soil pressure under the footing for this 5 ft 0 in. of material is

$$5(125) = 625 \text{ psf}$$

Therefore, the *effective allowable* soil pressure for the superimposed loads becomes

$$5,000 - 625 = 4,375 \text{ psf}$$

The required area of footing may be determined using service loads and allowable soil pressure or by modifying both the service loads and the allowable soil pressure with the ACI load factors. Using the service loads,

$$\text{Required } A = \frac{(225 + 175)(1,000)}{4,375} = 91.5 \text{ ft}^2$$

Use a 9 ft 6 in. square footing. This furnishes an actual area A of 90.3 ft², which is approximately 1% less than that required. However, because of the many assumptions and uncertainties involved, the dimensions selected will be satisfactory.

3. The factored soil pressure from superimposed loads may now be calculated.

$$p_u = \frac{P_u}{A} = \frac{[1.4(225) + 1.7(175)](1,000)}{90.3} = 6,783 \text{ psf}$$

4. The footing thickness is usually determined by shear strength requirements. Therefore, assume a total footing thickness and check the shear strength. The

297

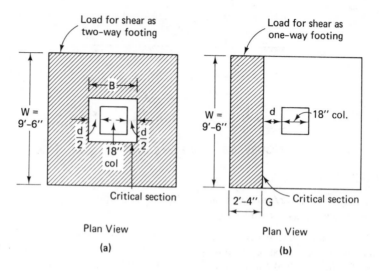

Load for shear as two-way footing

Load for shear as one-way footing

$W = 9'-6''$

B

$\dfrac{d}{2}$ $\dfrac{d}{2}$

18″ col

Critical section

Plan View

(a)

$W = 9'-6''$

d 18″ col.

$2'-4''$ G Critical section

Plan View

(b)

FIGURE 10-10 Footing shear analysis

thickness h will be assumed to be 24 in. Therefore, the effective depth, based on a 3-in. cover for bottom steel and No. 8 bars in each direction, is

$$d = 24 - 3 - 1 = 20 \text{ in.}$$

This constitutes an average effective depth which will be used for design calculations for both directions.

5. The shear strength of individual column footings is governed by the more severe of two conditions: two-way action (punching shear) or one-way action (beam shear). The location of the critical section for each type of behavior is depicted in Fig. 10–10. *For two-way action* [Fig. 10–10(a)],

$$B = \text{column width} + \left(\frac{d}{2}\right) 2$$

$$= 18 + 20 = 38 \text{ in.} = 3.17 \text{ ft}$$

The total factored shear acting on the critical section is

$$V_u = p_u(W^2 - B^2)$$

$$= 6.783(9.5^2 - 3.17^2)$$

$$= 544 \text{ kips}$$

The shear strength of the concrete is

$$V_c = \left(2 + \frac{4}{\beta_c}\right)\sqrt{f'_c}b_0 d$$

298

but should not exceed $4\sqrt{f'_c}b_0d$. Since $\beta_c = 1$, the maximum shear strength will be

$$V_c = 4\sqrt{f'_c}b_0d$$

$$= 4\sqrt{3,000}(38)(4)(20) = 666,030 \text{ lb}$$

$$= 666.0 \text{ kips}$$

$$\phi V_n = \phi V_c = 0.85(666.0) = 566 \text{ kips}$$

$$V_u < \phi V_n \qquad\qquad\qquad\qquad \text{(OK)}$$

For one-way action, the total factored shear acting on the critical section is

$$V_u = p_u WG$$

$$= 6.783(9.5)(2.33)$$

$$= 150.1 \text{ kips}$$

The shear strength of the concrete is

$$V_c = 2\sqrt{f'_c}b_wd$$

$$= 2\sqrt{3,000}(9.5)(12)(20)$$

$$= 249,761 \text{ lb}$$

$$= 249.8 \text{ kips}$$

$$\phi V_n = \phi V_c = 0.85(249.8) = 212.3 \text{ kips}$$

$$V_u < \phi V_n \qquad\qquad\qquad\qquad \text{(OK)}$$

The footing is satisfactory with respect to shear. We may check our assumption of step 2 with regard to weight of the footing and the soil on the footing. Assuming an earth weight of 100 lb/ft³:

$$150(2) + 100(3) = 600 \text{ psf}$$

Therefore, the assumption of 625 psf is slightly on the conservative (safe) side. No revision is warranted.

6. The critical section for bending moment may be taken at the face of column, as depicted in Fig. 10–11. Using the actual factored soil pressure and assuming the footing to act as a wide cantilever beam in both directions, the design moment may be computed.

$$M_u = p_u F\left(\frac{F}{2}\right)(W)$$

$$= 6,783(4)\left(\frac{4}{2}\right)(9.5)$$

$$= 515,508 \text{ ft-lb}$$

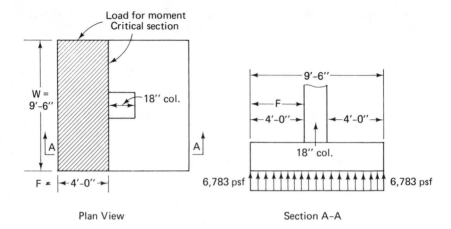

Plan View   Section A-A

FIGURE 10-11   Footing moment analysis

7. Design of steel:

$$\text{Required } \bar{k} = \frac{M_u}{\phi b d^2} = \frac{515.5(12)}{0.9(9.5)(12)(20)^2}$$

$$= 0.1507 \text{ ksi}$$

From Table A–6, the required $\rho = 0.0039$.

$$\rho_{\min} = \frac{200}{f_y} = \frac{200}{40,000} = 0.0050$$

Since $0.0039 < 0.0050$, use $\rho_{\min}$.

$$\text{Required } A_s = \rho b d$$

$$= 0.0050(9.5)(12)(20)$$

$$= 11.4 \text{ in.}^2$$

Since the footing is square and an average effective depth was used, the steel requirements in the other direction may be assumed to be identical. Therefore, use 20 No. 7 bars each way ($A_s = 12.0$ in.^{2} in each direction) and distribute the bars uniformly across the footing in each direction as shown in Fig. 10–12(a).

The required development length for the No. 7 footing bars is 18 in. if we neglect the modification factors. The development length furnished is 45 in., which is significantly in excess of that required (OK).

8. The design bearing strength cannot exceed

$$\phi(0.85 f'_c A_1)$$

except where the supporting surface is wider on all sides than the loaded area, for which case the design bearing strength cannot exceed

$$\phi(0.85f'_c A_1)\sqrt{\frac{A_2}{A_1}}$$

but in no case is the bearing strength to exceed

$$\phi(0.85f'_c A_1)(2)$$

Since the supporting surface is wider on all sides, the bearing strength for the footing may be computed as follows:

$$\sqrt{\frac{A_2}{A_1}} = \sqrt{\frac{90.3}{2.25}} = 6.3$$

Therefore, use 2.0.

$$\text{Footing bearing strength} = \phi(0.85f'_c A_1)(2.0)$$

$$= 0.70(0.85)(3.0)(18)^2(2.0)$$

$$= 1,157 \text{ kips}$$

The column bearing strength is computed as follows:

$$\phi(0.85)f'_c A_1 = 0.70(0.85)(4.0)(18)^2 = 771 \text{ kips}$$

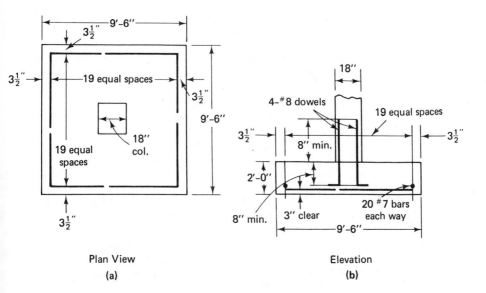

Plan View
(a)

Elevation
(b)

FIGURE 10-12   Design sketch for Example 10-3

The actual design bearing load

$$P_u = 1.4(225) + 1.7(175)$$

$$= 612.5 \text{ kips}$$

Since $612.5 < 771 < 1,157$, the entire column load can be transferred by concrete alone. However, the ACI Code requires a minimum dowel area of

$$\text{Required } A_s = 0.005 A_g$$

$$= 0.005(18)^2 = 1.62 \text{ in.}^2$$

Use a minimum of four bars. Four No. 6 bars, $A_s = 1.76 \text{ in.}^2$, is satisfactory. However, it is general practice in a situation such as this to use dowels of the same diameter as the column steel. Therefore, use 4 No. 8 dowels and place one in each corner ($A_s = 3.16 \text{ in.}^2$).

The development length for dowels into the column and footing must be adequate despite the fact that full load transfer can be made without dowels.

Since bars are in compression, the basic development length for dowels into the footing is:

$$0.02 \, d_b \frac{f_y}{\sqrt{f'_c}} = 0.02(1)\left(\frac{40,000}{\sqrt{3,000}}\right) = 14.6 \text{ in.}$$

but not less than

$$0.0003 \, d_b f_y = 0.0003(1)(40,000) = 12 \text{ in.}$$

The required development length ℓ_d equals the basic development length times any modification factors. We will use the modification factor for the case where the steel provided is in excess of the steel required:

$$\frac{\text{Required } A_s}{\text{Provided } A_s} = \frac{1.62}{3.16} = 0.51$$

$$\text{Required } \ell_d = 14.6(0.51) = 7.4 \text{ in.}$$

For bars in compression, ℓ_d must not be less than 8 in. Therefore, use ℓ_d of 8 in. into both the column and the footing, since f'_c for the column concrete is higher and the required ℓ_d for the bars into the column would be less than that into the footing. The actual anchorage used may be observed in Fig. 10–12(b). The dowels should be placed adjacent to the corner longitudinal bars. Generally these dowels are furnished with a 90° hook at their lower ends and placed on top of the main footing reinforcement. This will tie the dowel in place and reduce the possibility of the dowel being dislodged during construction. The hook cannot be considered effective in the required development length (ACI 12.5.3).

10-6 RECTANGULAR REINFORCED CONCRETE FOOTINGS

Rectangular footings generally are used where space limitations require it. The design of these footings is very similar to that of the square column footing with the one major exception that each direction must be investigated independently. Shear is checked for two-way action in the normal way, but for one-way action it is checked across the shorter side only. Bending moment must be considered separately for each direction. Each direction will generally have a different area of steel required. The reinforcing steel running in the long direction should be placed below the short-direction steel so that it may have the larger effective depth to carry the larger bending moments in that direction.

In rectangular footings, the *distribution* of the reinforcement is different than for square footings (ACI Code, Section 15.4.4). The reinforcement in the long direction should be uniformly distributed over the shorter footing width. A part of the required reinforcement in the short direction is placed in a band equal to the length of the short side of the footing. The portion of the total required steel that should go into this band is

$$\frac{2}{\beta + 1}$$

where β is the ratio of the long side to the short side of the footing. The remainder of the reinforcement is uniformly distributed in the outer portions of the footing. This distribution is depicted in Fig. 10–13. Other features of the design are similar to those for the square column footing.

EXAMPLE 10-4

Design a reinforced concrete footing to support an 18-in. square tied concrete column, as shown in Fig. 10–14. One dimension of the footing is limited to a maximum of 7 ft.

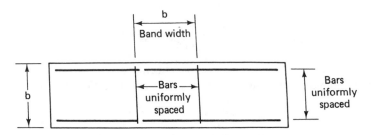

FIGURE 10-13 Rectangular footing plan

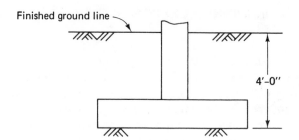

Finished ground line

4'-0"

FIGURE 10-14 Sketch for Example 10-4

1. Design data: service dead load = 175 kips, service live load = 175 kips, allowable soil pressure = 5,000 psf, f'_c for both footing and column = 3,000 psi, f_y for all steel = 40,000 psi, and longitudinal column steel consists of No. 8 bars.

2. Since the footing thickness is unknown at this point, one can use an average weight for the 4 ft 0 in. from the bottom of the footing to the finished ground line of 125 lb/ft³. The soil pressure under the footing for this 4 ft 0 in. of material is

$$4(125) = 500 \text{ psf}$$

Therefore, the *effective allowable* soil pressure for the superimposed loads becomes

$$5,000 - 500 = 4,500 \text{ psf}$$

Based on service loads, the required area of footing may be calculated.

$$\text{Required } A = \frac{(175 + 175)(1,000)}{4,500} = 77.8 \text{ ft}^2$$

Use a rectangular footing 7 ft 0 in. by 11 ft 6 in. This furnishes an actual area A of 80.5 ft².

3. The factored soil pressure from superimposed loads may now be calculated.

$$p_u = \frac{P_u}{A} = \frac{[1.4(175) + 1.7(175)](1,000)}{80.5} = 6,739 \text{ psf}$$

4. Assume that $h = 24$ in. Therefore, the effective depth, based on a 3-in. cover for bottom steel and 1-in.-diameter bars in each direction, will be

$$d = 24 - 3 - 1 = 20 \text{ in.}$$

This constitutes an *average* effective depth which will be used for design calculations for both directions.

304

5. First, we check the shear strength *for two-way action* [with reference to Fig. 10–15(a)].

$$B = \text{column width} + \left(\frac{d}{2}\right)2$$

$$= 18 + 20 = 38 \text{ in.} = 3.17 \text{ ft}$$

The total factored shear acting on critical section is

$$V_u = p_u(A - B^2)$$

$$= 6.739(80.5 - 3.17^2)$$

$$= 474.8 \text{ kips}$$

The shear strength of the concrete is

$$V_c = \left(2 + \frac{4}{\beta_c}\right)\sqrt{f_c'}b_0d$$

but must not exceed $4\sqrt{f_c'}b_0d$.

$$\beta_c = \frac{11.5}{7.0} = 1.64$$

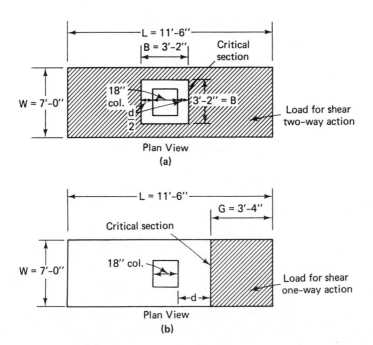

Plan View
(a)

Plan View
(b)

FIGURE 10-15 Footing shear analysis

Therefore,

$$V_c = \left(2 + \frac{4}{1.64}\right)\sqrt{f'_c}b_0d$$

$$= 4.44\sqrt{f'_c}b_0d$$

Therefore, the maximum shear strength will be

$$V_c = 4\sqrt{f'_c}b_0d$$

$$= 4\sqrt{3,000}\,(38)(4)(20)$$

$$= 666,030\text{ lb}$$

$$= 666.0\text{ kips}$$

$$\phi V_n = \phi V_c = 0.85(666.0) = 566\text{ kips}$$

$$V_u < \phi V_n \qquad\qquad\qquad\qquad\qquad\text{(OK)}$$

For one-way action, consider shear across the short side only. The critical section is at a distance equal to the effective depth of the member from the face of column [see Fig. 10–15(b)].

The total factored shear acting on the critical section is

$$V_u = p_u WG$$

$$= 6.739(7.0)(3.33)$$

$$= 157.1\text{ kips}$$

The shear strength of the concrete is

$$V_c = 2\sqrt{f'_c}b_wd$$

$$= 2\sqrt{3,000}\,(7.0)(12)(20)$$

$$= 184,034\text{ lb}$$

$$= 184.0\text{ kips}$$

$$\phi V_n = \phi V_c = 0.85(184.0) = 156.4\text{ kips}$$

$$V_u > \phi V_n$$

The footing is very slightly understrength in shear. We will consider it satisfactory. A check, at this point, on the assumption of step 2 will show that the sum of the weight of the footing and of the soil on the footing is

$$2(150) + 2(100) = 500\text{ psf} \qquad\qquad\qquad\text{(OK)}$$

6. For bending moment, each direction must be considered independently with the critical section taken at the face of the column. Using the actual factored soil pressure and assuming the footing to act as a wide cantilever beam in each direction, the design moment may be calculated. With reference to Fig. 10–16(b), for moment in the long direction,

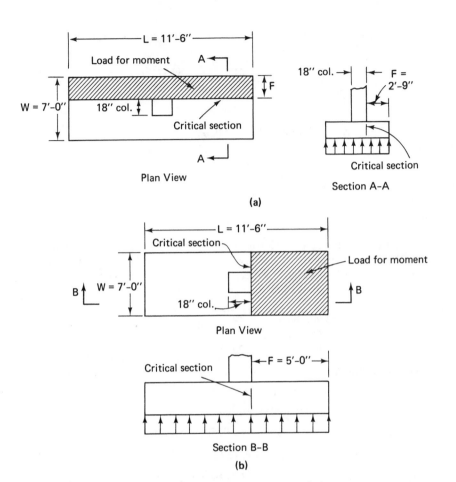

FIGURE 10-16 Footing moment analysis

$$M_u = p_u F\left(\frac{F}{2}\right)(W)$$

$$= 6,739(5)\left(\frac{5}{2}\right)(7.0)$$

$$= 589,663 \text{ ft-lb}$$

For moment in the short direction [Fig. 10–16(a)],

$$M_u = p_u F\left(\frac{F}{2}\right)(L)$$

$$= 6,739(2.75)\left(\frac{2.75}{2}\right)(11.5)$$

$$= 293,041 \text{ ft-lb}$$

7. We next determine the steel design. For the long direction, where $M_u = 589,663$ ft-lb:

$$\text{Required } \bar{k} = \frac{M_u}{\phi bd^2} = \frac{589.7(12)}{(0.9)(7)(12)(20)^2}$$

$$= 0.2340 \text{ ksi}$$

From Table A–6, the required $\rho = 0.0062$.

$$\rho_{\min} = 0.0050 \qquad \rho > \rho_{\min} \qquad \qquad \text{(OK)}$$

$$\text{Required } A_s = \rho bd$$

$$= 0.0062(7)(12)(20)$$

$$= 10.4 \text{ in.}^2$$

Use 14 No. 8 bars ($A_s = 11.1$ in.²). These bars will run in the long direction and will be distributed uniformly across the 7 ft 0 in. width. They will be placed in the bottom layer where they will have the advantage of slightly greater effective depth. For the short direction, where $M_u = 293,041$ ft-lb:

$$\text{Required } \bar{k} = \frac{M_u}{\phi bd^2} = \frac{293.0(12)}{(0.9)(11.5)(12)(20)^2}$$

$$= 0.0708 \text{ ksi}$$

From Table A–6, required $\rho = 0.0018$. But $\rho_{\min} = 0.0050$. Therefore, $\rho < \rho_{\min}$ and $\rho_{\min}$ should be used.

$$\text{Required } A_s = \rho bd$$

$$= 0.0050(11.5)(12)(20)$$

$$= 13.8 \text{ in.}^2$$

Use 18 No. 8 bars ($A_s = 14.2$ in.²). These bars will run in the short direction but will *not* be distributed uniformly across the 11 ft 6 in. side.

In rectangular footings, a portion of the total reinforcement required in the short direction is placed in a band centered on the column and having a width equal to the short side. The portion of the total required steel that goes into this band is

$$\frac{2}{\beta + 1}$$

where

$$\beta = \frac{\text{long-side dimension}}{\text{short-side dimension}} = \frac{11.5}{7.0} = 1.64$$

from which

$$\frac{2}{\beta + 1} = \frac{2}{1.64 + 1} = 0.757 = 75.7\%$$

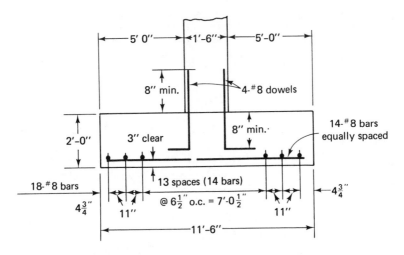

FIGURE 10-17 Design sketch for Example 10-4

Therefore, 75.7% of 18 No. 8 bars must be placed in a band width = 7 ft 0 in. The balance of the required bars will be distributed equally in the outer portions of the footing.

$$(0.757)(18) = 13.6 \text{ bars}$$

Use 14 bars in the 7 ft 0 in. band width. The bar arrangement is depicted in Fig. 10–17.

The required development length for the No. 8 footing bars is 23 in. if we neglect the modification factors. The minimum development length furnished (in the short direction) is 30 in., which is in excess of that required (OK).

8. Since the supporting surface is wider on all sides, the bearing strength for the footing may be computed as follows:

$$\phi(0.85 f'_c \, A_1)\sqrt{\frac{A_2}{A_1}}$$

$$\sqrt{\frac{A_2}{A_1}} = \sqrt{\frac{49.0}{2.25}} = 4.67$$

Therefore, use 2.0.

$$\text{Footing bearing strength} = \phi(0.85 f'_c A_1)(2.0)$$

$$= .70(0.85)(3.0)(18)^2(2.0)$$

$$= 1,157 \text{ kips}$$

The column bearing strength is computed as follows:

$$\phi(0.85)f'_c A_1 = 0.70(0.85)(3)(18)^2 = 578 \text{ kips}$$

Actual design bearing load

$$P_u = 1.4(175) + 1.7(175)$$
$$= 542.5 \text{ kips}$$

Since $542.5 < 578 < 1{,}156.7$, the entire column load can be transferred by concrete alone.

For proper connection between column and footing, use 4 No. 8 dowels (to match the column steel), one in each corner. The development length requirement for the No. 8 dowels is identical to that of Example 10–3.

10-7 ECCENTRICALLY LOADED FOOTINGS

Where footings are subject to eccentric vertical loads or to moments transmitted by the supported column, the design varies somewhat from that of the preceding sections. The soil pressure is no longer uniform across the footing width but may be assumed to vary linearly. The resultant force should be within the middle one-third of the footing base to ensure a positive contact surface between footing and soil.

With the soil pressure distribution known, the footing must be designed to resist all moments and shears, as were the concentrically loaded footings.

10-8 COMBINED FOOTINGS

This category includes footings that support more than one column or wall. The two-column type of combined footing, which is relatively common, generally results out of necessity. Two conditions that may lead to its use are (1) an exterior column that is immediately adjacent to a property line where it is impossible to utilize an individual column footing, and (2) two columns that are closely spaced causing their individual footings to be closely spaced. In these situations, a rectangular or trapezoidal shape combined footing would generally be used. The choice as to which shape to use is based on the difference in column loads as well as physical (dimensional) limitations. If the footing cannot be rectangular, a trapezoidal shape would then be selected.

The physical dimensions (except thickness) of the combined footing are generally established by the allowable soil pressure. Also, the centroid of the footing area should coincide with the line of action of the

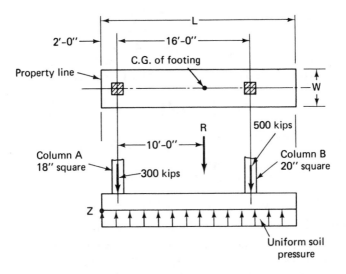

FIGURE 10-18   Sketch for Example 10-5

resultant of the two column loads. These dimensions are usually determined using service loads in combination with an allowable soil pressure.

EXAMPLE 10–5

Determine the shape and proportions of a combined footing subject to two column loads as shown in Fig. 10–18.

1. Design data: The service load on the footing from column *A* is 300 kips, from column *B* is 500 kips, and the allowable soil pressure is 6,000 psf.

2. Locate the resultant column load by a summation of moments at point *z* in Fig. 10–18.

$$\Sigma M_z = 300(2) + 500(18) = 800(x)$$

from which

$$x = 12 \text{ ft } 0 \text{ in.} \quad \text{(measured from } z\text{)}$$

3. Assuming a rectangular shape, establish the length of footing *L* so that the centroid of the footing area coincides with the line of action of resultant force *R*.

$$\text{Required } L = 12(2) = 24 \text{ ft } 0 \text{ in.}$$

4. Assume a footing thickness of 3 ft 0 in. Therefore, its weight = 150(3) = 450 psf and allowable soil pressure for superimposed loads = 6,000 − 450 = 5,550 psf. This neglects any soil *on* the footing.

311

5. The footing area required is

$$\frac{R}{5,550} = \frac{800,000}{5,550} = 144.1 \text{ ft}^2$$

6. With a length = 24 ft 0 in., the footing width W required is

$$\frac{144.1}{24} = 6 \text{ ft 0 in.}$$

7. The actual uniform soil pressure is

$$\frac{800,000}{6(24)} + 450 = 6,000 \text{ psf} \tag{OK}$$

EXAMPLE 10–6

Utilizing the design data from Example 10–5, determine the proportions of a combined footing if the footing length is limited to 22 ft.

1. The resultant column load is located as in Example 10–5 at a point 12 ft 0 in. from point z.

2. Assume a footing thickness of 3 ft 0 in. Therefore, its weight = 3(150) = 450 psf and the allowable soil pressure for superimposed loads is 6,000 − 450 = 5,550 psf.

3. The footing area required is then

$$\frac{800,000}{5,550} = 144.1 \text{ ft}^2$$

4. Assume a trapezoidal shape. The area A of a trapezoid is (see Fig. 10–19)

$$A = \frac{(b + b_1)L}{2}$$

from which

$$b + b_1 = \frac{2A}{L} = \frac{2(144.1)}{22} = 13.1 \text{ ft}$$

5. The center of gravity of the trapezoid and the resultant force of the column loads are to coincide. The location of the center of gravity, a distance c from point z, may be written

$$c = \frac{L(2b + b_1)}{3(b + b_1)} = \frac{L(b + b + b_1)}{3(b + b_1)} = 12 \text{ ft}$$

6. We now have two equations that contain b and b_1. The equation of step 5 may be easily solved for b by substitution of L = 22 ft and $(b + b_1)$ = 13.1 ft.

$$\frac{22(b + 13.1)}{3(13.1)} = 12$$

from which

$$b = \frac{12(3)(13.1)}{22} - 13.1 = 8.34 \text{ ft}$$

and

$$b_1 = 13.1 - 8.34 = 4.76 \text{ ft}$$

Use $b = 8$ ft 4 in. and $b_1 = 4$ ft 9 in.

$$\text{Actual } A = \frac{22(4.75 + 8.33)}{2} = 143.9 \text{ ft}^2$$

The footing is very slightly undersized. No revision is warranted.

The structural design of the rectangular and trapezoidal shape combined footings is generally based on a uniform soil pressure, despite the fact that loading combinations will almost always introduce some eccentricity with respect to the centroid of the footing. The determination of the footing thickness and reinforcement must be based on factored loads and soil pressure to be consistent with the ACI strength method approach. The assumed footing behavior, which will be briefly described, is a generally used approach to simplify the design of the footing.

In Fig. 10–19, it can be seen that the columns are positioned relatively close to the ends of the footing. Assuming the columns as the supports

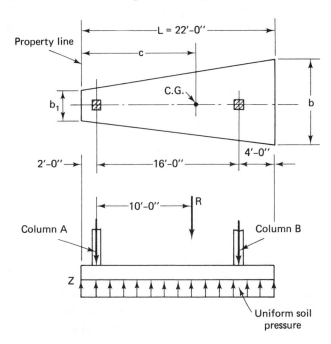

FIGURE 10-19 Sketch for Example 10–6

and the footing subjected to an upward uniformly distributed load caused by the uniform soil pressure; moments which create tension in the top of the footing will predominate in the longitudinal direction. Therefore the principal longitudinal reinforcement will be placed in the top of the footing equally distributed across the footing width. Somewhat smaller moments in the transverse direction will cause compression in the top of the footing. Transverse steel will be placed under each column in the bottom of the footing to distribute the column load in the transverse direction using the provisions for individual column footings. This in effect makes the combined footing act as a wide rectangular beam in the longitudinal direction which may then be designed using the ACI Code provisions for flexure.

Pertinent design considerations may be summarized as follows:

1. Main reinforcement (uniformly distributed) is placed in a longitudinal direction in the top of the footing, assuming the footing to be a longitudinal beam.

2. Shear should be checked considering both one-way shear at a distance d from the face of the column and two-way (punching) shear on a perimeter $d/2$ from the face of the column.

3. Stirrups or bent bars are frequently required to maintain an economical footing thickness. This assumes that the shear effect is uniform across the width of the footing.

4. Transverse reinforcement is generally uniformly placed in the bottom of the footing within a band having a width not greater than the column width plus twice the effective depth of the footing. The design treatment in the transverse direction is similar to the design treatment of the individual column footing assuming dimensions equal to the band width as previously described and the transverse footing width.

5. Longitudinal steel is also placed in the bottom of the footing in order to tie together and position the stirrups and transverse steel. Although the required steel areas may be rather small, the effects of cantilever moments in the vicinity of the columns should be checked.

10–9 CANTILEVER OR STRAP FOOTING

A third type of combined footing is generally termed a *cantilever* or *strap footing*. This is an economical type of footing when the proximity of a property line precludes the use of other types. For instance, an isolated column footing may be too large for the area available, and the nearest column is too distant to allow a rectangular or trapezoidal combined

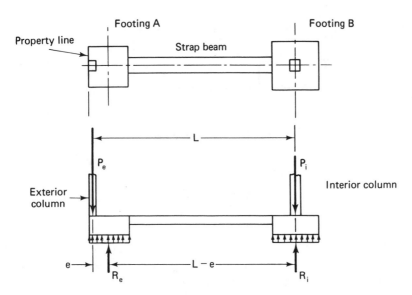

FIGURE 10-20 Cantilever footing

footing to be economical. The strap footing may be regarded as two individual column footings connected by a strap beam.

In Fig. 10–20 the exterior footing is placed eccentrically under the exterior column so that it does not violate the property-line limitations. This would produce a nonuniform pressure distribution under the footing which could lead to footing rotation. To balance this rotational, or overturning effect, the exterior footing is connected by a stiff beam, or strap, to the nearest interior footing and uniform soil pressures under the footings are assumed. The strap, which may be categorized as a flexural member, is subjected to both bending moment and shear, resulting from the forces P_e and R_e acting on the exterior footing. As shown in Fig. 10–20, the applied moment is counterclockwise and the shear will be positive because $R_e > P_e$. At the interior column, there is no eccentricity between the column load P_i and the resultant soil pressure force R_i. Therefore, we will assume that no moment is induced in the strap at the interior column.

We will define V as the vertical shear force necessary to keep the strap in equilibrium, as shown in Fig. 10–21. Then, with P_e known, V and R_e may be calculated using the principles of statics. A moment summation about R_e yields

$$P_e e = V(L - e)$$

$$V = \frac{P_e e}{L - e}$$

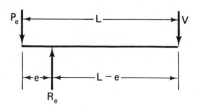

FIGURE 10-21 Strap beam

and, from a summation of vertical forces,

$$R_e = P_e + V$$

$$R_e = P_e + \frac{P_e e}{L - e}$$

Note that V acting downward on the strap beam also means that V is an uplift force on the interior footing. Therefore,

$$R_i = P_i - V$$

$$R_i = P_i - \frac{P_e e}{L - e}$$

In summary, R_e becomes greater than P_e by a magnitude of V while R_i becomes less than P_i by a magnitude equal to V.

The footing areas required are merely the reactions R_e and R_i based on service loads divided by the net allowable soil pressure. These values are based on an assumed trial e and may have to be recomputed until the trial e and the actual e are the same.

The structural design of the interior footing is simply the design of an isolated column footing subject to a load R_i. The exterior footing is generally considered as under one-way *transverse* bending similar to a wall footing with longitudinal steel furnished by extending the strap steel into the footing. The selection of footing thickness and reinforcement should be based on factored loads to be consistent with the ACI strength design approach.

The strap beam is assumed to be a flexural member with no bearing on the soil underneath. Many designers make a further simplifying assumption that the beam weight is carried by the underlying soil; hence the strap is designed as a rectangular beam subject to a constant shearing force and a linearly varying negative bending moment based on factored loads.

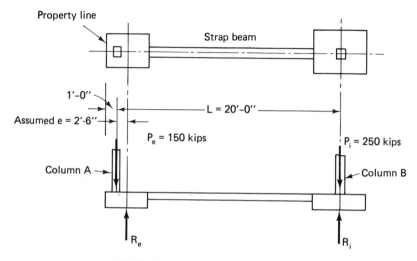

FIGURE 10-22   Sketch for Example 10-7

EXAMPLE 10–7

Determine the size of the exterior and interior footings of a strap footing for the conditions and design data furnished in Fig. 10–22.

1. Design data: the service load on the footing from column A is 150 kips, from column B is 250 kips, and the allowable soil pressure is 4,000 psf.

2. Assume that $e = 2$ ft 6 in. and footing thickness is 2 ft 0 in. Therefore, footing weight $= 2(150) = 300$ psf.

3. The allowable soil pressure for superimposed loads $= 4,000 - 300 = 3,700$ psf.

4. Determine the strap beam shear V.

$$V = \frac{P_e e}{L - e} = \frac{150(2.5)}{17.5} = 21.4 \text{ kips}$$

5. Footing reactions:

$$R_e = P_e + V = 150 + 21.4 = 171.4 \text{ kips}$$

$$R_i = P_i - V = 250 - 21.4 = 228.6 \text{ kips}$$

6. For the exterior footing, the required area is

$$\frac{171.4}{3.7} = 46.3 \text{ ft}^2$$

Use a footing 7 ft 0 in. by 7 ft 0 in. ($A = 49 \text{ ft}^2$). Note in Fig. 10–22 that actual e is then the same as assumed e, and no revision of calculations is necessary.

317

For the interior footing, the required area is

$$\frac{228.6}{3.7} = 61.8 \text{ ft}^2$$

Use a footing 8 ft 0 in. by 8 ft 0 in. ($A = 64$ ft²).

PROBLEMS

10–1. Design a plain concrete wall footing to carry a 12-in.-thick reinforced concrete wall. Service loads are 2.3 kips/ft dead load (includes the weight of the wall) and 2.3 kips/ft live load. Use $f'_c = 3,000$ psi and an allowable soil pressure of 4,000 psf.

10–2. Redesign the footing for the wall of Problem 10–1. Service loads are 8.0 kips/ft dead load and 8.0 kips/ft. live load.

10–3. Design a reinforced concrete footing for the wall of Problem 10-1 if the service loads are 6 kips/ft dead load (includes wall weight) and 15 kips/ft live load. Use $f_y = 40,000$ psi.

10–4. Design a square individual column footing to support an 18-in. square reinforced concrete tied column. Service loads are 200 kips dead load and 350 kips live load. Use $f'_c = 3,000$ psi and $f_y = 60,000$ psi. The allowable soil pressure is 3,500 psf and $w_e = 100$ lb/ft³. The column, reinforced with 8 No. 8 bars, has $f'_c = 5,000$ psi and $f_y = 60,000$ psi. Bottom of footing is to be 4 ft below grade.

10–5. Design a square individual column footing to support a 16-in. square reinforced concrete tied column. Service loads are 200 kips dead load and 160 kips live load. Both column and footing have $f'_c = 4,000$ psi and $f_y = 40,000$ psi. Allowable soil pressure = 5,000 psf and $w_e = 100$ lb/ft³. The bottom of the footing is to be 4 ft below grade. Column is reinforced with 8 No. 7 bars.

10–6. Design for load transfer from a 14-in. square tied column to a 13 ft 0 in. square footing. $P_u = 650$ kips, footing $f'_c = 3,000$ psi, column $f'_c = 5,000$ psi, $f_y = 60,000$ psi (all steel), and the column is reinforced with 8 No. 8 bars.

10–7. Redesign the footing of Problem 10–5 if there is a 7 ft 0 in. restriction on the width of the footing.

10–8. Footings are to be designed for two columns A and B spaced with $D = 16$ ft as shown. There are no dimensional limitations. The service load from column A is 100 kips and from column B 150 kips. The allowable soil pressure is 4,000 psf. Determine appropriate footing(s) size, type, and layout. Assume thickness and disregard reinforcing.

10–9. Same as Problem 10–8, except service loads are 700 kips from column A and 900 kips from column B and $D = 14$ ft.

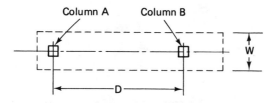

FIGURE P10-8 through P10-11

10–10. For the column layout shown, there is a width restriction W of 16 ft. D = 14 ft. Determine an appropriate footing size and layout if the service load from column A is 700 kips and from column B is 800 kips. Assume thickness and disregard reinforcing. The allowable soil pressure is 4,000 psf.

10–11. For the column layout shown, D = 24 ft. The footing for the 18-in. square column A must be placed flush with the left side of the column. Service loads are 200 kips for column A and 300 kips for column B. The allowable soil pressure is 4500 psf. Assume a footing thickness of 2 ft 6 in. and determine an appropriate size, type and layout for the footing.

11

Prestressed Concrete Fundamentals

11-1 INTRODUCTION

According to the ACI definition, prestressed concrete is a material that has had internal stresses induced to balance out, to a desired degree, stresses due to externally applied loads. Since tensile stresses are undesirable in concrete members, the object of prestressing is to create compressive stresses (prestress) at the same locations as the tensile stresses within the member so that the tensile stresses will be diminished or will disappear altogether. The diminishing or elimination of tensile stresses within the concrete will result in members that have fewer cracks or are crack-free at service load levels. This is one of the advantages of prestressed concrete over reinforced concrete, particularly in corrosive atmospheres. Prestressed concrete offers other advantages. Because beam cross sections are primarily in compression, diagonal tension stresses are reduced and the beams are stiffer at service loads. Additionally, sections can be smaller, resulting in less dead weight.

Despite the advantages, one must consider the higher unit cost of stronger materials, the need for expensive accessories, the necessity for close inspection and quality control, and, in the case of precasting, a higher initial investment in plant.

11-2 DESIGN APPROACH AND BASIC CONCEPTS

Since most of the advantages of prestressed concrete are at service load levels and since permissible stresses in the "green" concrete often control the amount of prestress force to be used, the major part of

Precast, Post-tensioned Bridge Girder Sections.

analysis and design calculations are made using service loads, permissible stresses, and basic assumptions as outlined in Sections 18.2 through 18.5 of the ACI Code. However, the strength requirements of the Code must also be met. Therefore, at some point, the design must be checked using appropriate load factors and strength reduction factors.

The normal method for applying prestress force to a concrete member is through the use of steel tendons. There are two basic methods of arriving at the final prestressed member, pretensioning and post-tensioning.

Pretensioning may be defined as a method of prestressing concrete in which the tendons are tensioned before the concrete is placed. This operation, which may be performed in a casting yard, is basically a five-step process:

1. The tendons are placed in a prescribed pattern on the casting bed between two anchorages. The tendons are then tensioned to a value not to exceed 80% of the specified tensile strength of the tendons or 94% of the specified yield strength (ACI Code, Section 18.5.1). The tendons are then anchored so that the load in them is maintained.

2. If the concrete forms are not already in place, they may then be assembled around the tendons.

3. The concrete is then placed in the forms and allowed to cure. Proper quality control must be exercised and curing may be accelerated

with the use of steam or other methods. The concrete will bond to the tendons.

4. When the concrete attains a prescribed strength, normally within 24 hours or less, the tendons are cut at the anchorages. Since the tendons are now bonded to the concrete, as they are cut from their anchorages the high prestress force must be transferred to the concrete. As the high tensile force of the tendon creates a compressive force on the concrete section, the concrete will tend to shorten slightly. The stresses that exist once the tendons have been cut are often called the stresses at *transfer*.

5. The prestressed member is then removed from the forms and moved to a storage area so that the casting bed can be prepared for further use.

Pretensioned members are usually manufactured at a casting yard or plant which is somewhat removed from the job site where the members will eventually be used. In this case, they are usually delivered to the job site ready to be set in place. Where a project is of sufficient magnitude to warrant it economically, a casting yard may be built on the job site, thus decreasing transportation costs and allowing larger members to be precast without the associated transportation problems.

Figure 11–1 depicts the various stages in the manufacture of a precast pretensioned member.

Post-tensioning may be defined as a method of prestressing concrete in which the tendons are tensioned *after* the concrete has cured. The operation is commonly a six-step process.

Tendon Tensioned between Anchorages

(a)

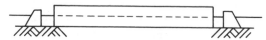

Forms Assembled and Concrete Placed in Forms

(b)

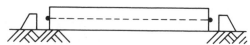

Tendons Cut and Compression Transferred
to Member

(c)

FIGURE 11-1   Pretensioned member

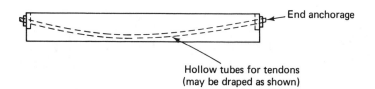

Hollow tubes for tendons
(may be draped as shown)

FIGURE 11-2 Post-tensioned member

1. Concrete forms are assembled with flexible hollow tubes (metal or plastic) placed in the forms and held at specified locations.

2. Concrete is then placed in the forms and allowed to cure to a prescribed strength.

3. Tendons are placed in the tubes (In some systems a complete tendon assembly is placed in the forms prior to the pouring of the concrete.)

4. The tendons are tensioned by jacking against an anchorage device or end plate which, in some cases, has been previously embedded in the end of the member. The anchorage device will incorporate some method for gripping the tendon and holding the load.

5. If the tendons are to be bonded, the space in the tubes around the tendons may be grouted using a pumped grout. Some members utilize unbonded tendons.

6. The end anchorages may be covered with a protective coating.

While post-tensioning is sometimes performed in a plant away from the project, it is most often done at the job site, particularly for units too large to be shipped assembled or for unusual applications (Fig. 11–2).

The tensioning of the tendons is normally done with hydraulic jacks. There are many patented devices available to accomplish the anchoring of the tendon ends to the concrete.

11–3 STRESS PATTERNS IN PRESTRESSED CONCRETE BEAMS

The stress pattern existing on the cross section of a prestressed concrete beam may be determined by superimposing the stresses due to the loads and forces acting on the beam at any particular time. For our purposes, the following sign convention will be adopted:

Tensile stresses are positive (+)

Compressive stresses are negative (−)

323

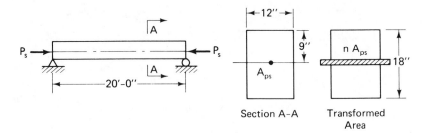

FIGURE 11-3 Sketches for Example 11-1

Since we will assume a crack-free cross section at service load level, the entire cross section will remain effective in carrying stress. Also, the entire concrete cross section will be utilized in the calculation of centroid and moment of inertia.

For purposes of explanation, we will consider a rectangular shape with tendons placed at the centroid of the section and we will investigate the induced stresses. Although the rectangular shape is used for some applications, it is often less economical than the more complex shapes.

EXAMPLE 11-1

For the section shown in Fig. 11-3, determine the stresses due to prestress immediately after transfer and also the stresses at midspan when the member is placed on a 20-ft simple span. Use f'_c = 5,000 psi and assume that the concrete has attained a strength of 4,000 psi at the time of transfer. Use a central prestressing force of 100 kips.

1. Compute the stress in the concrete at time of initial prestress. With the prestressing force P_s applied at the centroid of the section and assumed acting on the gross section A_c, the concrete stress will be uniform over the entire section.

$$f = \frac{P_s}{A_c} = \frac{-100}{12(18)} = -0.463 \text{ ksi}$$

2. Compute the stresses due to the beam dead load.
   Weight of beam:

$$w_{DL} = \frac{12(18)}{144}(0.150) = 0.225 \text{ kips/ft}$$

   Moment due to dead load:

$$M_{DL} = \frac{w_{DL}\,\ell^2}{8} = \frac{0.225(20)^2}{8} = 11.25 \text{ ft-kips}$$

Compute the moment of inertia I of the beam using the gross cross section and neglecting the transformed area of the tendons (since the steel is at the centroid, it will have no effect on the moment of inertia).

$$I_g = \frac{bh^3}{12} = \frac{12(18)^3}{12} = 5,832 \text{ in.}^4$$

Dead load stresses:

$$f = \frac{Mc}{I} = \frac{11.25(12)(9)}{5,832} = \pm 0.208 \text{ ksi}$$

3. Compute the stresses due to prestress plus dead load.

$$f_{\text{(initial prestress + DL)}} = -0.463 \pm 0.208$$

$$= -0.671 \text{ ksi} \quad \text{(compression, top)}$$

$$= -0.255 \text{ ksi} \quad \text{(compression, bottom)}$$

These results are shown on the stress summation diagram, Fig. 11–4.

It can be seen that the tensile stresses due to the DL moment in the bottom of the beam have been completely canceled out and that compression exists on the entire cross section. It is also evident that a limited additional positive moment may be carried by the beam without resulting in a net tensile stress in the bottom of the beam. This situation may be further improved upon by *lowering the location of the tendon* to induce additional compressive stresses in the bottom of the beam.

Example 11–1 reflects two stages of the prestress process. The transfer stage occurs in a pretensioned member when the tendons are cut at the ends of the member and the prestress force has been transferred to the beam as shown in Fig. 11–1. When the beam in this problem is removed from the forms (picked up by its end points), dead load stresses are introduced and in this second stage both the beam weight (dead load)

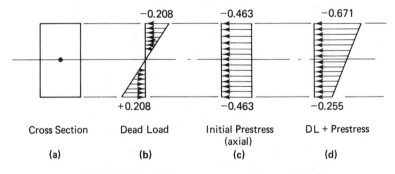

FIGURE 11-4 **Midspan stresses for Example 11-1**

and the prestress force are contributors to the stress pattern within the beam. This stage is important because it occurs early in the life of the beam (sometimes within 24 hours of casting) and the concrete stresses must be held within permissible values as specified in the ACI Code, Section 18.4.

If the prestress force were placed below the neutral axis in Example 11–1, negative bending moment would occur in the member at transfer, causing the beam to curve upward and pick up its dead load. Hence, for a simple beam such as this, where the prestress force is eccentric, the stresses due to the initial prestress would never exist alone without the counteracting stresses from the dead load moment.

11–4 PRESTRESSED CONCRETE MATERIALS

The application of the prestress force both strains or "stretches" the tendons and at the same time induces tensile stresses in them. If the tendon strain is reduced for some reason, the stress will also be reduced. In prestressed concrete this is known as a *loss of prestress*. This loss occurs after the prestress force has been introduced. Contributory factors to this loss are creep and shrinkage of the concrete, elastic compression of the concrete member, relaxation of the tendon stress, anchorage seating loss, and friction losses due to intended or unintended curvature in post-tensioning tendons. Estimates on magnitudes of these losses vary. However, it is generally known that an ordinary steel bar tensioned to its yield strength (40,000 psi) would lose its entire prestress by the time all stress losses had taken place. Therefore, for prestressed concrete applications, it is necessary to use very high strength steels, where the previously mentioned strain losses will result in a much smaller percentage change in original prestress force.

The most commonly used steel for pretensioned prestressed concrete is in the form of a seven-wire uncoated stress-relieved strand having a minimum tensile strength (f_{pu}) of 250,000 psi or 270,000 psi, depending on grade. The seven-wire strand is made up of seven cold-drawn wires. The center wire is straight and the six outside wires are laid helically around it. All six outside wires are the same diameter, and the center wire is slightly larger. This, in effect, guarantees that each of the outside wires will bear on and grip the center wire.

Prestressing steel does not exhibit the definite yield point characteristic found in the normal ductile steel used in reinforcing steel (see Fig. 11–5). The yield strength for prestressing wire and strand is a "specified yield strength" which is obtained from the stress–strain diagram at 1% strain, according to ASTM. Nevertheless, the specified yield point is not

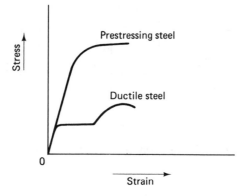

FIGURE 11-5 Comparative stress–strain curves

as important in prestressing steel as is the yield point in the ductile steels. It is a consideration when determining the ultimate strength of a beam.

In normal reinforced concrete members that are designed to ensure tension failures, the strength of the concrete is secondary in importance to the strength of the steel in the determination of the flexural strength of the member. In prestressed applications, concrete in the 4,000–6,000 psi range is commonly used. Some of the reasons for this are (1) volumetric changes for higher-strength concrete are smaller, which will result in smaller prestress losses; (2) bearing and development stresses are higher; and (3) higher-strength concrete is more easily obtained in precast work than in cast-in-place work because of better-quality control.

In addition, high early strength cement (type III) is normally used to obtain as rapid a turnover time as possible for optimum use of forms.

11-5 ANALYSIS OF RECTANGULAR PRESTRESSED CONCRETE BEAMS

The analysis of flexural stresses in a prestressed member should be performed for different stages of loading; namely, the initial service load stage, which includes dead load plus prestress before losses, the final service load stage, which includes dead load plus prestress plus live load after losses, and finally the ultimate strength stage, which involves load and understrength factors. Generally, checking of prestressed members is accomplished at the service load level based on unfactored loads. However, the ultimate strength of a member should also be checked, utilizing the same strength principles as for nonprestressed reinforced concrete members.

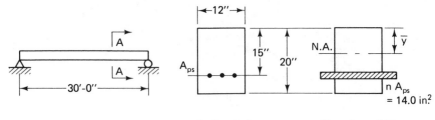

Section A-A   Transformed Area

FIGURE 11-6 Sketches for Example 11-2

EXAMPLE 11-2

For the beam of cross section shown in Fig. 11-6, analyze the flexural stresses at midspan at transfer and in service. Neglect losses. Use a prestressed steel area A_{ps} of 2.0 in.^{2}, $f'_c = 6,000$ psi, and assume that the concrete has attained a strength of 5,000 psi at the time of transfer. Initial prestress force = 250 kips. The service dead load = 0.25 kip/ft, which does not include the weight of the beam. The service live load = 1.0 kip/ft. Use $n = 7$.

Assume that the gross section is effective and use the transformed area (neglecting displaced concrete) for the moment of inertia.

1. Beam weight:

$$w_{DL} = \frac{20(12)}{144}(0.150) = 0.25 \text{ kips/ft}$$

Moment due to the beam weight:

$$M_{DL} = \frac{w_{DL}\,\ell^2}{8} = \frac{0.25(30)^2}{8} = 28.1 \text{ ft-kips}$$

Moment due to superimposed loads (DL + LL):

$$M_{DL+LL} = \frac{w_{(DL+LL)}\ell^2}{8} = \frac{1.25(30)^2}{8} = 140.6 \text{ ft-kips}$$

2. Location of the neutral axis:

$$\bar{y} = \frac{\Sigma(Ay)}{\Sigma A}$$

using the top of the section as the reference axis.

$$\bar{y} = \frac{12(20)(10) + 14(15)}{12(20) + 14} = 10.28 \text{ in.}$$

The eccentricity e of the strands from the neutral axis is

$$15 - 10.28 = 4.72 \text{ in.}$$

328

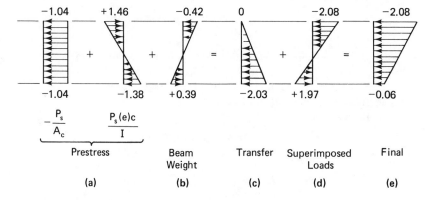

FIGURE 11-7 Midspan stress summary for Example 11–2

Moment of inertia about the neutral axis:

$$I = \frac{12(20)^3}{12} + 12(20)(0.28)^2 + 14(4.72)^2$$

$$I = 8{,}331 \text{ in.}^4$$

3. The stresses may now be calculated. These are summarized in Fig. 11–7.

   a. Initial prestress
      As a result of the eccentric prestressing force, the induced stress at the initial prestress stage will *not* be uniform but may be computed from

$$f = -\frac{P_s}{A_c} \pm \frac{Mc}{I}$$

$$= -\frac{P_s}{A_c} \pm \frac{P_s(e)c}{I}$$

where P_s/A_c represents the axial effect of the prestress force and $\frac{P_s(e)c}{I}$ the eccentric, or moment, effect.

$$-\frac{P_s}{A_c} = -\frac{250}{12(20)} = -1.04 \text{ ksi} \quad \text{(compression top and bottom)}$$

$$+\frac{P_s(e)c}{I} = \frac{250(4.72)(10.28)}{8{,}331} = +1.46 \text{ ksi} \quad \text{(tension in top)}$$

$$-\frac{P_s(e)c}{I} = -\frac{250(4.72)(9.72)}{8{,}331} = -1.38 \text{ ksi} \quad \text{(compression in bottom)}$$

   b. Beam weight

$$f = \pm \frac{Mc}{I}$$

329

$$f = +\frac{28.1(12)(9.72)}{8,331} = +0.39 \text{ ksi} \quad \text{(tension in bottom)}$$

$$f = -\frac{28.1(12)(10.28)}{8,331} = -0.42 \text{ ksi} \quad \text{(compression in top)}$$

Summarizing the initial service load stage at the time of transfer [prestress plus beam weight (DL)]:
Top of beam:

$$-1.04 + 1.46 - 0.42 = 0$$

Bottom of beam:

$$-1.04 - 1.38 + 0.39 = -2.03 \text{ ksi} \quad \text{(compression)}$$

The permissible stresses (ACI Code, Section 18.4) immediately after prestress transfer (before losses) are in terms of f'_{ci}, which is the compressive strength (psi) of concrete at the time of initial prestress.

$$\text{Compression} = 0.60 f'_{ci}$$
$$= 0.60(5,000) = 3,000 \text{ psi} = 3.0 \text{ ksi}$$
$$\text{Tension} = 3\sqrt{f'_{ci}}$$
$$= 3\sqrt{5,000} = 212 \text{ psi} = 0.212 \text{ ksi}$$

Since $2.03 < 3.0$ and $0 < 0.212$, the beam is satisfactory at this stage.

c. Superimposed loads (DL + LL)

$$f = \pm \frac{Mc}{I}$$

$$f = +\frac{140.6(12)(9.72)}{8331} = +1.97 \text{ ksi} \quad \text{(tension in bottom)}$$

$$f = -\frac{140.6(12)(10.28)}{8331} = -2.08 \text{ ksi} \quad \text{(compression in top)}$$

Summarizing the second service load stage when the beam has had service loads applied (prestress plus beam dead load plus superimposed loads):
Top of beam:

$$0 - 2.08 = -2.08 \text{ ksi} \quad \text{(compression)}$$

Bottom of beam:

$$-2.03 + 1.97 = -0.06 \text{ ksi} \quad \text{(compression)}$$

The permissible stresses in the concrete at the service load level (ACI Code, Section 18.4.2) are

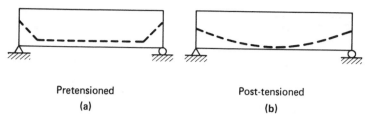

Pretensioned   Post-tensioned

(a)   (b)

FIGURE 11-8 Draped tendons

$$\text{Compression} = 0.45 f_c'$$

$$= 0.45 (6{,}000) = 2{,}700 \text{ psi} = 2.7 \text{ ksi}$$

$$\text{Tension} = 6\sqrt{f_c'}$$

$$= 6\sqrt{6{,}000} = 465 \text{ psi} = 0.465 \text{ ksi}$$

The permissible stresses given above assume that losses have been accounted for. Our analysis has neglected losses. Aside from this, the stresses in the beam would be satisfactory, since 2.08 < 2.70. Note in Fig. 11–7(c) that at transfer, no tensile stresses exist on the cross section.

The final stress distribution [Fig. 11–7(e)] in the beam of Example 11–2 shows that the entire beam cross section is under compression. This is the stress pattern that exists at *midspan* (since the applied moments calculated were midspan moments). It is evident that moments will decrease toward the supports of a simply supported beam and that the stress pattern will change drastically. For example, if the eccentricity of the tendon were constant in the beam in question, the net prestress [Fig. 11–7(a)] would exist at the beam ends, where moment due to beam weight and applied loads is zero. Since the tensile stress is undesirable, the location of the tendon is changed in the area of the end of the beam, so that eccentricity is decreased. This results in a *curved* or *draped* tendon within the member, as shown in Fig. 11–8. Post-tensioned members may have curved tendons accurately placed to satisfy design requirements, but in pretensioned members, because of the nature of the fabrication process, only approximate curves are formed by forcing the tendon up or down at a few points.

Although shear was not included in the beam analysis, the reader should be aware that the draping of the tendons, in addition to affecting the flexural stresses at the beam ends, produces a force acting vertically upward which has the effect of reducing the shear force due to dead and live loads. Prestressed concrete flexural members are available in numerous shapes suitable for various applications. A few of the more common ones are shown in Fig. 11–9.

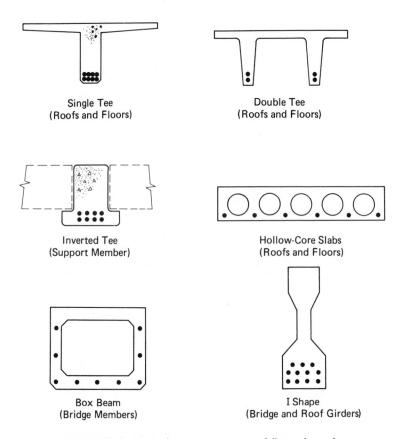

Single Tee
(Roofs and Floors)

Double Tee
(Roofs and Floors)

Inverted Tee
(Support Member)

Hollow-Core Slabs
(Roofs and Floors)

Box Beam
(Bridge Members)

I Shape
(Bridge and Roof Girders)

FIGURE 11-9 Typical precast prestressed flexural members

PROBLEMS

11–1. The *plain concrete* beam shown having a rectangular cross section 10 in. wide and 18 in. deep is simply supported on a single span of 20 ft. Assuming no loads other than the dead load of the beam itself, find the bending stresses that exist at midspan.

11–2. A beam identical to that of Problem 11–1 is prestressed with a force P_s of 185 kips. The tendons are located at the center of the beam. $A_{ps} = 1.23$ in.^{2}. Find the stresses in the beam at midspan at transfer. There is no load other than the weight of the beam. Draw complete stress diagrams.

11–3. The tendons in the beam of Problem 11–2 are placed 11 in. below the top of the beam at midspan. Assuming that no tension is allowed, determine the maximum service uniform load that the beam can carry in addition to its own

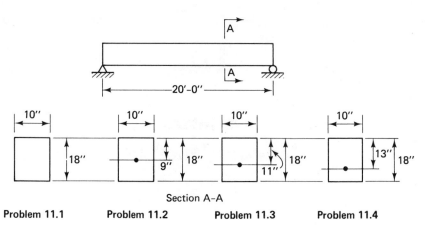

Section A-A

Problem 11.1   Problem 11.2   Problem 11.3   Problem 11.4

FIGURE P11-1 through P11-4

weight as governed by the stresses at midspan. Assume that f'_{ci} = 3,500 psi and f'_c = 5,000 psi.

11-4. Same as Problem 11-3, except that the tendons are 13 in. below the top of the beam at midspan and permissible stresses are as defined in the ACI Code, Section 18.4.

12

Composite
Bending Members

12-1 INTRODUCTION

Where reinforced concrete floor systems consisting of slabs and supporting beams are poured monolithically, the design of the beams assumes
that a portion of the slab is utilized as a compressive flange. Therefore,
the slab and beam behave as a unit in resisting the induced shears and
moments.

An alternative floor system is one in which the supporting beams are
erected prior to the placement of the slab. If the two elements are
properly interconnected, they, too, will resist induced shears and moments as a unit. The members resulting from such an interconnected
combination of beams and slabs are called *composite bending members*.
Common examples of composite bending members are either structural
steel beams or precast concrete beams in combination with a cast-in-
place concrete slab. The resulting floor system will have a cross section
as shown in Fig. 12-1.

A significant advantage of composite construction is the use of the
structural steel or precast concrete beams as supports for the slab forms.
Referring to Fig. 12-1, the interconnection between the two elements
must be so designed that as the total composite unit deflects, there is no
relative movement (slip) between the cast-in-place concrete slab and the
top of the precast beam (or top of the steel beam). The connection can,
therefore, be seen to be primarily a horizontal *shear* connection between
the two elements. For the steel beam, steel studs with heads are the

Precast Floor Components for a Parking Garage.

most common mechanical shear connector and are applied by welding to the flange as shown in Fig. 12–2(a).

Hooked studs [Fig. 12–2(b)], channel sections in short lengths [Fig. 12–2(c)], and spirals [Fig. 12–2(d)] have also been used to serve as shear connectors. For the precast concrete beam of Fig. 12–1(b), the interconnection between beam and slab is accomplished by intentionally roughening the top surface of the beam and extending vertical bars or stirrups from the beam into the slab. The detail must be such that the bars are fully anchored in both elements.

In the absence of shear connectors between the slab and steel beam, the resulting system is said to be *noncomposite*. The magnitude of shear transfer due to friction and/or bond between the two elements is highly indeterminate and unreliable and is therefore neglected. The supporting

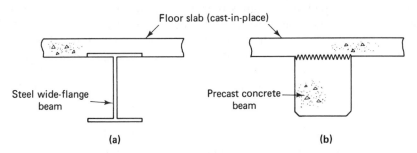

FIGURE 12-1 Typical composite cross sections

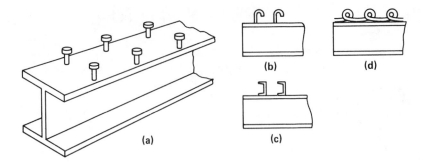

FIGURE 12-2 Shear connectors for composite beams

beam alone is then assumed to carry the total vertical loads from the slab. In some cases, if a limited number of shear connectors are installed which will transmit a limited, but known, amount of shear, the system is termed *partially composite*. Where both contact surfaces of the interconnecting elements are concrete, full transfer of horizontal shear forces may be assumed if the requirements of the ACI Code, Section 17.5.2, are satisfied.

Bridges having reinforced concrete decks supported on steel girders are commonly designed for composite action. In buildings, composite construction is most efficient with heavy loadings, relatively long spans, and beams that are spaced as far apart as practical with repetitive bay framing. Composite floor systems are stiffer and stronger than similar noncomposite floor systems with beams of the same size. In practice, this means that lighter and shallower beams may be used in composite systems. It also means that deflections of the steel or precast concrete section, if it alone must carry construction loads, may be excessive. Excessive dead load deflections as well as high dead load stresses can be prevented through the use of temporary shoring. This is a method of temporarily supporting the beam during the placing of the concrete slab. If shoring is not used, the beam itself must support all loads until the slab has developed its design strength, at which time the composite section will be available to resist all further loads. If temporary shoring is used because of strength and/or deflection considerations, it will be used only until the slab has developed sufficient strength; at which time the full composite section will be available to resist both dead and live loads. Tests to destruction have shown that there is no difference in the ultimate strength of similar members, whether constructed shored or unshored.

Figure 12–3 shows stress patterns in a composite beam constructed *without* temporary shoring. The steel beam alone supports the construction loads (form work, fresh concrete, construction live loads) and any other loads that may be put in place before the slab has gained sufficient strength. Figure 12–3(a) shows a typical stress diagram with the neutral axis at the center of the steel beam. Once the slab has gained strength, the *composite section* will resist further loads (primarily live loads) because of the mechanical interconnection provided between the beam and slab by the shear connectors. The stress pattern induced by these loads is shown in Fig. 12–3(b). Note that the neutral axis is much higher in the full composite section than it is for the steel beam alone. The load-carrying capacity of the composite beam is greater than that of the steel beam acting alone. Note also that the action of the full composite section is utilized only in resisting the loads placed *after* the slab has gained strength. Figure 12–3(d) shows the stress pattern that exists at ultimate moment capacity.

The stress patterns in a composite beam constructed *with* shoring are depicted in Fig. 12–4. The amount of shoring used will govern the amount of stress induced in the steel beam during construction. Sufficient shoring can be placed so that the construction load stresses will be minor. Thus the beam will remain virtually unstressed until after the slab has hardened and attained sufficient strength. After the temporary shores are removed, the composite section resists both dead and live loads. Over a long period of time, creep in the concrete slab due to sustained load will cause the steel beam to carry an additional portion of the sustained load while the full composite section resists the transient live loads.

The ultimate load stress patterns shown in Figs. 12–3(d) and 12–4(b),

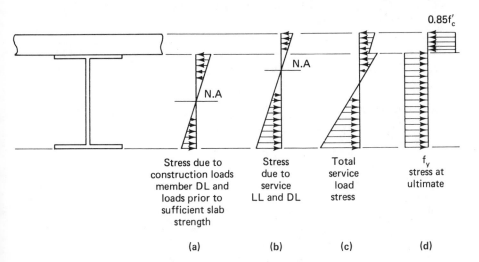

Stress due to construction loads member DL and loads prior to sufficient slab strength	Stress due to service LL and DL	Total service load stress	f_y stress at ultimate
(a)	(b)	(c)	(d)

FIGURE 12-3 Composite beam stresses without shoring

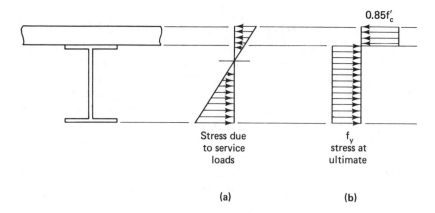

FIGURE 12-4 Composite beam stresses with shoring

are identical, assuming the same cross section for each condition, since the magnitude of the total tensile force, $A_s f_y$, is fixed. As may be observed, the neutral axis is portrayed as being located in the slab. This is reasonable for composite beams of ordinary proportions that may occur in buildings and implies that the full steel beam will be in tensile yielding.

It can be seen, then, that a composite beam must be checked at several stages with varying load and available strength combinations. The ACI Code, Chapter 17, contains general guidelines for composite flexural members consisting of concrete elements. The American Institute of Steel Construction has published in its Specification[1] provisions for the designing of composite steel–concrete beams. Additionally, the AISC *Manual* contains tables and examples of table usage for the foregoing. The following calculations will lend an understanding of the behavioral aspects of a simple composite steel–concrete beam.

12–2 FLEXURAL STRENGTH OF A COMPOSITE STEEL–CONCRETE BEAM

The same assumptions as previously used for the determination of M_R (ultimate moment strength) will be used to find M_R for a composite beam. Among these are: (1) the steel in tension will yield, (2) the maximum concrete strain will be 0.003, and (3) the stress–strain assumptions must be compatible. The AISC specification rules for effective flange width are exactly the same as for T-beams under the ACI Code. In addition to the M_R calculations, the allowable service load moment will be calculated, since the AISC provisions are in terms of working stresses.

EXAMPLE 12–1

Find the ultimate moment strength (M_R) for the fully composite steel–concrete beam shown in Fig. 12–5. Assume shoring so that the full capacity of the composite section is available to resist the design moment. The beam is a W16×40 and the slab is 5 in. thick. The steel is ASTM A36 (yield stress $F_y = 36$ ksi) and $f'_c = 3,000$ psi. From the AISC *Manual*, the necessary properties of the W16×40 are $A = 11.8$ in.², depth $d = 16.01$ in., and $I_x = 518$ in.⁴.

Determine the location of the neutral axis. The steel beam is assumed to be fully yielded (F_y exists on the entire cross section).

$$N_T = AF_y = 11.8(36) = 424.8 \text{ kips}$$

$$N_T = N_C = 0.85f'_c ab$$

$$a = \frac{N_T}{0.85f'_c b} = \frac{424.8}{0.85(3)(78)} = 2.14 \text{ in.}$$

Therefore,

$$c = \frac{a}{\beta_1} = \frac{2.14}{0.85} = 2.51 \text{ in.}$$

Check the assumption of a fully yielded beam by determining the strain at the top of the top flange ϵ_f as shown in Fig. 12–6.

$$\frac{\epsilon_f}{2.49} = \frac{0.003}{2.51}$$

$$\epsilon_f = \frac{2.49}{2.51}(0.003) \approx 0.003$$

$$\epsilon_y = \frac{F_y}{E_s} = \frac{36,000}{29,000,000} = 0.0012$$

Since $0.003 > 0.0012$, the top of the flange has yielded and the assumption is valid.

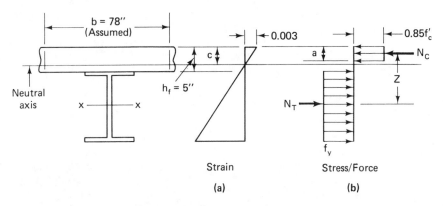

FIGURE 12-5   Sketch for Example 12–1

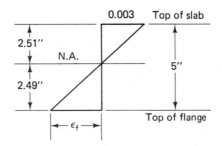

0.003   Top of slab

2.51″

N.A.

5″

2.49″

ϵ_f

Top of flange

FIGURE 12-6   Concrete strain diagram for Example 12–1

Find M_R.

$$M_R = \phi N_T Z \quad \text{or} \quad \phi N_c Z$$

$$Z = \frac{d}{2} + \left(h_f - \frac{a}{2} \right) = 8.00 + \left(5.00 - \frac{2.14}{2} \right)$$

$$= 11.93 \text{ in.}$$

$$M_R = 0.90(424.8)(11.93)\frac{1}{12} = 380 \text{ ft-kips}$$

The AISC Specification, under which the preceding composite steel–concrete beam would be be designed, is in terms of working stresses (allowable stresses). Among the design assumptions are the following:

1. The maximum concrete stress = $0.45 f'_c$. Concrete tension stresses are neglected.

2. The maximum steel stress = $0.66 F_y$, where F_y is the yield stress of the steel beam.

3. The compression area of the concrete on the compression side of the neutral axis is to be treated as an *equivalent area of steel* by dividing it by the modular ratio n (see Chapter 7).

4. The properties of the composite section are computed in accordance with elastic theory.

EXAMPLE 12–2

For the composite steel–concrete beam of Example 12–1, find the allowable service moment as governed by the AISC Specification provisions.

To determine the allowable service moment, the concrete slab will be transformed into an equivalent steel area. The neutral axis location will be determined, the moment of inertia will be calculated, the governing allowable stress will be determined, and the allowable moment will be found.

From Table A–5, $n = 9$. Therefore, the width of the equivalent steel area may be found as

$$\frac{\text{Slab width}}{n} = \frac{78}{9} = 8.67 \text{ in.}$$

The transformed steel section is shown in Fig. 12–7(a).

A trial calculation for the neutral-axis location will show that it will fall in the slab (equivalent steel area). Therefore, a reference axis is established at the top of the section and $\bar{y}$ is found. Note that concrete in tension, below the neutral axis, is not included in the calculation, since it is not effective.

$$\bar{y} = \frac{\Sigma (Ay)}{\Sigma A} = \frac{\dfrac{(8.67)\bar{y}^2}{2} + 11.8\left(\dfrac{16.01}{2} + 5.0\right)}{8.67\bar{y} + 11.8}$$

$$8.67\bar{y}^2 + 11.8\bar{y} = 4.33\bar{y}^2 + 153.5$$

$$4.33\bar{y}^2 + 11.8\bar{y} = 153.5$$

$$\bar{y}^2 + 2.73\bar{y} = 35.45$$

Using a method of solution for this expression similar to that in Chapter 3, it is found that

$$\bar{y} = 4.74 \text{ in.}$$

The moment of inertia of the effective transformed area may now be found.

$$I = \Sigma I_c + \Sigma Ad^2$$

$$= \frac{1}{3}(8.67)(4.74)^3 + 518 + 11.8\left(\frac{16.01}{2} + 0.26\right)^2$$

$$= 1,632 \text{ in.}^4$$

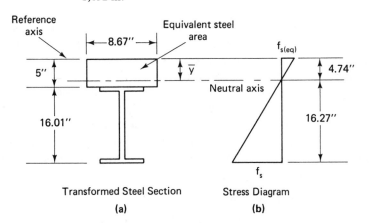

Transformed Steel Section

(a)

Stress Diagram

(b)

FIGURE 12-7 Sketches for Example 12-2

Determine which allowable stress will govern. The allowable concrete stress is $0.45f'_c = 1.35$ ksi. The allowable steel stress is $0.66F_y = 23.8$ ksi. In Fig. 12–7(b), $f_{s(eq)}$ is the maximum compressive stress on the equivalent steel area and f_s is the maximum tensile stress in the steel beam. The computed concrete stress f_c may be found (see Chapter 7 for background) as

$$f_c = \frac{f_{s(eq)}}{n}$$

If we assume that f_s at the bottom of the beam is equal to the allowable steel stress, 23.8 ksi, then, from Fig. 12–7(b),

$$f_{s(eq)} = \frac{4.74}{16.27}(23.8) = 6.92 \text{ ksi}$$

from which

$$f_c = \frac{6.92}{9} = 0.77 \text{ ksi}$$

Since 0.77 ksi < 1.35 ksi (the concrete allowable), it can be seen that the steel will reach its allowable stress first and that Fig. 12–7(b) does represent the stress diagram at allowable service moment. Utilization of the flexure formula will result in

$$\text{Allowable moment} = \frac{(\text{allowable steel stress})(I)}{c}$$

$$= \frac{23.8(1,632)}{16.27(12)} = 199 \text{ ft-kips}$$

12–3 SHEAR AND DEFLECTIONS

For composite steel–concrete beams, the AISC *Manual* contains guidelines for determining the specific number of stud shear connectors necessary based on the shear that can be carried by each stud. Both full and partial composite designs are covered. For composite members utilizing precast reinforced concrete beams, the ACI Code, Section 17.5, requires full transfer of the shear force between elements. Full transfer may be assumed when *all* of the following criteria are satisfied:

    1. The contact surfaces are clean, free of laitance, and intentionally roughened (roughened with a full amplitude of approximately 1/4 in.).

    2. Specified minimum ties are provided.

    3. Web members are designed to resist total vertical shear.

    4. All shear reinforcement is fully anchored into all interconnected elements.

If the previous criteria are not satisfied, horizontal shear must be investigated in accordance with the ACI Code, Section 17.5.4 or 17.5.5, to verify full composite action.

The AISC guidelines for deflections suggest minimum depth/span ratios for composite steel–concrete beams, where the depth is taken as full depth from the top of the slab to the bottom of the beam. If rational calculations for deflections are performed, the moment of inertia as determined for stress (see Example 12–2) may be used for live load deflections, but for long-time (sustained load) deflections, to approximate creep effects, the moment of inertia should be determined using $2n$ in place of n for the concrete-to-steel transformation.

The ACI Code guidelines for deflection control of composite reinforced concrete members (Section 9.5.5) follow the Code provisions for deflections previously discussed. For shored construction, the composite member may be considered equivalent to a cast-in-place member. The minimum-depth values of ACI Table 9.5(a) apply (with the normal restrictions). For unshored construction, if the depth of the precast member meets the requirements of ACI Table 9.5(a), deflections occurring after composite action begins need not be calculated. It may be necessary to check the long-time deflection of the precast member prior to the beginning of effective composite action.

PROBLEMS

12–1. Calculate M_R for the cross section shown. The beam is a W21×68 of A36 steel and $f'_c = 3,000$ psi. Assume full composite action.

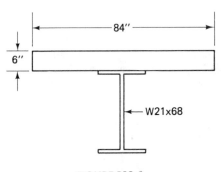

FIGURE P12-1

12–2. For the composite beam of Problem 12–1, find the allowable service moment strength.

12–3. Calculate M_R for the typical composite beam of the floor system shown. The steel beams are W12×40 of A36 steel, are spaced 6 ft 0 in. on center, and are on a simple span of 20 ft. Assume full composite action and an effective slab width of 72 in. $f'_c = 3,500$ psi.

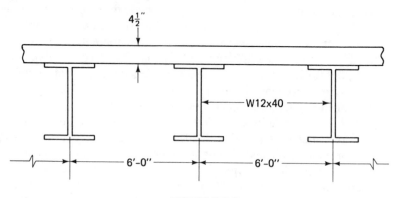

FIGURE P12-3

12–4. The superimposed service loads for the floor system of Problem 12–3 are 50 psf dead load and 350 psf live load. Develop and sketch complete service load stress diagrams and determine if the composite beam is adequate. Assume full shoring.

12–5. Repeat Problem 12–4 assuming no shoring. The following additional loads are applicable: 10 psf load due to forms and 75 psf construction live load. Draw stress diagrams for all loading stages through initial occupancy. Forms are removable.

REFERENCE

1. *Manual of Steel Construction*, 7th ed., American Institute of Steel Construction, Wrigley Bldg., 400 N. Michigan Ave., Chicago, Illinois 60611.

13

Detailing Reinforced Concrete Structures

13-1 INTRODUCTION

The contract documents package for a typical building as developed by an architect/engineer's office commonly includes both drawings and specifications. The drawings typically concern the following areas: site, architectural, structural, mechanical, and electrical. The specifications supplement and amplify the drawings. The contract documents package is the product that results from what may be categorized as the planning and design phase of a project.

The next sequential phase may be categorized as the *construction phase*. It includes many subcategories for reinforced concrete structures, two of which are *detailing* and *fabricating* of the reinforcing steel. As described in the ACI *Manual of Standard Practice for Detailing Reinforced Concrete Structures* (ACI 315–74),[1] detailing consists of the preparation of placing drawings, reinforcing bar details, and bar lists which are used for the fabrication and placement of the reinforcement in a structure. Fabricating consists of the actual shopwork required for the reinforcing steel, such as cutting, bending, bundling, and tagging.

Most bar fabricators not only supply the reinforcing steel but also prepare the placing drawings and bar lists, fabricate the bars, and deliver to the project site. In some cases, the bar fabricator may also act as the placing subcontractor.

It is general practice in the United States for all reinforced concrete used in building projects to be designed, detailed, and fabricated in

345

Reactor Containment Foundation Mat—Seabrook Station, New Hampshire.

accordance with the ACI Code (ACI 318–77). In addition, the Concrete Reinforcing Steel Institute regularly publishes its *Manual of Standard Practice*[2] which contains the latest recommendations of the reinforcing steel industry for standardization of materials and practices.

Techniques have also been developed which make use of electronic computers and other data-processing equipment to facilitate the generation of bar lists and other components of the detailing process. This not only aids in standardization and accuracy of the documents produced, but can also be readily incorporated into the stock control system and the shopwork planning of the reinforcing steel fabricator.

13–2 PLACING DRAWINGS

The *placing drawing* (commonly called a *shop drawing*) consists of a plan view with sufficient sections to clarify and define bar placement. As such, it is the guide that the ironworkers will use as they place the reinforcing steel on the job. Additionally, the placing drawing will contain typical views of beams, girders, joists, columns, and other members as necessary. Frequently, tabulations called *schedules* are used to list similar members, which vary in size, shape, and reinforcement details. A bar list and/or bending details may or may not be shown on the placing drawing, as some fabricators not only have their own preferred format but prefer the list to be prepared as a separate entity.

The preparation of the placing drawing is based on the complete set of contract documents and generally contains only the information necessary for bar fabrication and placing. Building dimensions are not shown unless they are necessary to locate the steel properly.

A typical placing drawing of a beam and girder floor system may be observed in Fig. 13–2. This may be contrasted to the engineering drawing of the same system in Fig. 13–1, which in reality is part of the structural plans of the contract documents. The engineering drawing, as may be observed, shows the floor plan view locating and identifying the structural elements along with views of typical beams, girders, and slabs and their accompanying schedules. The placing drawing supplements the engineering drawing by furnishing all the information necessary for bar fabrication and placement. The precise size, shape, dimensions, and location of each bar are furnished utilizing a marking system which will be discussed later in this chapter. Information relative to bar supports may also be included. Figure 13–2 illustrates a placing drawing which includes placing data for bar supports as well as a bending details schedule.

Placing drawings, in addition to controlling the placement of the steel in the forms, serve as the basis for ordering the steel. Therefore, a proper interpretation of the contract documents by the fabricator is absolutely essential. Generally, all placing drawings are submitted to the architect or engineer for checking and approval before shop fabrication will begin.

D BAR MARKS

wo identification systems are required. The
n of the various structural members, while
ntification of the individual bars within the
em for the structural members may consist
l indentification for each beam, girder and
nated numerically as used in Figs. 13–1 and
n of alphabetical and numerical coordinates
lumns are numbered consecutively in one
cutively in the other. A coordinate system
undation-engineering drawing of Fig. 13–3,
a coordinate designation such as B2 or C3.
ablished on the architectural and structural
e detailer unless the detailer requires a more

Footings, as may be observed in Fig. 13–3, are generally designated

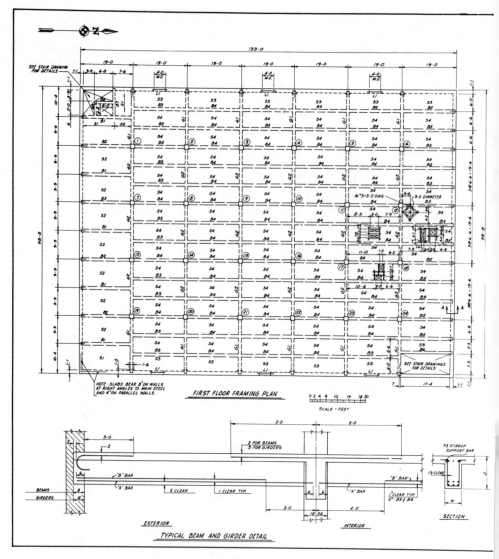

FIGURE 13-1 Engineering drawing for beam and girder framing
Courtesy—American Concrete Institute—Ref. 1

with an F prefix followed by a number, such as F1 and F3, without regard to a coordinate system. Footing piers or pedestals may be identified using the coordinate system such as B2 or D4, or may be designated with a P prefix followed by a number, such as P1 or P3. Beams, joists, girders, lintels, slabs, and walls are generally given

348

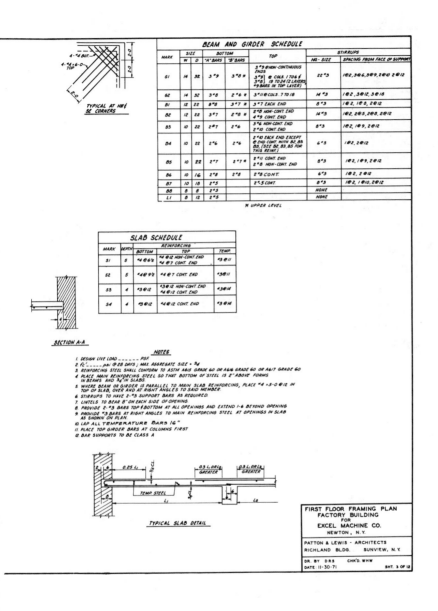

designations which indicate the specific floor in the building, the type of member, and an identifying number. For example, 1G2 indicates a first floor girder numbered 2; RB4 indicates a roof beam numbered 4.

Generally, the alphabetical designations for the various structural members are as follows:

349

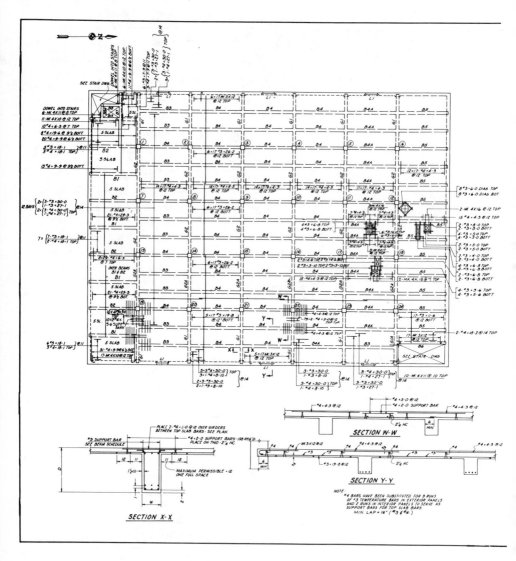

FIGURE 13-2 Placing drawing for beam and girder framing
Courtesy—American Concrete Institute—Ref. 1

B: beams	L: lintels
J: joists	P: piers
W: walls	G: girders
F: footings	S: slabs

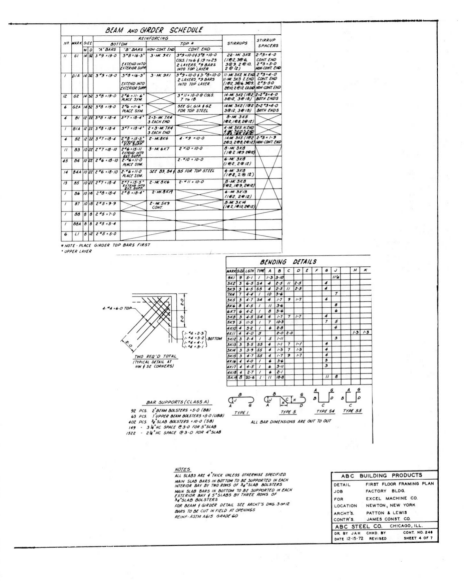

Along with a marking system established for the structural members, a system of identifying and marking the reinforcing bars must be established.

In buildings, only bent bars are furnished with a mark number or designation. The straight bar has its own identification by virtue of its size and length. Numerous systems are in use throughout the industry,

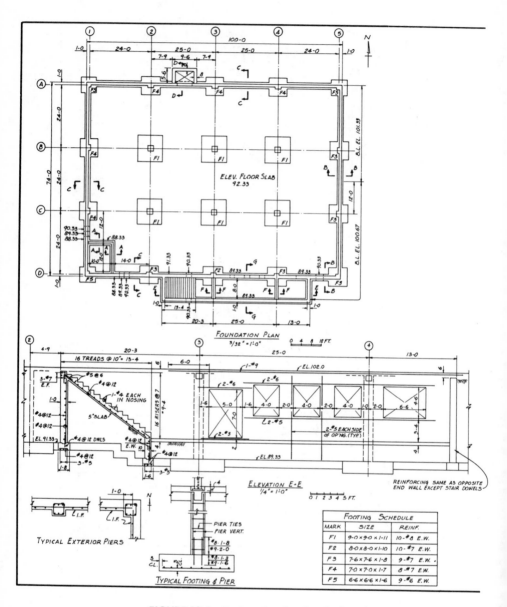

FIGURE 13-3 Engineering drawing for foundations
Courtesy—American Concrete Institute—Ref. 1

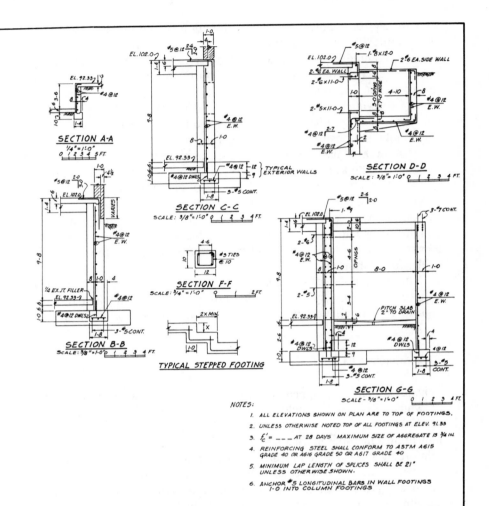

SECTION A-A
1/4" = 1'-0"
0 1 2 3 4 5 FT.

SECTION B-B
SCALE: 3/8" = 1'-0" 0 1 2 3 4 FT.

SECTION C-C
SCALE: 3/8" = 1'-0" 0 1 2 3 4 FT.

} TYPICAL EXTERIOR WALLS

SECTION F-F
SCALE: 3/4" = 1'-0" 0 1 2 FT.

TYPICAL STEPPED FOOTING

SECTION D-D
SCALE: 3/8" = 1'-0" 0 1 2 3 4 FT.

SECTION G-G
SCALE - 3/8" = 1'-0" 0 1 2 3 4 FT.

NOTES:

1. ALL ELEVATIONS SHOWN ON PLAN ARE TO TOP OF FOOTINGS.

2. UNLESS OTHERWISE NOTED TOP OF ALL FOOTINGS AT ELEV. 91.33

3. f'_c = ___ AT 28 DAYS MAXIMUM SIZE OF AGGREGATE IS 3/4 IN.

4. REINFORCING STEEL SHALL CONFORM TO ASTM A615
   GRADE 40 OR A616 GRADE 50 OR A617 GRADE 40

5. MINIMUM LAP LENGTH OF SPLICES SHALL BE 21"
   UNLESS OTHERWISE SHOWN.

6. ANCHOR #5 LONGITUDINAL BARS IN WALL FOOTINGS
   1'-0 INTO COLUMN FOOTINGS

PIER SCHEDULE			
PIER	SIZE	VERTICAL	TIES
B2, B3, B4, C2, C3, C4	16 × 16	8 - #9	#3@12
D3, D4	16 × 16	8 - #8	#3@12
A1 THRU A5 B1, B5, C1, C5 D2, D5	16 × 16	6 - #8	#3@12
D1	16 × 16	6 - #8	#3@12

**FOUNDATIONS & PIERS
OFFICE BUILDING
FOR
BAILEY-JONES CO.
EASTON, PA.**

R.A. SMITH & ASSOCIATES
ARCHITECTS-ENGINEERS
BATH, PA.

DR. BY: W.C.R.
DATE: 11-30-71   SHEET 1 OF 10

with the system choice generally a function of the building type, size, and complexity, as well as the standards of each fabricator.

One commonly used system is the use of an arbitrary letter such as K followed by consecutive numbers throughout the project without regard to bar location or shape. This system may be observed in the placing drawing of Fig. 13–2. The letter is prefixed with the size of the bar. For example, 6K7 represents a No. 6 bar whose shape and dimensions may be observed in the bending details schedule and whose location, in this case, may be established by the beam and girder schedule as being the top reinforcing at the noncontinuous end of the B3 members.

An alternative system is to use many letters rather than one arbitrary letter. Column bars may be designated with a C and footing bars with an F. For example, a 7F5 would be a No. 7 footing bar, whose shape and dimensions would be established in a bending details schedule and whose location would be observed in a footing schedule, typical footing details, or the foundation plan.

Other acceptable systems are currently being used. Of primary importance in any system that is chosen is that it should be simple, logical, easy to understand, and not lead to ambiguity or confusion.

13–4 SCHEDULES

Schedules generally appear both on engineering drawings and placing drawings. Typical schedules may be observed in Fig. 13–1, 13–2, and 13–3. On the engineering drawings (Fig. 13–1), it is a tabular form indicating a member mark number, concrete dimensions, and the member reinforcing steel size and location. Specific design details pertinent to the member reinforcing may also be furnished in the schedule. On the engineering drawing, the schedule must be correlated with a typical section and plan view to be meaningful. Schedules are generally used for the typical members, among which are slabs, beams, girders, joists, columns, footings, and piers. Typical footing and pier schedules may be observed in Fig. 13–3, and typical beam, girder, and slab schedules may be observed in Fig. 13–1.

Similar schedules are used on the placing drawings, but added information is furnished. The placing drawing schedule is more detailed and generally indicates the number of bars, member mark number, and physical dimensions of member reinforcing steel. Additionally, it indicates size and length of straight bars, mark numbers (if bent bars), location of bars, and spacing, along with all specific notes and comments relative to the reinforcing bars. This schedule must also be worked together with typical sections and plan views. The schedules are generally accompanied by a bar list, indicating bending details which may or may not be presented on the placing drawing. There is no standard format for either engineering drawing or placing drawing schedules. They are merely a convenient technique of presenting information for a group of similar items, such as groups of beams, girders, columns, and footings. The schedule format will vary somewhat to conform to the requirements of a particular job.

13-5 FABRICATING STANDARDS

The fabrication process consists of cutting, bending, bundling, and tagging the reinforcing steel. Our discussion will primarily be limited to the cutting and bending because of their effect on a member's structural capacity. Bending, which includes the making of standard hooks, is generally accomplished in accordance with the requirements of the ACI Code (ACI 318-77), the provisions of which have been discussed in Chapter 5.

In the fabricating shop, bars to be bent are first cut to length as stipulated and then sent to a special bending department, where they are bent as designated in the bending details schedule or bar lists. The common types of bent bars have been standardized throughout the industry, and applicable configurations are generally incorporated onto the placing drawing or bar lists in conjunction with the bending details (see Fig. 13-2). Each configuration, sometimes called a bar type, has a designation such as 7, 8, or 9, and S1 or T1, with each dimension designated by a letter. Typical bar bends may be observed in Fig. 13-4. In addition, standards have been established with respect to the details of the hooks and bends. The ACI Code (Sections 7.1, 7.2, and 7.3) establishes minimum requirements and is graphically portrayed in Fig. 13-5. This table also shows the extra length of bar needed for the hook (A or G), which must be added to the sum of all other detailed dimensions to arrive at the total length of bar. It is common practice to show all bar dimensions as *out to out* (meaning outside to outside) of bar. The ACI Code also stipulates that bars must be bent cold unless indicated otherwise by the engineer/architect. Field bending of bars partially embedded in concrete is not allowed unless specifically permitted by the engineer/architect (ACI 7.3).

Straight bars are cut to the prescribed length from longer stock-length bars which are received in the fabricating shop from the mills. Tolerances in fabrication of reinforcing steel are generally standardized and may be observed in Fig. 13-6. For instance, the cutting tolerance for straight bars is the specified length ± 1 in. unless special tolerances are called for. Due consideration for these tolerances must be made by both the engineer and contractor in the design and construction phases.

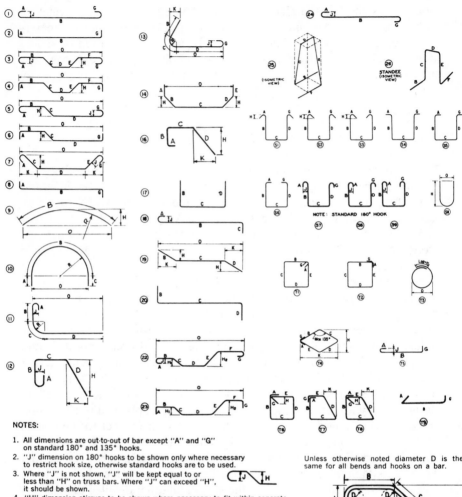

NOTES:

1. All dimensions are out-to-out of bar except "A" and "G" on standard 180° and 135° hooks.
2. "J" dimension on 180° hooks to be shown only where necessary to restrict hook size, otherwise standard hooks are to be used.
3. Where "J" is not shown, "J" will be kept equal to or less than "H" on truss bars. Where "J" can exceed "H", it should be shown.
4. "H" dimension stirrups to be shown where necessary to fit within concrete.
5. Where bars are to be bent more accurately than standard bending tolerances, bending dimensions which require closer fabrication should have limits indicated.
6. Figures in circles show types.
7. For recommended diameter "D", of bends, hooks, etc., see tables, page 6-4.
8. Type S1-S9, S11, T1-T9 apply to bar sizes #3, #4 and #5 only, except for 135° Seismic Stirrup/Tie Hooks which include #6 bars.

Unless otherwise noted diameter D is the same for all bends and hooks on a bar.

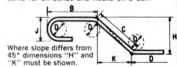

Where slope differs from 45° dimensions "H" and "K" must be shown.

ENLARGED VIEW SHOWING
BAR BENDING DETAILS

FIGURE 13-4   Typical bar bends
Courtesy—Concrete Reinforcing Steel Institute—Ref. 2

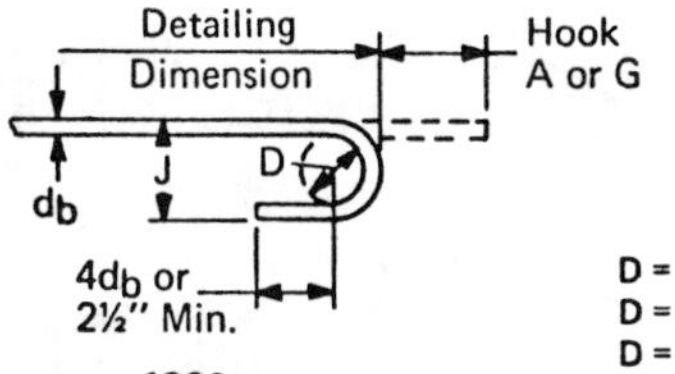

Detailing
Dimension
Hook
A or G
d_b
J
D
$4d_b$ or
2½″ Min.

180°

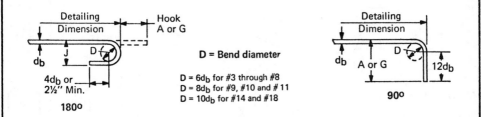

D = Bend diameter

D = $6d_b$ for #3 through #8
D = $8d_b$ for #9, #10 and # 11
D = $10d_b$ for #14 and #18

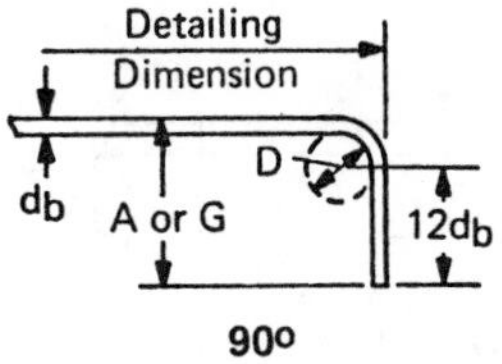

Detailing
Dimension
D
d_b
A or G
$12d_b$

90°

Bar size	Dimensions of standard 180-deg hooks, all grades			Dimensions of standard 90-deg hooks, all grades	
	A or G	J	D	A or G	D
#3	5″	3″	2¼″	6″	2¼″
#4	6	4	3	8	3
#5	7	5	3¾	10	3¾
#6	8	6	4½	1′-0″	4½
#7	10	7	5¼	1-2	5¼
#8	11	8	6	1-4	6
#9	1′-3″	11¼	9	1-7	9
#10	1-5	1′-0¾″	10¼	1-10	10¼
#11	1-7	1-2¼	11¼	2-0	11¼
#14	2-2	1-8½	17	2-7	17
#18	2-11	2-3	22¾	3-5	22¾

NOTE: When available depth is limited, #3 through #11 Grade 40 bars having 180-deg hooks may be bent with $D=5d_b$ and correspondingly smaller A and J dimensions.

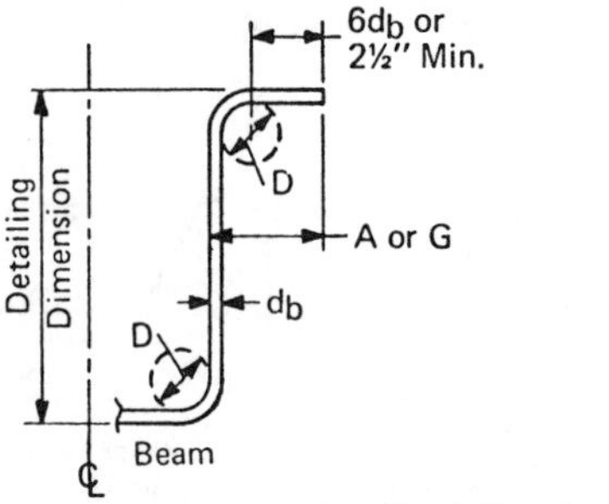

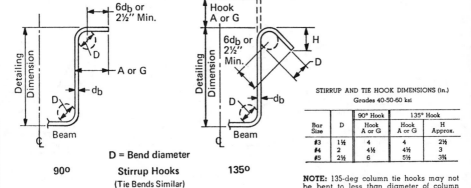

$6d_b$ or
2½″ Min.
D
A or G
d_b
D
Detailing
Dimension
Beam
℄

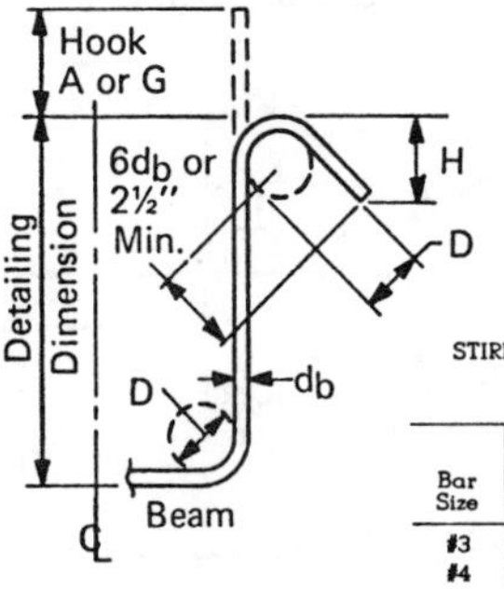

Hook
A or G
H
$6d_b$ or
2½″
Min.
D
Detailing
Dimension
D
d_b
Beam
℄

D = Bend diameter

90° **Stirrup Hooks** 135°

(Tie Bends Similar)

STIRRUP AND TIE HOOK DIMENSIONS (in.)

Grades 40-50-60 ksi

Bar Size	D	90° Hook	135° Hook	
		Hook A or G	Hook A or G	H Approx.
#3	1½	4	4	2½
#4	2	4½	4½	3
#5	2½	6	5½	3¾

NOTE: 135-deg column tie hooks may not be bent to less than diameter of column vertical bar enclosed in hook.

HOOKS AND BENDS OF WELDED WIRE FABRIC

Inside diameter of bends in welded wire fabric, plain or deformed, for stirrups and ties shall be at least four wire diameters for wire larger than D6 or W6 and two wire diameters for all other wires. Bends with inside diameter of less than eight wire diameters shall not be less than four wire diameters from nearest welded intersection.

FIGURE 13-5 Standard hook details
Courtesy—American Concrete Institute—Ref. 1

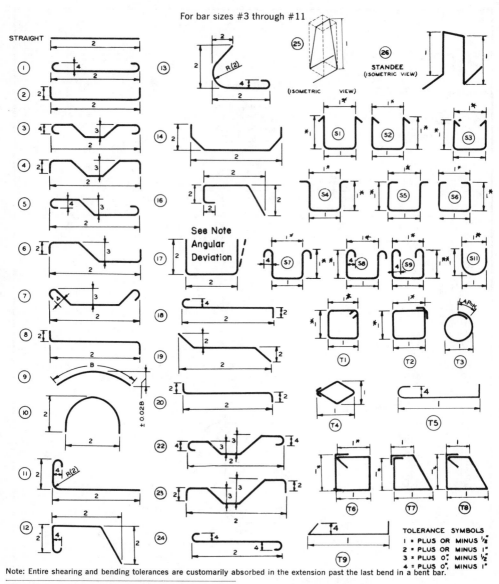

For bar sizes #3 through #11

Note: Entire shearing and bending tolerances are customarily absorbed in the extension past the last bend in a bent bar.

*Dimensions on this line are to be within tolerance shown but are not to differ from the opposite parallel dimension more than ½".
Angular Deviation — maximum ±2½° or ±½"/ft., but not less than ½", on all 90° hooks and bends.
Tolerances for Types S1-S9, S11, T1-T9 apply to bar sizes #3, #4 and #5 only, except for 135° Seismic Stirrup/Tie Hooks which include #6 bars.

FIGURE 13-6 Standard fabricating tolerances
Courtesy—Concrete Reinforcing Steel Institute—Ref. 2

ABC STEEL PRODUCTS CO.
CHICAGO, ILLINOIS

GRADE _As Noted_ _ _ _ _ _

ORDER NO. _B-32701_ _ _ _ _

PROJECT _Grant Office Bldg._ _

CUSTOMER _Stanford Const. Co._ _ _

LOCATION _Tulsa, Okla._ _ _ _ _

MATERIAL FOR _Footings & Dowels_

DRG. NO. _A_ _ _ _ _ _ _ _ _ _

SHEET _1_ . OF _1_ _ _ _ _ _

DATE _3-12-69_ REV. _7-29-69_

MADE BY _A.R.H._ CH'K BY _B.R.T._

For typical bend types refer to _ _ _ _ _ _ _ _ _ _

Item	Grade	No. Pieces	Size	Length	Mark	Type	A	B	C	D	E	F	G	H	J	K	R	O
1	STRAIGHT																	
2	60	16	11	11-0														
3																		
4		16	9	4-8														
5																		
6		64	7	11-6														
7		80	7	7-6														
8																		
9		4	6	15-0														
10		48	6	11-6														
11		4	6	11-0														
12		8	6	10-6														
13		32	6	9-6														
14		184	6	7-6														
15																		
16	60	19	4	3-6														
17																		
18	HEAVY BENDING																	
19																		
20	60	24	11	6-11	D1100	2	2-0	4-11										
21																		
22		36	10	6-0	D1000	2	1-10	4-2										
23																		
24		6	9	6-0	D900	2	1-7	4-5										
25		8	9	5-6	D901	2	1-7	3-11										
26																		
27		131	6	3-7	D600	2	1-0	2-7										
28																		
29	60	95	5	3-2	D500	2	10	2-4										
30																		
31	LIGHT BENDING																	
32																		
33	40	16	3	10-5	T300	T3												3-0
34																		
35																		
36																		
37																		
38																		
39																		

FIGURE 13-7 Typical bar list
Courtesy—Concrete Reinforcing Steel Institute—Ref. 4

13–6 BAR LISTS

The bar list serves several purposes. It is used for fabrication including cutting, bending and shipping as well as placement and inspection. It represents a bill of materials indicating complete descriptions of the various bar items. The information in a bar list is obtained from the placing drawing or while the placing drawing is being prepared. A typical bar list form may be observed in Fig. 13–7, and its similarity to the bending details schedule of Fig. 13–2 is apparent.

The bar list generally includes both straight and bent bars and indicates all bar dimensions and bends as well as grade of steel and number of pieces. A bar list of this type may be used in addition to a bending detail schedule on the placing drawing. However, some fabricators prefer not to use a bending detail schedule. With this system the detailers will utilize sketches until the drawing is complete and then transfer the information to separate bar lists. Two bar lists are prepared, one for straight bars and one for bent bars.

13–7 EXTRAS

Bars are sold on the basis of weight. To a base price are added various extra charges (extras) dependent principally on the amount of effort required to produce the final product. Among these extras are *bending extras* and *special fabrication extras*.

Bent bars are generally classified as *heavy bending* or *light bending*. Extra charges are made for all shop bending, with the charge a function of the classification. Due to the increased amount of handling and number of bends per pound of steel, light bending charges per pound are appreciably more than the charges for heavy bending. According to ACI, *heavy bending* is defined as bar sizes Nos. 4–18, which are bent at not more than six points, radius bent to one radius, and bending not otherwise defined. *Light bending* includes all No. 3 bars; all stirrups and column ties; and all bars Nos. 4–18 which are bent at more than six points, bent in more than one plane, radius bent with more than one radius in any one bar, or a combination of radius and other bending. Special fabrication includes fabrication of bars specially suited to conditions for a given project. This may include special tolerances and variations from minimum standards as well as unusual bends and spirals.

13–8 BAR SUPPORTS AND BAR PLACEMENT

Bar supports are used to hold the bars firmly at their designated locations before and during the placing of concrete. These supports may be of metal, plastic, precast concrete, or other approved materials. Standard

SYMBOL	BAR SUPPORT ILLUSTRATION	TYPE OF SUPPORT	STANDARD SIZES
SB	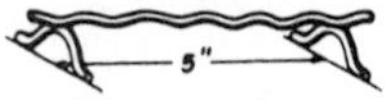	Slab Bolster	¾, 1, 1½, and 2 inch heights in 5 ft. and 10 ft. lengths
SBU*		Slab Bolster Upper	Same as SB
BB		Beam Bolster	1, 1½, 2; over 2″ to 5″ heights in increments of ¼″ in lengths of 5 ft.
BBU*	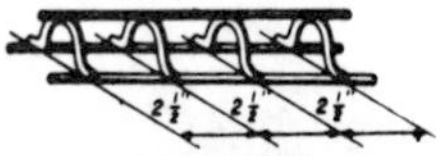	Beam Bolster Upper	Same as BB
BC		Individual Bar Chair	¾, 1, 1½, and 1¾″ heights
JC		Joist Chair	4, 5, and 6 inch widths and ¾, 1. and 1½ inch heights
HC		Individual High Chair	2 to 15 inch heights in increments of ¼ in.
HCM*	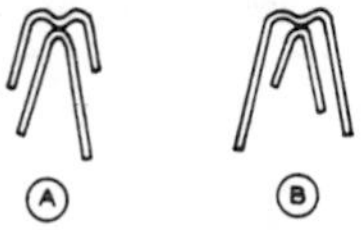	High Chair for Metal Deck	2 to 15 inch heights in increments of ¼ in.
CHC		Continuous High Chair	Same as HC in 5 foot and 10 foot lengths
CHCU*	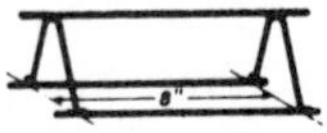	Continuous High Chair Upper	Same as CHC
CHCM*	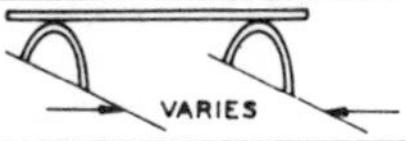	Continuous High Chair for Metal Deck	Up to 5 inch heights in increments of ¼ in.
JCU**	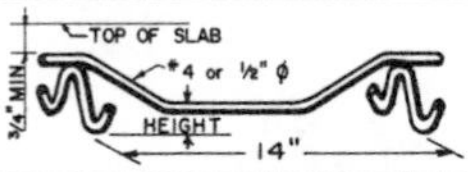	Joist Chair Upper	14″ Span. Heights —1″ through +3½″ vary in ¼″ increments

*Available in Class A only, except on special order.
**Available in Class A only, with upturned or end bearing legs.

FIGURE 13-8 **Standard types and sizes of wire bar supports**
Courtesy—Concrete Reinforcing Steel Institute—Ref. 2

specifications for steel wire bar supports as well as the placing of bar supports have been established by the Concrete Reinforcing Steel Institute (CRSI). Standard types and sizes of wire bar supports are shown in Fig. 13–8. It is a general practice that unless the engineers drawings or specifications show otherwise, bar supports will be furnished in accordance with CRSI standards. More detailed information relative to bar supports may be found in the CRSI publications *Placing Reinforcing Bars*[3] and *Manual of Standard Practice*.[2]

REFERENCES

1. *Manual of Standard Practice for Detailing Reinforced Concrete Structures* (ACI 315–74), The American Concrete Institute, P.O. Box 19150, Redford Station, Detroit, Michigan 48219.

2. *Manual of Standard Practice*, 1976 ed., Concrete Reinforcing Steel Institute, 180 North LaSalle Street, Chicago, Illinois 60601.

3. *Placing Reinforcing Bars*, 3rd ed., Concrete Reinforcing Steel Institute, 180 North LaSalle Street, Chicago, Illinois 60601.

4. *Reinforcing Bar Detailing*, 1970, Concrete Reinforcing Steel Institute, 180 North LaSalle Street, Chicago, Illinois 60601.

A

Tables and Charts

TABLE A–1
Reinforcing Steel

Reinforcing Steel Summary*

Type of Steel and ASTM Specification Number	Sizes Available	Grade	Minimum Tensile Strength (psi)	f_y (psi)	ϵ_y
Billet steel, A615	Nos. 3–11	40	70,000	40,000	0.00138
	Nos. 3–11 Nos. 14 and 18	60	90,000	60,000	0.00207
Axle steel, A617	Nos. 3–11	40	70,000	40,000	0.00138
		60	90,000	60,000	0.00207
Rail steel, A616	Nos. 3–11	50	80,000	50,000	0.00172
		60	90,000	60,000	0.00207
Low alloy steel, A706	Nos. 3–11 Nos. 14 and 18	60	80,000 (Min.: 1.25 f_y)	60,000 (Max.: 78,000)	0.00207

* The nominal dimensions of a deformed bar (diameter, area, perimeter) are equivalent to those of a plain round bar having the same weight per foot as the deformed bar.

TABLE A–1
Reinforcing Steel (cont.)

Properties of Steel Reinforcing Bars
Unit Weight and Nominal Diameter

Bar Number	2	3	4	5	6	7	8	9	10	11	14	18
Unit weight per foot (lb)	0.167	0.376	0.668	1.043	1.502	2.044	2.670	3.400	4.303	5.313	7.65	13.60
Diameter (in.)	0.250	0.375	0.500	0.625	0.750	0.875	1.000	1.128	1.270	1.410	1.693	2.257

Areas of Multiples of Reinforcing Bars (in.^{2})

Number of Bars	Bar Number									
	2	3	4	5	6	7	8	9	10	11
1	0.05	0.11	0.20	0.31	0.44	0.60	0.79	1.00	1.27	1.56
2	0.10	0.22	0.40	0.62	0.88	1.20	1.58	2.00	2.54	3.12
3	0.15	0.33	0.60	0.93	1.32	1.80	2.37	3.00	3.81	4.68
4	0.20	0.44	0.80	1.24	1.76	2.40	3.16	4.00	5.08	6.24
5	0.25	0.55	1.00	1.55	2.20	3.00	3.93	5.00	6.35	7.80
6	0.30	0.66	1.20	1.86	2.64	3.60	4.74	6.00	7.62	9.36
7	0.35	0.77	1.40	2.17	3.08	4.20	5.53	7.00	8.89	10.9
8	0.40	0.88	1.60	2.48	3.52	4.80	6.32	8.00	10.2	12.5
9	0.45	0.99	1.80	2.79	3.96	5.40	7.11	9.00	11.4	14.0
10	0.50	1.10	2.00	3.10	4.40	6.00	7.90	10.0	12.7	15.6
11	0.55	1.21	2.20	3.41	4.84	6.60	8.69	11.0	14.0	17.2
12	0.60	1.32	2.40	3.72	5.28	7.20	9.48	12.0	15.2	18.7
13	0.65	1.43	2.60	4.03	5.72	7.80	10.3	13.0	16.5	20.3
14	0.70	1.54	2.80	4.34	6.16	8.40	11.1	14.0	17.8	21.8
15	0.75	1.65	3.00	4.65	6.60	9.00	11.8	15.0	19.0	23.4
16	0.80	1.76	3.20	4.96	7.04	9.60	12.6	16.0	20.3	25.0
17	0.85	1.87	3.40	5.27	7.48	10.2	13.4	17.0	21.6	26.5
18	0.90	1.98	3.60	5.58	7.92	10.8	14.2	18.0	22.9	28.1
19	0.95	2.09	3.80	5.89	8.36	11.4	15.0	19.0	24.1	29.6
20	1.00	2.20	4.00	6.20	8.80	12.0	15.8	20.0	25.4	31.2

Bar Spacing (in)	Bar Number									
	2	3	4	5	6	7	8	9	10	11
2	0.30	0.66	1.20	1.86						
2½	0.24	0.53	0.96	1.49	2.11					
3	0.20	0.44	0.80	1.24	1.76	2.40	3.16	4.00		
3½	0.17	0.38	0.69	1.06	1.51	2.06	2.71	3.43	4.35	
4	0.15	0.33	0.60	0.93	1.32	1.80	2.37	3.00	3.81	4.68
4½	0.13	0.29	0.53	0.83	1.17	1.60	2.11	2.67	3.39	4.16
5	0.12	0.26	0.48	0.74	1.06	1.44	1.90	2.40	3.05	3.74
5½	0.11	0.24	0.44	0.68	0.96	1.31	1.72	2.18	2.77	3.40
6	0.10	0.22	0.40	0.62	0.88	1.20	1.58	2.00	2.54	3.12
6½	0.09	0.20	0.37	0.57	0.81	1.11	1.46	1.85	2.34	2.88
7	0.09	0.19	0.34	0.53	0.75	1.03	1.35	1.71	2.18	2.67
7½	0.08	0.18	0.32	0.50	0.70	0.96	1.26	1.60	2.03	2.50
8	0.08	0.16	0.30	0.46	0.66	0.90	1.18	1.50	1.90	2.34
9	0.07	0.15	0.27	0.41	0.59	0.80	1.05	1.33	1.69	2.08
10	0.06	0.13	0.24	0.37	0.53	0.72	0.95	1.20	1.52	1.87
11	0.05	0.12	0.22	0.34	0.48	0.65	0.86	1.09	1.39	1.70
12	0.05	0.11	0.20	0.31	0.44	0.60	0.79	1.00	1.27	1.56
13	0.04	0.10	0.18	0.29	0.41	0.55	0.73	0.92	1.17	1.44
14	0.04	0.09	0.17	0.27	0.38	0.51	0.68	0.86	1.09	1.34
15	0.04	0.09	0.16	0.25	0.35	0.48	0.64	0.80	1.02	1.25
16	0.04	0.08	0.15	0.23	0.33	0.45	0.59	0.75	0.95	1.17
17	0.03	0.08	0.14	0.22	0.31	0.42	0.56	0.71	0.90	1.10
18	0.03	0.07	0.13	0.21	0.29	0.40	0.53	0.67	0.85	1.04

TABLE A-4
Design Constants

f'_c (psi)	$f_y = 40,000$ psi			$f_y = 50,000$ psi			$f_y = 60,000$ psi		
	ρ_{min}	$0.75\rho_b$*	Recommended Design ρ	ρ_{min}	$0.75\rho_b$*	Recommended Design ρ	ρ_{min}	$0.75\rho_b$*	Recommended Design ρ
3,000	0.005	0.0278	0.0135	0.004	0.0206	0.0108	0.0033	0.0161	0.0090
4,000	0.005	0.0372	0.0180	0.004	0.0275	0.0144	0.0033	0.0214	0.0120
5,000	0.005	0.0436	0.0225	0.004	0.0324	0.0180	0.0033	0.0252	0.0150
6,000	0.005	0.0490	0.0270	0.004	0.0364	0.0216	0.0033	0.0283	0.0180

*See Eq. (2–3) for ρ_b.

	f'_c(psi)			
	3,000	3,500	4,000	5,000
E_c (psi)*	3,120,000	3,370,000	3,605,000	4,030,000
n†	9	9	8	7
$\sqrt{f'_c}$ (ksi)	0.055	0.059	0.063	0.071
$2\sqrt{f'_c}$ (ksi)	0.110	0.118	0.126	0.141
$4\sqrt{f'_c}$ (ksi)	0.219	0.237	0.253	0.283
$7.5\sqrt{f'_c}$ (ksi)‡	0.411	0.444	0.474	0.530
$8\sqrt{f'_c}$ (ksi)	0.438	0.474	0.506	0.566

* E_c for normal-weight concrete = $57,000\sqrt{f'_c}$.

† Nearest whole number.

‡ Modulus of rupture (f_r).

TABLE A-6
Coefficient of Resistance ($\bar{k}$) vs. Reinforcement Ratio (ρ)
($f'_c = 3,000$ psi; $f_y = 40,000$ psi)
units of $\bar{k}$ are (ksi)

ρ	$\bar{k}$	ρ	$\bar{k}$	ρ	$\bar{k}$
0.0010	0.0397	0.0056	0.2142	0.0102	0.3754
0.0011	0.0436	0.0057	0.2178	0.0103	0.3787
0.0012	0.0476	0.0058	0.2214	0.0104	0.3821
0.0013	0.0515	0.0059	0.2251	0.0105	0.3854
0.0014	0.0554	0.0060	0.2287	0.0106	0.3887
0.0015	0.0593	0.0061	0.2323	0.0107	0.3921
0.0016	0.0632	0.0062	0.2359	0.0108	0.3954
0.0017	0.0671	0.0063	0.2395	0.0109	0.3987
0.0018	0.0710	0.0064	0.2431	0.0110	0.4020
0.0019	0.0749	0.0065	0.2467	0.0111	0.4053
0.0020	0.0788	0.0066	0.2503	0.0112	0.4086
0.0021	0.0826	0.0067	0.2539	0.0113	0.4119
0.0022	0.0865	0.0068	0.2575	0.0114	0.4152
0.0023	0.0903	0.0069	0.2611	0.0115	0.4185
0.0024	0.0942	0.0070	0.2646	0.0116	0.4218
0.0025	0.0980	0.0071	0.2682	0.0117	0.4251
0.0026	0.1019	0.0072	0.2717	0.0118	0.4283
0.0027	0.1057	0.0073	0.2753	0.0119	0.4316
0.0028	0.1095	0.0074	0.2788	0.0120	0.4348
0.0029	0.1134	0.0075	0.2824	0.0121	0.4381
0.0030	0.1172	0.0076	0.2859	0.0122	0.4413
0.0031	0.1210	0.0077	0.2894	0.0123	0.4445
0.0032	0.1248	0.0078	0.2929	0.0124	0.4478
0.0033	0.1286	0.0079	0.2964	0.0125	0.4510
0.0034	0.1324	0.0080	0.2999	0.0126	0.4542
0.0035	0.1362	0.0081	0.3034	0.0127	0.4574
0.0036	0.1399	0.0082	0.3069	0.0128	0.4606
0.0037	0.1437	0.0083	0.3104	0.0129	0.4638
0.0038	0.1475	0.0084	0.3139	0.0130	0.4670
0.0039	0.1512	0.0085	0.3173	0.0131	0.4702
0.0040	0.1550	0.0086	0.3208	0.0132	0.4733
0.0041	0.1587	0.0087	0.3243	0.0133	0.4765
0.0042	0.1625	0.0088	0.3277	0.0134	0.4797
0.0043	0.1662	0.0089	0.3311	0.0135	0.4828	recmd.
0.0044	0.1699	0.0090	0.3346	0.0136	0.4860
0.0045	0.1736	0.0091	0.3380	0.0137	0.4891
0.0046	0.1774	0.0092	0.3414	0.0138	0.4923
0.0047	0.1811	0.0093	0.3449	0.0139	0.4954
0.0048	0.1848	0.0094	0.3483	0.0140	0.4985
0.0049	0.1885	0.0095	0.3517	0.0141	0.5016
ρ_{min} 0.0050	0.1922	0.0096	0.3551	0.0142	0.5047
0.0051	0.1958	0.0097	0.3585	0.0143	0.5078
0.0052	0.1995	0.0098	0.3619	0.0144	0.5109
0.0053	0.2032	0.0099	0.3653	0.0145	0.5140
0.0054	0.2069	0.0100	0.3686	0.0146	0.5171
0.0055	0.2105	0.0101	0.3720	0.0147	0.5202

TABLE A-6
Coefficient of Resistance ($\bar{k}$) vs. Reinforcement Ratio (ρ)
($f'_c = 3,000$ psi; $f_y = 40,000$ psi)
units of $\bar{k}$ are (ksi)

ρ	$\bar{k}$	ρ	$\bar{k}$	ρ	$\bar{k}$
0.0148	0.5233	0.0195	0.6607	0.0242	0.7843
0.0149	0.5263	0.0196	0.6635	0.0243	0.7867
0.0150	0.5294	0.0197	0.6662	0.0244	0.7892
0.0151	0.5325	0.0198	0.6690	0.0245	0.7917
0.0152	0.5355	0.0199	0.6718	0.0246	0.7941
0.0153	0.5386	0.0200	0.6745	0.0247	0.7966
0.0154	0.5416	0.0201	0.6773	0.0248	0.7990
0.0155	0.5446	0.0202	0.6800	0.0249	0.8015
0.0156	0.5477	0.0203	0.6827	0.0250	0.8039
0.0157	0.5507	0.0204	0.6854	0.0251	0.8063
0.0158	0.5537	0.0205	0.6882	0.0252	0.8088
0.0159	0.5567	0.0206	0.6909	0.0253	0.8112
0.0160	0.5597	0.0207	0.6936	0.0254	0.8136
0.0161	0.5627	0.0208	0.6963	0.0255	0.8160
0.0162	0.5657	0.0209	0.6990	0.0256	0.8184
0.0163	0.5686	0.0210	0.7016	0.0257	0.8208
0.0164	0.5716	0.0211	0.7043	0.0258	0.8232
0.0165	0.5746	0.0212	0.7070	0.0259	0.8255
0.0166	0.5775	0.0213	0.7097	0.0260	0.8279
0.0167	0.5805	0.0214	0.7123	0.0261	0.8303
0.0168	0.5835	0.0215	0.7150	0.0262	0.8326
0.0169	0.5864	0.0216	0.7176	0.0263	0.8350
0.0170	0.5893	0.0217	0.7203	0.0264	0.8373
0.0171	0.5923	0.0218	0.7229	0.0265	0.8397
0.0172	0.5952	0.0219	0.7255	0.0266	0.8420
0.0173	0.5981	0.0220	0.7282	0.0267	0.8443
0.0174	0.6010	0.0221	0.7308	0.0268	0.8467
0.0175	0.6039	0.0222	0.7334	0.0269	0.8490
0.0176	0.6068	0.0223	0.7360	0.0270	0.8513
0.0177	0.6097	0.0224	0.7386	0.0271	0.8536
0.0178	0.6126	0.0225	0.7412	0.0272	0.8559
0.0179	0.6155	0.0226	0.7438	0.0273	0.8582
0.0180	0.6184	0.0227	0.7463	0.0274	0.8605
0.0181	0.6212	0.0228	0.7489	0.0275	0.8627
0.0182	0.6241	0.0229	0.7515	0.0276	0.8650
0.0183	0.6269	0.0230	0.7540	0.0277	0.8673
0.0184	0.6298	0.0231	0.7566	0.0278	0.8695
0.0185	0.6326	0.0232	0.7591	*Max*	
0.0186	0.6355	0.0233	0.7617		
0.0187	0.6383	0.0234	0.7642		
0.0188	0.6411	0.0235	0.7667		
0.0189	0.6439	0.0236	0.7693		
0.0190	0.6467	0.0237	0.7718		
0.0191	0.6495	0.0238	0.7743		
0.0192	0.6523	0.0239	0.7768		
0.0193	0.6551	0.0240	0.7793		
0.0194	0.6579	0.0241	0.7818		

TABLE A-7
Coefficient of Resistance ($\bar{k}$) vs. Reinforcement Ratio (ρ)
($f'_c = 3,000$ psi; $f_y = 60,000$ psi)
units of $\bar{k}$ are (ksi)

ρ	$\bar{k}$	ρ	$\bar{k}$	ρ	$\bar{k}$	ρ	$\bar{k}$
0.0010	0.0593	0.0029	0.1681	0.0048	0.2717	0.0067	0.3703
0.0011	0.0651	0.0030	0.1736	0.0049	0.2771	0.0068	0.3754
0.0012	0.0710	0.0031	0.1792	0.0050	0.2824	0.0069	0.3804
0.0013	0.0768	0.0032	0.1848	0.0051	0.2876	0.0070	0.3854
0.0014	0.0826	0.0033	0.1903	0.0052	0.2929	0.0071	0.3904
0.0015	0.0884	0.0034	0.1958	0.0053	0.2982	0.0072	0.3954
0.0016	0.0942	0.0035	0.2014	0.0054	0.3034	0.0073	0.4004
0.0017	0.1000	0.0036	0.2069	0.0055	0.3086	0.0074	0.4053
0.0018	0.1057	0.0037	0.2123	0.0056	0.3139	0.0075	0.4103
0.0019	0.1115	0.0038	0.2178	0.0057	0.3191	0.0076	0.4152
0.0020	0.1172	0.0039	0.2233	0.0058	0.3243	0.0077	0.4202
0.0021	0.1229	0.0040	0.2287	0.0059	0.3294	0.0078	0.4251
0.0022	0.1286	0.0041	0.2341	0.0060	0.3346	0.0079	0.4299
0.0023	0.1343	0.0042	0.2395	0.0061	0.3397	0.0080	0.4348
0.0024	0.1399	0.0043	0.2449	0.0062	0.3449	0.0081	0.4397
0.0025	0.1456	0.0044	0.2503	0.0063	0.3500	0.0082	0.4445
0.0026	0.1512	0.0045	0.2557	0.0064	0.3551	0.0083	0.4494
0.0027	0.1569	0.0046	0.2611	0.0065	0.3602	0.0084	0.4542
0.0028	0.1625	0.0047	0.2664	0.0066	0.3653	0.0085	0.4590

TABLE A-7
Coefficient of Resistance ($\bar{k}$) vs. Reinforcement Ratio (ρ)
(f'_c = 3,000 psi; f_y = 60,000 psi)
units of $\bar{k}$ are (ksi)

ρ	$\bar{k}$	ρ	$\bar{k}$	ρ	$\bar{k}$	ρ	$\bar{k}$
0.0086	0.4638	0.0105	0.5522	0.0124	0.6355	0.0143	0.7137
0.0087	0.4686	0.0106	0.5567	0.0125	0.6397	0.0144	0.7176
0.0088	0.4733	0.0107	0.5612	0.0126	0.6439	0.0145	0.7216
0.0089	0.4781	0.0108	0.5657	0.0127	0.6482	0.0146	0.7255
0.0090	0.4828	0.0109	0.5701	0.0128	0.6524	0.0147	0.7295
0.0091	0.4875	0.0110	0.5746	0.0129	0.6565	0.0148	0.7334
0.0092	0.4923	0.0111	0.5790	0.0130	0.6607	0.0149	0.7373
0.0093	0.4970	0.0112	0.5835	0.0131	0.6649	0.0150	0.7412
0.0094	0.5016	0.0113	0.5879	0.0132	0.6690	0.0151	0.7451
0.0095	0.5063	0.0114	0.5923	0.0133	0.6731	0.0152	0.7489
0.0096	0.5109	0.0115	0.5967	0.0134	0.6773	0.0153	0.7528
0.0097	0.5156	0.0116	0.6010	0.0135	0.6814	0.0154	0.7566
0.0098	0.5202	0.0117	0.6054	0.0136	0.6854	0.0155	0.7604
0.0099	0.5248	0.0118	0.6097	0.0137	0.6895	0.0156	0.7642
0.0100	0.5294	0.0119	0.6140	0.0138	0.6936	0.0157	0.7680
0.0101	0.5340	0.0120	0.6184	0.0139	0.6976	0.0158	0.7718
0.0102	0.5386	0.0121	0.6227	0.0140	0.7017	0.0159	0.7756
0.0103	0.5431	0.0122	0.6269	0.0141	0.7057	0.0160	0.7793
0.0104	0.5477	0.0123	0.6312	0.0142	0.7097	0.0161	0.7830

TABLE A-8
Coefficient of Resistance ($\bar{k}$) vs. Reinforcement Ratio (ρ)
($f'_c = 4,000$ psi; $f_y = 40,000$ psi)
units of $\bar{k}$ are (ksi)

ρ	$\bar{k}$	ρ	$\bar{k}$	ρ	$\bar{k}$	ρ	$\bar{k}$
0.0010	0.0398	0.0057	0.2204	0.0104	0.3906	0.0151	0.5504
0.0011	0.0437	0.0058	0.2241	0.0105	0.3941	0.0152	0.5536
0.0012	0.0477	0.0059	0.2278	0.0106	0.3976	0.0153	0.5569
0.0013	0.0516	0.0060	0.2315	0.0107	0.4011	0.0154	0.5602
0.0014	0.0555	0.0061	0.2352	0.0108	0.4046	0.0155	0.5635
0.0015	0.0595	0.0062	0.2390	0.0109	0.4080	0.0156	0.5667
0.0016	0.0634	0.0063	0.2427	0.0110	0.4115	0.0157	0.5700
0.0017	0.0673	0.0064	0.2464	0.0111	0.4150	0.0158	0.5733
0.0018	0.0712	0.0065	0.2501	0.0112	0.4185	0.0159	0.5765
0.0019	0.0752	0.0066	0.2538	0.0113	0.4220	0.0160	0.5798
0.0020	0.0791	0.0067	0.2574	0.0114	0.4254	0.0161	0.5830
0.0021	0.0830	0.0068	0.2611	0.0115	0.4289	0.0162	0.5863
0.0022	0.0869	0.0069	0.2648	0.0116	0.4323	0.0163	0.5895
0.0023	0.0908	0.0070	0.2685	0.0117	0.4358	0.0164	0.5927
0.0024	0.0946	0.0071	0.2721	0.0118	0.4392	0.0165	0.5959
0.0025	0.0985	0.0072	0.2758	0.0119	0.4427	0.0166	0.5992
0.0026	0.1024	0.0073	0.2795	0.0120	0.4461	0.0167	0.6024
0.0027	0.1063	0.0074	0.2831	0.0121	0.4495	0.0168	0.6056
0.0028	0.1102	0.0075	0.2868	0.0122	0.4530	0.0169	0.6088
0.0029	0.1140	0.0076	0.2904	0.0123	0.4564	0.0170	0.6120
0.0030	0.1179	0.0077	0.2941	0.0124	0.4598	0.0171	0.6152
0.0031	0.1217	0.0078	0.2977	0.0125	0.4632	0.0172	0.6184
0.0032	0.1256	0.0079	0.3013	0.0126	0.4666	0.0173	0.6216
0.0033	0.1294	0.0080	0.3049	0.0127	0.4701	0.0174	0.6248
0.0034	0.1333	0.0081	0.3086	0.0128	0.4735	0.0175	0.6279
0.0035	0.1371	0.0082	0.3122	0.0129	0.4768	0.0176	0.6311
0.0036	0.1410	0.0083	0.3158	0.0130	0.4802	0.0177	0.6343
0.0037	0.1448	0.0084	0.3194	0.0131	0.4836	0.0178	0.6375
0.0038	0.1486	0.0085	0.3230	0.0132	0.4870	0.0179	0.6406
0.0039	0.1524	0.0086	0.3266	0.0133	0.4904	0.0180	0.6438
0.0040	0.1562	0.0087	0.3302	0.0134	0.4938	0.0181	0.6469
0.0041	0.1600	0.0088	0.3338	0.0135	0.4971	0.0182	0.6501
0.0042	0.1638	0.0089	0.3374	0.0136	0.5005	0.0183	0.6532
0.0043	0.1676	0.0090	0.3409	0.0137	0.5038	0.0184	0.6563
0.0044	0.1714	0.0091	0.3445	0.0138	0.5072	0.0185	0.6595
0.0045	0.1752	0.0092	0.3481	0.0139	0.5105	0.0186	0.6626
0.0046	0.1790	0.0093	0.3517	0.0140	0.5139	0.0187	0.6657
0.0047	0.1828	0.0094	0.3552	0.0141	0.5172	0.0188	0.6688
0.0048	0.1866	0.0095	0.3588	0.0142	0.5206	0.0189	0.6720
0.0049	0.1904	0.0096	0.3623	0.0143	0.5239	0.0190	0.6751
0.0050	0.1941	0.0097	0.3659	0.0144	0.5272	0.0191	0.6782
0.0051	0.1979	0.0098	0.3694	0.0145	0.5305	0.0192	0.6813
0.0052	0.2016	0.0099	0.3729	0.0146	0.5338	0.0193	0.6844
0.0053	0.2054	0.0100	0.3765	0.0147	0.5372	0.0194	0.6875
0.0054	0.2091	0.0101	0.3800	0.0148	0.5405	0.0195	0.6905
0.0055	0.2129	0.0102	0.3835	0.0149	0.5438	0.0196	0.6936
0.0056	0.2166	0.0103	0.3870	0.0150	0.5471	0.0197	0.6967

374

TABLE A-8
Coefficient of Resistance $(\bar{k})$ vs. Reinforcement Ratio (ρ)
$(f'_c = 4,000 \text{ psi}; f_y = 40,000 \text{ psi})$
units of $\bar{k}$ are (ksi)

ρ	$\bar{k}$	ρ	$\bar{k}$	ρ	$\bar{k}$
0.0198	0.6998	0.0244	0.8359	0.0334	1.0735
0.0199	0.7028	0.0245	0.8388	0.0336	1.0784
0.0200	0.7059	0.0246	0.8416	0.0338	1.0832
0.0201	0.7089	0.0248	0.8473	0.0340	1.0880
0.0202	0.7120	0.0250	0.8530	0.0342	1.0928
0.0203	0.7150	0.0252	0.8586	0.0344	1.0976
0.0204	0.7181	0.0254	0.8642	0.0346	1.1023
0.0205	0.7211	0.0256	0.8698	0.0348	1.1071
0.0206	0.7242	0.0258	0.8754	0.0350	1.1118
0.0207	0.7272	0.0260	0.8810	0.0352	1.1165
0.0208	0.7302	0.0262	0.8865	0.0354	1.1212
0.0209	0.7332	0.0264	0.8920	0.0356	1.1258
0.0210	0.7362	0.0266	0.8975	0.0358	1.1305
0.0211	0.7393	0.0268	0.9030	0.0360	1.1351
0.0212	0.7423	0.0270	0.9085	0.0362	1.1397
0.0213	0.7453	0.0272	0.9139	0.0364	1.1443
0.0214	0.7483	0.0274	0.9194	0.0366	1.1488
0.0215	0.7512	0.0276	0.9248	0.0368	1.1534
0.0216	0.7542	0.0278	0.9302	0.0370	1.1579
0.0217	0.7572	0.0280	0.9355	0.0372	1.1624
0.0218	0.7602	0.0282	0.9409		
0.0219	0.7632	0.0284	0.9462		
0.0220	0.7661	0.0286	0.9516		
0.0221	0.7691	0.0288	0.9568		
0.0222	0.7720	0.0290	0.9621		
0.0223	0.7750	0.0292	0.9674		
0.0224	0.7779	0.0294	0.9726		
0.0225	0.7809	0.0296	0.9779		
0.0226	0.7838	0.0298	0.9831		
0.0227	0.7868	0.0300	0.9882		
0.0228	0.7897	0.0302	0.9934		
0.0229	0.7926	0.0304	0.9986		
0.0230	0.7955	0.0306	1.0037		
0.0231	0.7985	0.0308	1.0088		
0.0232	0.8014	0.0310	1.0139		
0.0233	0.8043	0.0312	1.0190		
0.0234	0.8072	0.0314	1.0240		
0.0235	0.8101	0.0316	1.0291		
0.0236	0.8130	0.0318	1.0341		
0.0237	0.8158	0.0320	1.0391		
0.0238	0.8187	0.0322	1.0441		
0.0239	0.8216	0.0324	1.0490		
0.0240	0.8245	0.0326	1.0540		
0.0241	0.8273	0.0328	1.0589		
0.0242	0.8302	0.0330	1.0638		
0.0243	0.8331	0.0332	1.0687		

TABLE A-9
Coefficient of Resistance ($\bar{k}$) vs. Reinforcement Ratio (ρ)
(f'_c = 4,000 psi; f_y = 60,000 psi)
units of $\bar{k}$ are (ksi)

ρ	$\bar{k}$	ρ	$\bar{k}$	ρ	$\bar{k}$	ρ	$\bar{k}$
0.0010	0.0595	0.0036	0.2091	0.0062	0.3516	0.0088	0.4870
0.0011	0.0654	0.0037	0.2148	0.0063	0.3570	0.0089	0.4921
0.0012	0.0712	0.0038	0.2204	0.0064	0.3623	0.0090	0.4971
0.0013	0.0771	0.0039	0.2259	0.0065	0.3676	0.0091	0.5022
0.0014	0.0830	0.0040	0.2315	0.0066	0.3729	0.0092	0.5072
0.0015	0.0889	0.0041	0.2371	0.0067	0.3782	0.0093	0.5122
0.0016	0.0946	0.0042	0.2427	0.0068	0.3835	0.0094	0.5172
0.0017	0.1005	0.0043	0.2482	0.0069	0.3888	0.0095	0.5222
0.0018	0.1063	0.0044	0.2538	0.0070	0.3941	0.0096	0.5272
0.0019	0.1121	0.0045	0.2593	0.0071	0.3993	0.0097	0.5322
0.0020	0.1179	0.0046	0.2648	0.0072	0.4046	0.0098	0.5372
0.0021	0.1237	0.0047	0.2703	0.0073	0.4098	0.0099	0.5421
0.0022	0.1294	0.0048	0.2758	0.0074	0.4150	0.0100	0.5471
0.0023	0.1352	0.0049	0.2813	0.0075	0.4202	0.0101	0.5520
0.0024	0.1410	0.0050	0.2868	0.0076	0.4254	0.0102	0.5569
0.0025	0.1467	0.0051	0.2922	0.0077	0.4306	0.0103	0.5618
0.0026	0.1524	0.0052	0.2977	0.0078	0.4358	0.0104	0.5667
0.0027	0.1581	0.0053	0.3031	0.0079	0.4410	0.0105	0.5716
0.0028	0.1638	0.0054	0.3086	0.0080	0.4461	0.0106	0.5765
0.0029	0.1695	0.0055	0.3140	0.0081	0.4513	0.0107	0.5814
0.0030	0.1752	0.0056	0.3194	0.0082	0.4564	0.0108	0.5862
0.0031	0.1809	0.0057	0.3248	0.0083	0.4615	0.0109	0.5911
0.0032	0.1866	0.0058	0.3302	0.0084	0.4666	0.0110	0.5959
0.0033	0.1922	0.0059	0.3356	0.0085	0.4718	0.0111	0.6008
0.0034	0.1979	0.0060	0.3409	0.0086	0.4768	0.0112	0.6056
0.0035	0.2035	0.0061	0.3463	0.0087	0.4819	0.0113	0.6104

TABLE A-9
Coefficient of Resistance ($\bar{k}$) vs. Reinforcement Ratio (ρ)
($f'_c = 4{,}000$ psi; $f_y = 60{,}000$ psi)
units of $\bar{k}$ are (ksi)

ρ	$\bar{k}$	ρ	$\bar{k}$	ρ	$\bar{k}$	ρ	$\bar{k}$
0.0114	0.6152	0.0140	0.7362	0.0165	0.8459	0.0190	0.9489
0.0115	0.6200	0.0141	0.7407	0.0166	0.8501	0.0191	0.9529
0.0116	0.6248	0.0142	0.7452	0.0167	0.8544	0.0192	0.9568
0.0117	0.6295	0.0143	0.7497	0.0168	0.8586	0.0193	0.9608
0.0118	0.6343	0.0144	0.7542	0.0169	0.8628	0.0194	0.9648
0.0119	0.6390	0.0145	0.7587	0.0170	0.8670	0.0195	0.9687
0.0120	0.6438	0.0146	0.7632	0.0171	0.8712	0.0196	0.9726
0.0121	0.6485	0.0147	0.7675	0.0172	0.8754	0.0197	0.9765
0.0122	0.6532	0.0148	0.7720	0.0173	0.8796	0.0198	0.9804
0.0123	0.6579	0.0149	0.7765	0.0174	0.8837	0.0199	0.9843
0.0124	0.6626	0.0150	0.7809	0.0175	0.8879	0.0200	0.9882
0.0125	0.6673	0.0151	0.7853	0.0176	0.8920	0.0201	0.9921
0.0126	0.6720	0.0152	0.7897	0.0177	0.8961	0.0202	0.9960
0.0127	0.6766	0.0153	0.7941	0.0178	0.9003	0.0203	0.9998
0.0128	0.6813	0.0154	0.7984	0.0179	0.9044	0.0204	1.0037
0.0129	0.6859	0.0155	0.8028	0.0180	0.9085	0.0205	1.0075
0.0130	0.6905	0.0156	0.8072	0.0181	0.9116	0.0206	1.0113
0.0131	0.6951	0.0157	0.8115	0.0182	0.9166	0.0207	1.0152
0.0132	0.6998	0.0158	0.8158	0.0183	0.9207	0.0208	1.0190
0.0133	0.7044	0.0159	0.8202	0.0184	0.9248	0.0209	1.0227
0.0134	0.7089	0.0160	0.8245	0.0185	0.9288	0.0210	1.0265
0.0135	0.7135	0.0161	0.8288	0.0186	0.9328	0.0211	1.0303
0.0136	0.7181	0.0162	0.8331	0.0187	0.9369	0.0212	1.0341
0.0137	0.7226	0.0163	0.8373	0.0188	0.9409	0.0213	1.0378
0.0138	0.7272	0.0164	0.8416	0.0189	0.9449	0.0214	1.0416
0.0139	0.7317						

TABLE A-10
Coefficient of Resistance ($\bar{k}$) vs. Reinforcement Ratio (ρ)
($f'_c = 5,000$ psi; $f_y = 60,000$ psi)
units of $\bar{k}$ are (ksi)

ρ	$\bar{k}$	ρ	$\bar{k}$	ρ	$\bar{k}$	ρ	$\bar{k}$
0.0010	0.0596	0.0055	0.3172	0.0100	0.5576	0.0145	0.7810
0.0011	0.0655	0.0056	0.3227	0.0101	0.5628	0.0146	0.7857
0.0012	0.0714	0.0057	0.3282	0.0102	0.5679	0.0147	0.7905
0.0013	0.0773	0.0058	0.3338	0.0103	0.5731	0.0148	0.7952
0.0014	0.0832	0.0059	0.3393	0.0104	0.5782	0.0149	0.8000
0.0015	0.0890	0.0060	0.3448	0.0105	0.5833	0.0150	0.8047
0.0016	0.0949	0.0061	0.3502	0.0106	0.5884	0.0151	0.8094
0.0017	0.1008	0.0062	0.3557	0.0107	0.5935	0.0152	0.8142
0.0018	0.1066	0.0063	0.3612	0.0108	0.5986	0.0153	0.8189
0.0019	0.1125	0.0064	0.3667	0.0109	0.6037	0.0154	0.8236
0.0020	0.1183	0.0065	0.3721	0.0110	0.6088	0.0155	0.8283
0.0021	0.1241	0.0066	0.3776	0.0111	0.6138	0.0156	0.8329
0.0022	0.1300	0.0067	0.3830	0.0112	0.6189	0.0157	0.8376
0.0023	0.1358	0.0068	0.3884	0.0113	0.6239	0.0158	0.8423
0.0024	0.1416	0.0069	0.3938	0.0114	0.6290	0.0159	0.8469
0.0025	0.1474	0.0070	0.3992	0.0115	0.6340	0.0160	0.8516
0.0026	0.1531	0.0071	0.4047	0.0116	0.6390	0.0161	0.8562
0.0027	0.1589	0.0072	0.4100	0.0117	0.6440	0.0162	0.8609
0.0028	0.1647	0.0073	0.4154	0.0118	0.6490	0.0163	0.8655
0.0029	0.1704	0.0074	0.4208	0.0119	0.6540	0.0164	0.8701
0.0030	0.1762	0.0075	0.4262	0.0120	0.6590	0.0165	0.8747
0.0031	0.1819	0.0076	0.4315	0.0121	0.6640	0.0166	0.8793
0.0032	0.1877	0.0077	0.4369	0.0122	0.6690	0.0167	0.8839
0.0033	0.1934	0.0078	0.4422	0.0123	0.6739	0.0168	0.8885
0.0034	0.1991	0.0079	0.4476	0.0124	0.6789	0.0169	0.8930
0.0035	0.2048	0.0080	0.4529	0.0125	0.6838	0.0170	0.8976
0.0036	0.2105	0.0081	0.4582	0.0126	0.6888	0.0171	0.9022
0.0037	0.2162	0.0082	0.4635	0.0127	0.6937	0.0172	0.9067
0.0038	0.2219	0.0083	0.4688	0.0128	0.6986	0.0173	0.9112
0.0039	0.2276	0.0084	0.4741	0.0129	0.7035	0.0174	0.9158
0.0040	0.2332	0.0085	0.4794	0.0130	0.7084	0.0175	0.9203
0.0041	0.2389	0.0086	0.4847	0.0131	0.7133	0.0176	0.9248
0.0042	0.2445	0.0087	0.4899	0.0132	0.7182	0.0177	0.9293
0.0043	0.2502	0.0088	0.4952	0.0133	0.7231	0.0178	0.9338
0.0044	0.2558	0.0089	0.5005	0.0134	0.7280	0.0179	0.9383
0.0045	0.2614	0.0090	0.5057	0.0135	0.7328	0.0180	0.9428
0.0046	0.2670	0.0091	0.5109	0.0136	0.7377	0.0181	0.9473
0.0047	0.2726	0.0092	0.5162	0.0137	0.7425	0.0182	0.9517
0.0048	0.2782	0.0093	0.5214	0.0138	0.7473	0.0183	0.9562
0.0049	0.2838	0.0094	0.5266	0.0139	0.7522	0.0184	0.9606
0.0050	0.2894	0.0095	0.5318	0.0140	0.7570	0.0185	0.9651
0.0051	0.2950	0.0096	0.5370	0.0141	0.7618	0.0186	0.9695
0.0052	0.3005	0.0097	0.5422	0.0142	0.7666	0.0187	0.9739
0.0053	0.3061	0.0098	0.5473	0.0143	0.7714	0.0188	0.9783
0.0054	0.3117	0.0099	0.5525	0.0144	0.7762	0.0189	0.9827

TABLE A–10
Coefficient of Resistance ($\bar{k}$) vs. Reinforcement Ratio (ρ)
($f'_c = 5,000$ psi; $f_y = 60,000$ psi)
units of $\bar{k}$ are (ksi)

ρ	$\bar{k}$	ρ	$\bar{k}$	ρ	$\bar{k}$	ρ	$\bar{k}$
0.0190	0.9871	0.0206	1.0563	0.0222	1.1233	0.0238	1.1881
0.0191	0.9915	0.0207	1.0605	0.0223	1.1274	0.0239	1.1921
0.0192	0.9959	0.0208	1.0648	0.0224	1.1315	0.0240	1.1961
0.0193	1.0002	0.0209	1.0690	0.0225	1.1356	0.0241	1.2000
0.0194	1.0046	0.0210	1.0732	0.0226	1.1397	0.0242	1.2040
0.0195	1.0090	0.0211	1.0775	0.0227	1.1438	0.0243	1.2079
0.0196	1.0133	0.0212	1.0817	0.0228	1.1478	0.0244	1.2119
0.0197	1.0176	0.0213	1.0859	0.0229	1.1519	0.0245	1.2158
0.0198	1.0220	0.0214	1.0901	0.0230	1.1560	0.0246	1.2197
0.0199	1.0263	0.0215	1.0942	0.0231	1.1600	0.0247	1.2236
0.0200	1.0306	0.0216	1.0984	0.0232	1.1641	0.0248	1.2275
0.0201	1.0349	0.0217	1.1026	0.0233	1.1681	0.0249	1.2314
0.0202	1.0392	0.0218	1.1067	0.0234	1.1721	0.0250	1.2353
0.0203	1.0435	0.0219	1.1109	0.0235	1.1761	0.0251	1.2392
0.0204	1.0478	0.0220	1.1150	0.0236	1.1801	0.0252	1.2431
0.0205	1.0520	0.0221	1.1192	0.0237	1.1841		

Number of Bars in One Layer	Bar Number							
	3 & 4	5	6	7	8	9	10	11
2	6.0	6.0	6.5	6.5	7.0	7.5	8.0	8.0
3	7.5	8.0	8.0	8.5	9.0	9.5	10.5	11.0
4	9.0	9.5	10.0	10.5	11.0	12.0	13.0	14.0
5	10.5	11.0	11.5	12.5	13.0	14.0	15.5	16.5
6	12.0	12.5	13.5	14.0	15.0	16.5	18.0	19.5
7	13.5	14.5	15.0	16.0	17.0	18.5	20.5	22.5
8	15.0	16.0	17.0	18.0	19.0	21.0	23.0	25.0
9	16.5	17.5	18.5	20.0	21.0	23.0	25.5	28.0
10	18.0	19.0	20.5	21.5	23.0	25.5	28.0	31.0

* Tabulated values based on No. 3 stirrups, minimum clear distance of 1 in., and a 1½-in. cover.

Basic Development Length (B.D.L.) for Tension Bars*
(In.)

Bar Number	f_y (ksi)	f'_c (psi)			
		3,000	4,000	5,000	6,000
3	40	6	6	6	6
	50	8	8	8	8
	60	9	9	9	9
4	40	8	8	8	8
	50	10	10	10	10
	60	12	12	12	12
5	40	10	10	10	10
	50	13	13	13	13
	60	15	15	15	15
6	40	13	12	12	12
	50	16	15	15	15
	60	19	18	18	18
7	40	18	15	14	14
	50	22	19	18	18
	60	26	23	21	21
8	40	23	20	18	16
	50	29	25	22	20
	60	35	30	27	24
9	40	29	25	23	21
	50	37	32	28	26
	60	44	38	34	31
10	40	37	32	29	26
	50	46	40	36	33
	60	56	48	43	39
11	40	46	39	35	32
	50	57	49	44	40
	60	68	59	53	48
14	40	62	54	48	44
	50	78	67	60	55
	60	93	81	72	66
18	40	80	70	62	57
	50	100	87	78	71
	60	121	104	93	85

* ℓ_d = B.D.L. × modification factors ≥ 12 in. For B.D.L. expressions, see ACI Code, Section 12.2.2.

Bar Number	f_y (ksi)	Type	f'_c(psi)			
			3,000	4,000	5,000	6,000
3	40	All	3.0	3.4	3.8	4.2
	60	All	4.4	5.1	5.7	6.3
4	40	All	3.9	4.6	5.1	5.6
	60	All	5.9	6.8	7.6	8.4
5	40	All	4.9	5.7	6.4	7.0
	60	All	7.4	8.5	9.5	10.5
6	40	All	6.3	6.8	7.6	8.4
	60	Top	7.9	8.5	9.5	10.5
		Other	9.5	10.2	11.5	12.5
7	40	All	8.6	8.6	8.9	9.8
	60	Top				
		Other	13.0	13.0	13.4	14.6
8	40	All	11.4	11.4	11.4	11.4
	60	Top				
		Other	17.1	17.1	17.1	17.1
9	40	All	14.4	14.4	14.4	14.4
	60	Top				
		Other	21.6	21.6	21.6	21.6
10	40	All	18.3	18.3	18.3	18.3
	60	Top				
		Other	24.4	24.4	24.4	24.4
11	40	All	22.5	22.5	22.5	22.5
	60	Top				
		Other	26.2	26.2	26.2	26.2

* The value of ℓ_e is larger of $0.04A_b \, \xi$ or $0.0004d_b \, \xi \sqrt{f'_c}$.

TABLE A-14
Preferred Maximum Number of Column Bars in One Row

Recommended Spiral or Tie Bar Number	Core Size (in.) = Column Size − 2 × cover	Circular Area (in²)	Bar Number							Square Area (in²)	Bar Number						
			5	6	7	8	9	10	11		5	6	7	8	9	10	11
3*	9	63.6	8	7	7	6	—	—	—	81	8	8	8	8	4	4	4
	10	78.5	10	9	8	7	6	—	—	100	12	8	8	8	8	4	4
	11	95.0	11	10	9	8	7	6	—	121	12	12	8	8	8	8	4
	12	113.1	12	11	10	9	8	7	6	144	12	12	12	8	8	8	8
	13	132.7	13	12	11	10	8	7	6	169	16	12	12	12	8	8	8
	14	153.9	14	13	12	11	9	8	7	196	16	16	12	12	12	8	8
	15	176.7	15	14	13	12	10	9	8	225	16	16	16	12	12	12	8
4	16	201.1	16	15	14	12	11	9	8	256	20	16	16	16	12	12	8
	17	227.0	18	16	15	13	12	10	9	289	20	20	16	16	12	12	8
	18	254.5	19	17	15	14	12	11	10	324	20	20	16	16	16	12	12
	19	283.5	20	18	16	15	13	11	10	361	24	20	20	16	16	12	12
	20	314.2	21	19	17	16	14	12	11	400	24	24	20	20	16	12	12
	21	346.4	22	20	18	17	15	13	11	441	28	24	24	20	20	16	12
	22	380.1	23	21	19	18	15	14	12	484	28	24	24	20	20	16	12
5	23	415.5	24	22	21	19	16	14	13	529	28	28	24	24	20	16	16
	24	452.4	25	23	21	20	17	15	13	576	32	28	24	24	20	16	16
	25	490.9	26	24	22	20	18	16	14	625	32	28	28	24	20	20	16
	26	530.9	28	25	23	21	19	16	14	676	32	32	28	24	24	20	16
	27	572.6	29	26	24	22	19	17	15	729	36	32	28	28	24	20	16

* # 4 tie for # 11 or larger longitudinal reinforcement.

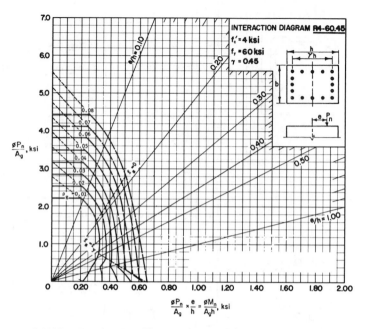

CHART A-15   Reprinted by permission of the American Concrete Institute from ACI Publication SP-17A(78): Design Handbook, Volume 2, Columns.

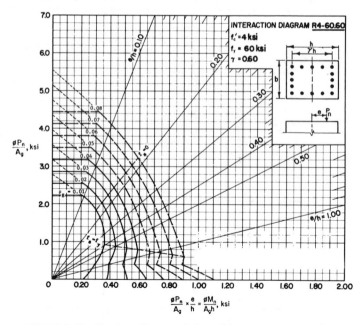

CHART A-16   Reprinted by permission of the American Concrete Institute from ACI Publication SP-17A(78): Design Handbook, Volume 2, Columns.

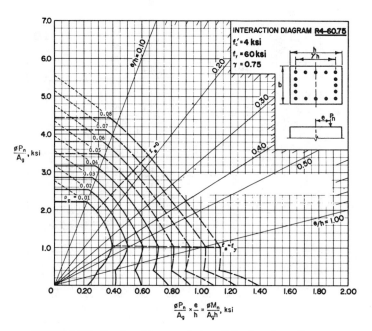

CHART A-17 Reprinted by permission of the American Concrete Institute from ACI Publication SP-17A(78): Design Handbook, Volume 2, Columns.

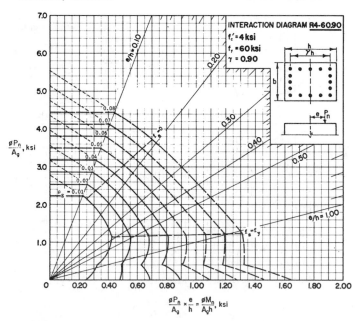

CHART A-18 Reprinted by permission of the American Concrete Institute from ACI Publication SP-17A(78): Design Handbook, Volume 2, Columns.

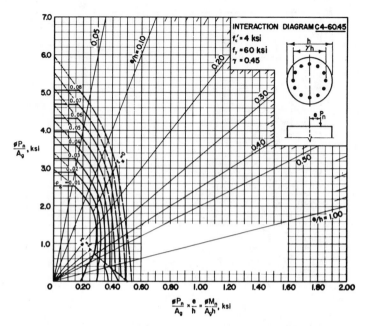

CHART A-19 Reprinted by permission of the American Concrete Institute from ACI Publication SP-17A(78): Design Handbook, Volume 2, Columns.

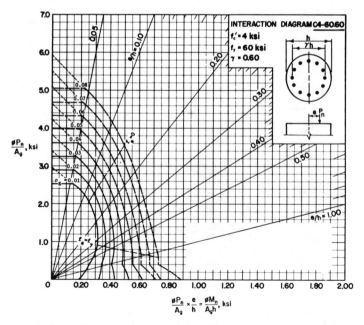

CHART A-20 Reprinted by permission of the American Concrete Institute from ACI Publication SP-17A(78): Design Handbook, Volume 2, Columns.

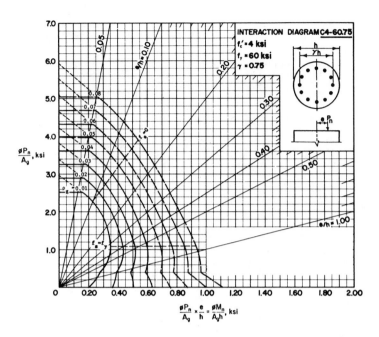

CHART A-21 Reprinted by permission of the American Concrete Institute from ACI Publication SP-17A(78): Design Handbook, Volume 2, Columns.

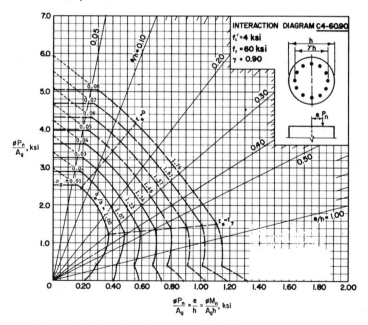

CHART A-22 Reprinted by permission of the American Concrete Institute from ACI Publication SP-17A(78): Design Handbook, Volume 2, Columns.

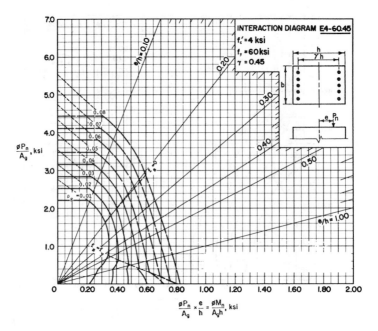

CHART A-23 Reprinted by permission of the American Concrete Institute from ACI Publication SP-17A(78): Design Handbook, Volume 2, Columns.

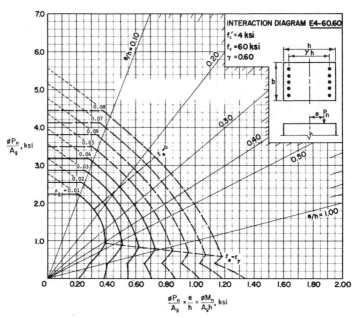

CHART A-24 Reprinted by permission of the American Concrete Institute from ACI Publication SP-17A(78): Design Handbook, Volume 2, Columns.

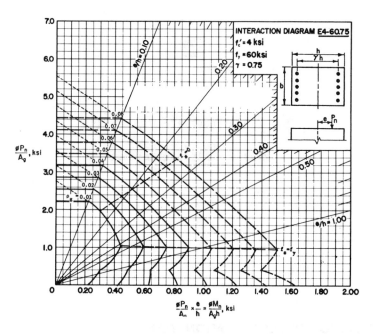

CHART A-25 Reprinted by permission of the American Concrete Institute from ACI Publication SP-17A(78): Design Handbook, Volume 2, Columns.

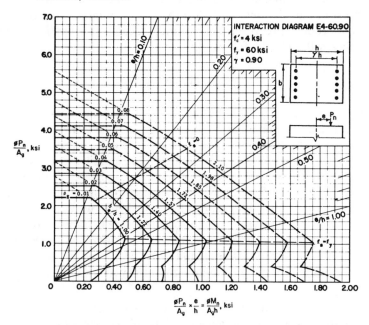

CHART A-26 Reprinted by permission of the American Concrete Institute from ACI Publication SP-17A(78): Design Handbook, Volume 2, Columns.

B

Supplementary Aids
and Guidelines

B-1 PRECISION FOR COMPUTATIONS FOR
REINFORCED CONCRETE

The widespread availability and use of electronic calculators for even
the simplest of calculations has led to the use of numbers that represent
a very high order of precision. For instance, a calculator having an eight-
digit display will yield the following:

$$\frac{8}{0.7} = 11.428571$$

It should be recognized that the numerator and denominator, each of
one-figure precision, resulted in a number that *indicates* eight-figure
precision. Since the quotient cannot be expected to be more precise than
the numbers that produced it, the result as represented above is a fallacy
and may lead one into a false sense of security associated with numbers
of very high precision. It is illogical to calculate, for instance, a required
steel area to four-figure precision when the loads were to two-figure
precision and the bars to be chosen have areas tabulated to two
(sometimes three)-figure precision. Likewise, the involved mathematical
expressions developed in this book and those presented by the ACI code
should be thought of in a similar light. They deal with a material,
concrete, which:

    1. Is made on site or plant-made and subject to varying amounts of
quality control and which will vary from the design strength.

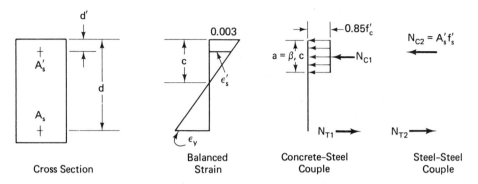

FIGURE B-1   Balanced condition—doubly reinforced beam

2. Is placed in forms that may or may not produce the design dimensions.

3. Contains reinforcing steel of a specified minimum strength, but which may vary above that strength.

In addition, the reinforced concrete member has reinforcing steel which may or may not be placed at the design location, and the design itself is generally based on loads that may be only "best estimates."

With the foregoing in mind, the following has been suggested by the Concrete Reinforcing Steel Institute as a rough guide for numerical precision in reinforced concrete calculations (and, admittedly, not always adhered to in this book):

1. Loads to the nearest 1 psf; 10 lb/ft; 100-lb concentration.
2. Span lengths to about 0.1 ft.
3. Total loads and reactions to 0.1 kip or three-figure precision.
4. Moments to the nearest 0.1 ft-kip or three-figure precision.
5. Individual bar areas to 0.01 in.²
6. Concrete sizes to the ½ in.
7. Effective beam depth to 0.1 in.
8. Column loads to nearest 1.0 kip.

B–2 A$_{s(max)}$ FOR DOUBLY REINFORCED BEAMS

A check on ductility of a doubly reinforced concrete beam can be made by determining $A_{s(max)}$ as the steel area which is 75% of the steel area required to produce the balanced condition.

Find the steel area required to produce the balanced condition A_{sb}. From Fig. B–1,

$$\frac{c}{0.003} = \frac{d}{0.003 + \epsilon_y}$$

$$c = \frac{0.003d}{0.003 + f_y/E_s} = \frac{0.003 \, dE_s}{0.003E_s + f_y}$$

Substituting $E_s = 29,000,000$ psi and noting that basic units are pounds and inches, at the balanced condition:

$$c = \frac{87,000d}{87,000 + f_y}$$

By equating the resultant forces at the balanced condition:

$$N_{Tb} = N_{Cb}$$

$$A_{sb}f_y = (0.85f_c')\beta_1 cb + A_s'f_s'$$

$$A_{s\,(max)} = 0.75A_{sb}$$

from which, by substitution,

$$A_{s\,(max)} = 0.75 \frac{(0.85f_c') \, \beta_1 cb}{f_y} + \frac{0.75A_s'f_s'}{f_y}$$

However, the ACI Code, Section 10.3.3, states that the 0.75 factor need not be applied to that portion of the tension steel that is equalized by compression steel. Adjusting for the preceding and substituting for c:

$$A_{s\,(max)} = \frac{0.75(0.85f_c')\beta_1 (87,000)db}{f_y(87,000 + f_y)} + \frac{A_s'f_s'}{f_y}$$

but, from Section 2-8, for a singly reinforced beam,

$$\rho_{max} = 0.75\rho_b = \frac{0.75(0.85)f_c'\beta_1}{f_y}\left[\frac{87,000}{f_y + 87,000}\right]$$

Therefore,

$$A_{s\,(max)} = \rho_{max}db + \frac{A_s'f_s'}{f_y} \tag{B-1}$$

If $f_s' = f_y$ at the balanced condition, this reduces to

$$A_{s\,(max)} = \rho_{max}db + A_s' \tag{B-2}$$

Note that ρ_{max} (or $0.75 \, \rho_b$) for a singly reinforced beam may be found in Table A-4.

Calculate f'_s *at the balanced condition.*

$$f'_s = \epsilon'_s E_s = \frac{c - d'}{c}(0.003)E_s = \left(1 - \frac{d'}{c}\right)(87,000)$$

Substitute for c:

$$f'_s = \left[1 - \frac{d'(87,000 + f_y)}{87,000d}\right](87,000)$$

$$f'_s = \left[1 - \frac{d'}{d}\left(\frac{87,000 + f_y}{87,000}\right)\right](87,000) \leqslant f_y \qquad \text{(B-3)}$$

For various values of f_y, the value of (d'/d) may be found for which $f'_s = f_y$. If the actual (d'/d) is less than this value, then $f'_s = f_y$. The values of (d'/d) for which $f'_s = f_y$ are

$$0.37 \text{ for } f_y = 40,000 \text{ psi}$$

$$0.184 \text{ for } f_y = 60,000 \text{ psi}$$

Procedure

1. Calculate the actual (d'/d).

2. For $(d'/d) \leqslant 0.37$ or 0.184 (as appropriate), use Eq. (B-2) for $A_{s\,(\text{max})}$.

3. For $(d'/d) > 0.37$ or 0.184 (as appropriate), use Eq. (B-3) for f'_s and Eq. (B-1) for $A_{s\,(\text{max})}$.

B-3 FLOW DIAGRAMS

The step-by-step procedures for analysis and design of reinforced concrete members may be presented in the form of flow diagrams (see Figs. B-2 to B-5). To aid the reader in grasping the overall calculation approach, which may sometimes include cycling steps, flow diagrams for the analysis and design of rectangular beams and T-beams are presented here. These flow diagrams represent, on an elementary level, the type of organization required to develop computer programs to aid in analysis/design calculations.

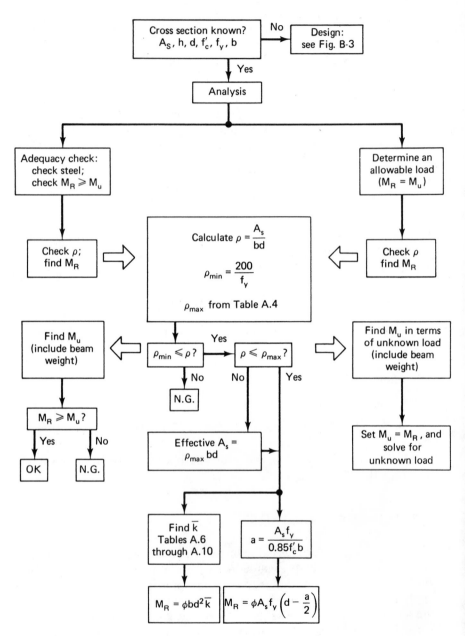

FIGURE B-2 Rectangular beam analysis for moment (tension steel only)

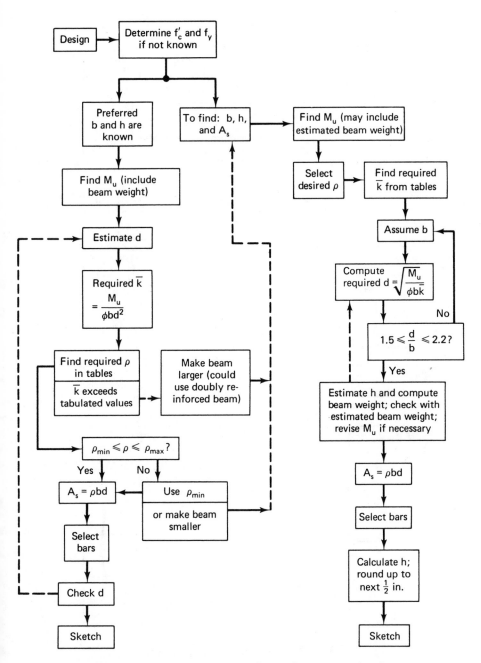

FIGURE B-3 Rectangular beam design for moment (tension steel only)

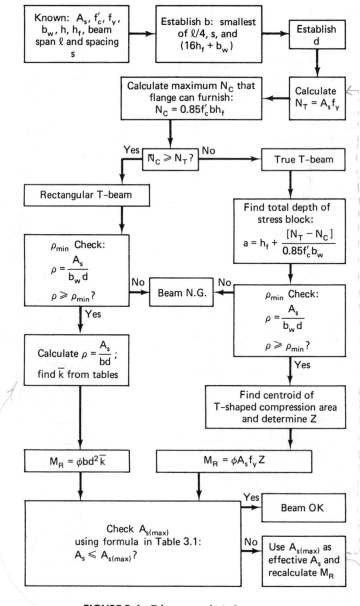

FIGURE B-4 T-beam analysis for moment

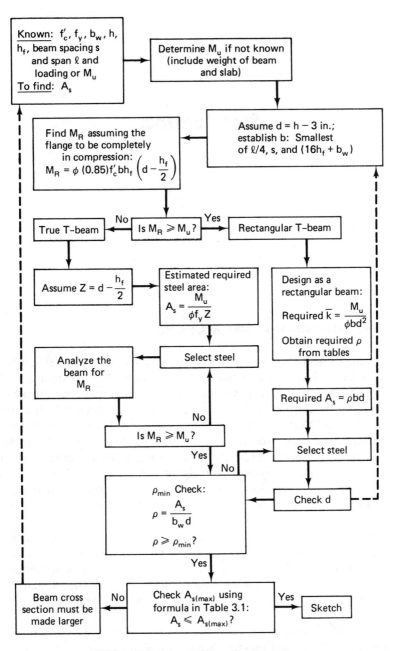

FIGURE B-5 T-beam design for moment

C

Metrication

C–1 THE INTERNATIONAL SYSTEM OF UNITS

The United States Customary System of weights and measures has been used as the primary unit system in this book. This system actually developed from the English system (British), which had been introduced in the original thirteen colonies when they were under British rule. Despite the fact that the English system was spread to many parts of the world during the past three centuries, it was widely recognized that there was a need for a single international coordinated measurement system. As a result, a second system of weights and measures, known as the metric system, was developed by a commission of French scientists and was adopted by France as the legal system of weights and measures in 1799.

Although the metric system was not accepted with enthusiasm at first, adoption by other nations occurred steadily after France made its use compulsory in 1840. In the United States, an Act of Congress in 1866 made it lawful throughout the land to employ the weights and measures of the metric system in all contracts, dealings, or court proceedings.

By 1900 a total of 35 nations, including the major nations of continental Europe and most of South America, had officially accepted the metric system. In 1971 the Secretary of Commerce, in transmitting to Congress the results of a three-year study authorized by the Metric Study Act of 1968, recommended that the United States change to predominant use of the metric system through a coordinated national program. The Congress responded by enacting the Metric Conversion Act of 1975 that established a U.S. Metric Board to carry out the planning, coordination, and public education that would facilitate a voluntary conversion to a modern metric system. Today, with the exception of a few small countries, the entire world is using the metric system or is changing to its use. Because of the many versions of the metric system that developed, an International General Conference on Weights and Measures in 1960 adopted an extensive revision and simplification of the system. The name *Le Systéme Internationale d'Unités* (International System of Units) with the international abbreviation SI was adopted for this modernized metric system. Further improvements in and additions to SI were made by the General Conference in 1964, 1968, 1971, and 1975.

The American Society of Civil Engineers (ASCE) resolved in 1970 to actively support conversion to SI and to adopt the revised edition of the American Society for Testing and Materials (ASTM) *Metric Practice* guide. The authors recognize that the introduction of and conversion to the SI in the building design and construction field in the United States is inevitable and will undoubtedly occur in the not-so-distant future. The application of SI units to building design and construction, when fully implemented, should serve to speed up calculations and activities requiring measuring.

As a means of introducing the SI metric system into reinforced concrete design, this appendix furnishes various design aids, conversion tables, and example problems utilizing the SI metric units.

Table C–1 presents equivalencies between U.S. Customary and SI systems for diameter, area, and weight (mass) of standard bar sizes.

C–2 SI STYLE AND USAGE

The SI consists of a limited number of *base* units that establish fundamental quantities and a large number of *derived* units, which come from the base units, to describe other quantities.

The SI base units pertinent to reinforced concrete design are listed in Table C–2, and the SI derived units pertinent to reinforced concrete design are listed in Table C–3.

Since the orders of magnitude of many quantities cover wide ranges of numerical values, SI *prefixes* have been established to deal with decimal-point placement. SI prefixes representing steps of 1000 are recommended to indicate orders of magnitude. Those recommended for use in reinforced concrete design are listed in Table C–4.

TABLE C–1
Properties of Steel Reinforcing Bars

	U.S. Customary			Metric (SI)		
Bar Number	Nominal Diameter (in.)	Nominal Area (in.²)	Nominal Weight (lb/ft)	Nominal Diameter (mm)	Nominal Area (mm²)	Nominal Mass (kg/m)
3	0.375	0.11	0.376	9.52	71	0.560
4	0.500	0.20	0.668	12.70	129	0.994
5	0.625	0.31	1.043	15.88	200	1.552
6	0.750	0.44	1.502	19.05	284	2.235
7	0.875	0.60	2.044	22.22	387	3.042
8	1.000	0.79	2.670	25.40	510	3.973
9	1.128	1.00	3.400	28.65	645	5.060
10	1.270	1.27	4.303	32.26	819	6.404
11	1.410	1.56	5.313	35.81	1006	7.907
14	1.693	2.25	7.650	43.00	1452	11.38
18	2.257	4.00	13.600	57.33	2581	20.24

From Table C–4 it is clear that one has a choice of ways to represent numbers. It is preferable to use numbers between 1 and 1000, whenever possible, by selecting the appropriate prefix. For example, 18 m is preferred to 0.018 km or 18 000 mm. A brief note must be made at this point concerning the presentation of numbers with many digits. It is common practice in the United States to separate digits into groups of three by means of commas. To avoid confusion with the widespread European practice of using a comma on the line as the decimal marker, the method of setting off groups of three digits with a gap, as shown in Table C–4, is recommended international practice. Note that this method is used on both the left and right sides of the decimal marker for any

TABLE C–2
SI Units and Symbols

Quantity	Unit	SI Symbol
Length	meter	m
Mass	kilogram	kg
Time	second	s
Angle*	radian	rad

* It is also permissible to use the arc degree *and its decimal submultiples* when the radian is not convenient.

TABLE C-3
SI Derived Units

Quantity	Unit	SI Symbol	Formula
Acceleration	meter per second squared	—	m/s^2
Area	square meter	—	m^2
Density (mass per unit volume)	kilogram per cubic meter	—	kg/m^3
Force	newton	N	$kg \cdot m/s^2$
Pressure or stress	pascal	Pa	N/m^2
Volume	cubic meter	—	m^3
Section modulus	meter to third power	—	m^3
Moment of inertia	meter to fourth power	—	m^4
Moment of force, torque	newton meter	—	$N \cdot m$
Force per unit length	newton per meter	—	N/m
Mass per unit length	kilogram per meter	—	kg/m
Mass per unit area	kilogram per square meter	—	kg/m^2

string of five or more digits. A group of four digits on either side of the decimal marker need not be separated.

A significant difference between SI and other measurement systems is the use of explicit and distinctly separate units for mass and force. The SI base unit kilogram (kg) denotes the base unit of mass, which is

TABLE C-4
SI Prefixes

Prefix	SI Symbol	Factor
mega	M	$1\ 000\ 000 = 10^6$
kilo	k	$1000 = 10^3$
milli	m	$0.001 = 10^{-3}$
micro	μ	$0.000\ 001 = 10^{-6}$

the quantity of matter of an object. This is a constant quantity which is independent of gravitational attraction. The derived SI unit newton (N) denotes the absolute derived unit of force (mass times acceleration: kg·m/s²). The term *weight* should be avoided since it is confused with mass and also because it describes a particular force that is related solely to gravitational acceleration, which varies on the surface of the earth. For the conversion of mass to force, the recommended value for the acceleration of gravity in the United States may be taken as $g = 9.81$ m/s².

As an example, we will consider the "weight" of reinforced concrete, which, in the design of bending members, must be considered as a load or force per unit length of span. In the U.S. Customary System, reinforced concrete weighs 150 lb/ft.³. This is equivalent in the SI to

$$150 \frac{\text{lb}}{\text{ft}^3} \times \left(\frac{3.2808 \text{ ft}}{1 \text{ m}}\right)^3 \times \frac{1 \text{ kg}}{2.2046 \text{ lb}} = 2403 \frac{\text{kg}}{\text{m}^3}$$

This represents a mass per unit volume (density) where the unit volume is 1 cubic meter, m³. To utilize the density of the concrete to obtain a force per cubic meter, Newton's law must be applied:

$$F = \text{mass times acceleration of gravity}$$

$$= mg$$

$$= 2403\,(9.81)$$

$$= 23\,573 \frac{\text{kg·m}}{\text{s}^2\text{m}^3}$$

$$= 23\,573 \text{ N/m}^3$$

$$= 23.57 \text{ kN/m}^3$$

To determine the dead load per unit length of a beam of dimensions $b = 500$ mm and $h = 1000$ mm:

$$\left(\frac{500}{1000}\right)\left(\frac{1000}{1000}\right)(23.57) = 11.79 \text{ kN/m}$$

Thus the load or force per unit length (1 meter) equals 11.79 kN.

C–3 CONVERSION FACTORS

Table C–5 contains conversion factors for the conversion of the U.S. Customary System units to SI units for quantities frequently used in reinforced concrete design.

TABLE C–5
Conversion Factors: U.S. Customary to SI Units

	Multiply		By		To Obtain
Length	inches	×	25.4	=	millimeters
	feet	×	0.3048	=	meters
	yards	×	0.9144	=	meters
	miles (statute)	×	1.609	=	kilometers
Area	square inches	×	645.2	=	square millimeters
	square feet	×	0.0929	=	square meters
	square yards	×	0.8361	=	square meters
Volume	cubic inches	×	16 387.	=	cubic millimeters
	cubic feet	×	0.028 32	=	cubic meters
	cubic yards	×	0.7646	=	cubic meters
	gallons (U.S. liquid)	×	0.003 785	=	cubic meters
Force	pounds	×	4.448	=	newtons
	kips	×	4448.	=	newtons
Force per unit length	pounds per foot	×	14.594	=	newtons per meter
	kips per foot	×	14 594.	=	newtons per meter
Load per unit volume	pounds per cubic foot	×	0.157 14	=	kilonewtons per cubic meter
Bending moment or torque	inch-pounds	×	0.1130	=	newton meters
	foot-pounds	×	1.356	=	newton meters
	inch-kips	×	113.0	=	newton meters
	foot-kips	×	1356.	=	newton meters
	inch-kips	×	0.1130	=	kilonewton meters
	foot-kips	×	1.356	=	kilonewton meters
Stress, pressure, loading (force per unit area)	pounds per square inch	×	6895.	=	pascals
	pounds per square inch	×	6.895	=	kilopascals
	pounds per square inch	×	0.006 895	=	megapascals
	kips per square inch	×	6.895	=	megapascals
	pounds per square foot	×	47.88	=	pascals
	pounds per square foot	×	0.047 88	=	kilopascals
	kips per square foot	×	47.88	=	kilopascals
	kips per square foot	×	0.047 88	=	megapascals
Mass	pounds	×	0.454	=	kilograms
Mass per unit volume (density)	pounds per cubic foot	×	16.02	=	kilograms per cubic meter
	pounds per cubic yard	×	0.5933	=	kilograms per cubic meter
Moment of inertia	inches^4	×	416 231.	=	millimeters^4
Mass per unit length	pounds per foot	×	1.488	=	kilograms per meter
Mass per unit area	pounds per square foot	×	4.882	=	kilograms per square meter

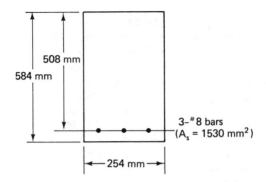

FIGURE C-1   Cross section for Example C-1

Reference is also made to the ACI Code (318–77), Appendix D, which furnishes a tabulation of SI metric equivalents of U.S. Customary Units as well as equation conversions for use in the SI system. Other reference sources which contain more complete treatment of the many aspects of metrication in the design and construction field are listed at the end of this appendix.

EXAMPLE C–1

Find M_R for the beam of cross section shown in Fig. C–1. Use the analysis procedure of Chapter 2.

1.
$$f_y = 275.8 \text{ MPa}$$
$$f_c' = 20.68 \text{ MPa}$$
$$b = 254 \text{ mm}, d = 508 \text{ mm}$$
$$h = 584 \text{ mm}, A_s = 1530 \text{ mm}^2$$

2. To be found: M_R.

3.
$$\rho = \frac{A_s}{bd} = \frac{1530}{254(508)} = 0.0118$$

4.
$$0.75\rho_b = 0.0278 = \rho_{\max}$$
$$0.0118 < 0.0278 \qquad \text{(OK)}$$

From the ACI Code, Appendix D, with f_y in units of MPa:

$$\rho_{\min} = \frac{1.4}{f_y} = \frac{1.4}{275.8} = 0.0050 \qquad \text{(OK)}$$

5.
$$a = \frac{A_s f_y}{0.85 f_c' b} = \frac{1530(275.8)}{0.85(20.68)(254)} = 94.5 \text{mm}$$

6.
$$Z = d - \frac{a}{2} = 508 - \frac{94.5}{2} = 460.75 \text{ mm}$$

7.
$$M_n = A_s f_y Z$$

8. $M_R = \phi M_n = 0.9(A_s f_y Z)$. All the quantities needed for the M_R calculation have been determined, but some conversion is required to make prefixes compatible. Rather than set up prefix conversion for each lengthy calculation, it is suggested, for situations such as this, that quantities be substituted in units of meters and newtons. Results will be in the same units. This method also lends itself to the use of numerical values expressed in powers-of-ten notation. For the M_R calculation:

$$A_s = 0.001\ 530 \text{ m}^2, \ f_y = 275\ 800\ 000 \text{ N/m}^2, \ Z = 0.460\ 75 \text{ m}$$

$$\text{from which } M_R = 0.9(0.001\ 530)(275\ 800\ 000)(0.460\ 75)$$

$$= 174\ 982 \text{ N·m}$$

$$= 175 \text{ kN·m}$$

Note that the final M_R is simply changed to kN·m. The kilo prefix is the most appropriate prefix for the majority of flexural problems which are presented in this book using the U.S. Customary System of units (see "Bending moment or Torque" in Table C–5 for comparison with ft-kips).

EXAMPLE C–2

Design a rectangular reinforced concrete beam for a simple span of 10 m to carry service loads of 14.6 kN/m dead load (does not include the load of the beam) and 29.2 kN/m live load. The maximum width of beam desired is 406 mm. Use $f_c' = 20.68$ MPa and $f_y = 413.7$ MPa. Assume a No. 3 stirrup and sketch the design. (See Chapter 2 for design procedure.)

1.
$$w_u = 1.4 w_{DL} + 1.7 w_{LL}$$

$$= 1.4(14.6) + 1.7(29.2)$$

$$= 70.08 \text{ kN/m}$$

$$M_u = \frac{w_u \ell^2}{8} = \frac{70.08(10)^2}{8} = 876 \text{ kN·m}$$

2. Assume that $\rho = 0.0090$ (Table A–4). Note that the given f_c' and f_y correspond to 3,000 psi and 60,000 psi, respectively.

3. From Table A–7,

$$\text{Required } \overline{k} = 0.4828 \text{ ksi}$$

Converting to the SI system,

$$\overline{k} = 0.4828(6.895) = 3.329 \text{ MPa}$$

4. Assume that b = 406 mm.

$$\text{Required } d = \sqrt{\frac{M_u}{\phi b \overline{k}}}$$

Substituting quantities in terms of meters and newtons,

$$\text{Required } d = \sqrt{\frac{876\,000}{0.9(0.406)(3\,329\,000)}}$$

$$= 0.849\,\text{m} = 849 \text{ mm}$$

$$\frac{d}{b} \text{ ratio} = \frac{849}{406} = 2.09 \qquad\qquad \text{(OK)}$$

5. Estimate the total beam depth for purposes of the beam dead load. Assume a No. 11 main bar, a No. 3 stirrup, and a minimum cover of 38 mm.

$$h = 849 + 17.9 + 9.5 + 38 = 914.4 \text{ mm}$$

Use an estimated beam depth of 950 mm. From Section C–2, the dead load due to reinforced concrete is 23.57 kN/m³. Therefore, the dead load of the beam per meter length is

$$0.406\,(0.950)(23.57) = 9.09 \text{ kN/m}$$

6. The additional M_u due to the beam dead load is

$$M_u = \frac{1.4 w_{DL}\,\ell^2}{8} = \frac{1.4(9.09)(10)^2}{8}$$

$$= 159.1 \text{ kN·m}$$

$$\text{Total } M_u = 876 + 159.1 = 1035.1 \text{ kN·m}$$

7. Using ρ, $\overline{k}$, and b as before, the effective depth required is

$$\text{Required } d = \sqrt{\frac{1\,035\,100}{0.9(0.406)(3\,329\,000)}}$$

$$= 0.922 \text{ m} = 922 \text{ mm}$$

$$\frac{d}{b} = \frac{922}{406} = 2.27 \qquad\qquad \text{(OK)}$$

8. $$\text{Required } A_s = \rho b d$$

$$= 0.0090\,(406)(922)$$

$$= 3369 \text{ mm}^2$$

9. Use 4 No. 11 bars.

$$A_s = 1006(4) = 4024 \text{ mm}^2$$

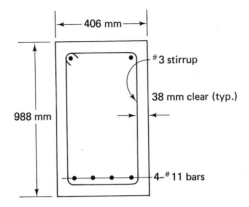

FIGURE C-2 Design sketch for Example C–2

The minimum beam width for 4 No. 11 bars, from Table A–11, is

$$14(25.4) = 356 \text{ mm} < 406 \text{ mm} \qquad \text{(OK)}$$

10. The total beam depth h may be taken as the effective depth required plus minimum concrete cover plus stirrup diameter plus one-half main steel diameter.

$$\text{Required } h = 922 + 38 + 9.5 + \frac{35.81}{2}$$

$$= 987.4 \text{ mm}$$

$$\text{Use } h = 988 \text{ mm.}$$

11. The design sketch is shown in Fig. C–2.

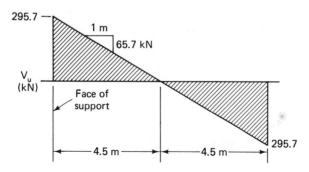

FIGURE C-3 Shear force diagram for Example C–3

EXAMPLE C–3

A simply supported rectangular reinforced concrete beam 330 mm wide and having an effective depth of 508 mm carries a total design load (w_u) of 65.7 kN/m on a 9.00 m clear span. (The given load includes the dead load of the beam.) Design the web reinforcement if f'_c = 20.68 MPa and f_y = 275.8 MPa (see Fig. C–3).

1. Draw a shear force diagram.

$$V_u = \frac{w_u \ell}{2} = \frac{65.7(9)}{2} = 295.7 \text{ kN}$$

2. Draw the required V_s diagram.

$$\text{Required } V_s = \frac{V_u}{\phi} - V_c$$

$$V_c = (0.17\sqrt{f'_c})b_w d \qquad \text{(ACI Code, Appendix D)}$$

$$= 0.17\sqrt{20.68}(330)(508)$$

$$= 129\ 599 \text{ N} = 129.6 \text{ kN}$$

Stirrups must be provided if $V_u > \frac{1}{2}\phi V_c$.

$$\frac{1}{2}\phi V_c = \frac{0.85(129.6)}{2} = 55.1 \text{ kN}$$

Stirrups are required, since 295.7 > 55.1. At the face of the support:

$$\text{Required } V_s = \frac{295.7}{0.85} - 129.6 = 218.3 \text{ kN}$$

The total uniform load of 65.7 kN/m is the slope of the V_u diagram. Therefore, the slope of the V_s diagram will be the slope of the V_u diagram divided by ϕ.

$$\text{Slope of } V_s \text{ diagram} = \frac{65.7}{0.85} = 77.3 \text{ kN/m}$$

From the face of the support, the point where the V_s diagram passes through zero may be determined:

$$V_s = 0 \text{ at } \frac{218.3}{77.3} = 2.82 \text{ m}$$

The V_s diagram is shown in Fig. C–4.

3. We next determine the length of the span over which stirrups are required, referencing from the face of the support:

$$\frac{295.7 - 55.1}{65.7} = 3.66 \text{ m}$$

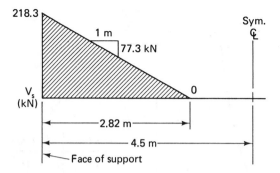

FIGURE C-4 V_s **diagram for Example C–3**

4. Assume a No. 3 stirrup ($A_v = 142$ mm²). The spacing at the critical section, based on an effective depth $d = 508$ mm, may now be found.

$$V_s = 218.3 - 0.508(77.3) = 179 \text{ kN}$$

Using basic units of m and N yields

$$\text{Required } s = \frac{A_v f_y d}{V_s} = \frac{(0.000\ 142)(275\ 800\ 000)(0.508)}{179\ 000}$$

$$= 0.111 \text{ m} = 111 \text{ mm}$$

If the required spacing were less than 100 mm, a larger-diameter bar would be selected for the stirrups.

5. For the ACI Code maximum spacing requirements, compare V_s at the critical section with $0.33\sqrt{f'_c}b_w d$ (ACI Code, Appendix D).

$$0.33\sqrt{f'_c}b_w d = 0.33\sqrt{20.68}(330)(508)$$

$$= 251\ 575 \text{ N} = 251.6 \text{ kN}$$

$$V_s = 179 \text{ kN at critical section}$$

Since $179 < 251.6$, the maximum spacing will be $d/2$ or 610 mm, whichever is smaller.

$$\frac{d}{2} = \frac{508}{2} = 254 \text{ mm}$$

A second criterion of the ACI Code is

$$\text{Maximum } s = \frac{A_v f_y}{0.34 b_w} = \frac{142(275.8)}{0.34(330)} \quad \text{(ACI Code, Appendix D)}$$

$$= 349.1 \text{ mm}$$

Therefore, of the foregoing criteria, the maximum spacing = 254 mm.

409

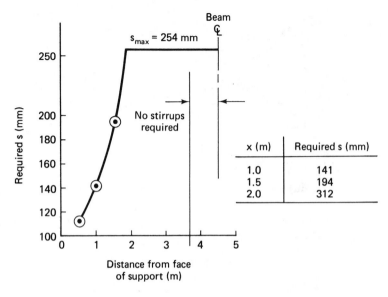

FIGURE C-5 Stirrup spacing requirements—Example C-3

6. Next determine the spacing requirements based on shear strength. At the critical section, the required spacing is 111 mm. The maximum spacing is 254 mm. At other points along the span, the spacing required may be determined as follows:

$$V_s = V_{s(\max)} - \text{slope}\,(x)$$

$$= 218.3 - 77.3(x)$$

$$\text{Required } s = \frac{A_v f_y d}{V_s}$$

Using basic units of m and N yields

$$\text{Required } s = \frac{(0.000\ 142)(275\ 800\ 000)(0.508)}{218\ 300 - 77\ 300(x)}$$

The results for several arbitrary values of x are shown tabulated and plotted in Fig. C–5.

7. Utilizing Fig. C–5, the stirrup pattern shown in Fig. C–6 may be developed. Stirrups have been placed the full length of the span, which is both common practice and conservative.

410

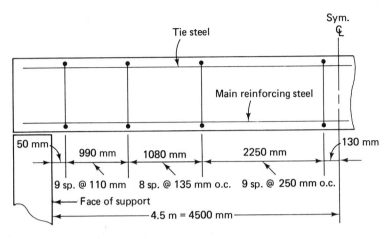

FIGURE C-6 Stirrup spacing—Example C-3

REFERENCES

1. Milton, Hans J., *Recommended Practice for the Use of Metric (SI) Units in Building Design and Construction*, NBS Technical Note 938, U.S. Department of Commerce/National Bureau of Standards, U.S. Government Printing Office, Washington, D.C., 1977.

2. McGannon, Harold E., *Measuring Things . . . Cubits to Modern Metrics . . .*, United States Steel Corporation, 1976.

3. American National Standard, ANSI Z 210.1–1976/ASTM E 380–76/IEEE Std 268–1976, *Metric Practice*, 1976 rev. ed., American Society for Testing and Materials, 1916 Race Street, Philadelphia, Pennsylvania 19013.

Index